INORGANIC NANOWIRES

Nanomaterials and Their Applications

Series Editor: M. Meyyappan

Inorganic Nanowires: *Applications, Properties, and Characterization*
M. Meyyappan, Mahendra Sunkara

Inorganic Nanoparticles: *Synthesis, Applications, and Perspectives*
Claudia Altavilla, Enrico Ciliberto, Editors

INORGANIC NANOWIRES
Applications, Properties, and Characterization

M. Meyyappan

NASA Research Center, Moffett Field
California, USA

Mahendra Sunkara

University of Louisville
Kentucky, USA

CRC Press
Taylor & Francis Group
Boca Raton   London   New York

CRC Press is an imprint of the
Taylor & Francis Group, an **informa** business

Cover Image: Vertical surround-gate transistor (Courtesy of H.T. Ng) and vertical zinc oxide nanowires (Courtesy of H.T. Ng and Pho Nguyen). The image at the bottom depicts a laser mode profile along a nanowire laser, showing waves being emitted from the top end and into the substrate (Courtesy of Cun-Zheng Ning).

CRC Press
Taylor & Francis Group
6000 Broken Sound Parkway NW, Suite 300
Boca Raton, FL 33487-2742

© 2010 by Taylor and Francis Group, LLC
CRC Press is an imprint of Taylor & Francis Group, an Informa business

No claim to original U.S. Government works

Printed in the United States of America on acid-free paper
10 9 8 7 6 5 4 3 2 1

International Standard Book Number: 978-1-4200-6782-8 (Hardback)

Library of Congress Cataloging-in-Publication Data

Meyyappan, M.
    Inorganic nanowires : applications, properties, and characterization / M. Meyyappan, Mahendra K. Sunkara.
        p. cm. -- (Nanomaterials and their applications)
    Includes bibliographical references and index.
    ISBN 978-1-4200-6782-8 (hardcover : alk. paper)
    1. Nanowires. 2. Inorganic compounds. 3. Materials. I. Sunkara, Mahendra K. II. Title.

TK7874.85.M493 2010
620'.5--dc22
                                             2009034836

Visit the Taylor & Francis Web site at
http://www.taylorandfrancis.com

and the CRC Press Web site at
http://www.crcpress.com

Contents

Preface

Ever since the U.S. National Nanotechnology Initiative was unveiled in 2000, nanotechnology as a field of research has exploded due to its potential impact on electronics, computing, memory and data storage, communications, materials and manufacturing, health and medicine, energy, transportation, national security, and other economic sectors. Fundamental to the success in any of these sectors is the development of novel nanostructured materials. It is a well-established fact that properties do change at the nanoscale relative to their bulk counterparts. There is no shortage of candidates for nanostructured materials for any of the above applications. Carbon nanotubes (CNTs) certainly top any list, followed by nanoparticles, powders, quantum dots, conducting organic molecules, and biomolecules. In the last five to seven years, inorganic nanowires (INWs) have received much attention since semiconducting, metallic, oxide, and other nanowires exhibit properties that are interesting and appropriate enough for the applications listed above and they can potentially compete with any of the other nanostructured material candidates, including CNTs in some cases.

There is a long history to inorganic materials synthesized in the form of cylindrical wires; R.S. Wagner demonstrated submicron-scale wires of semiconducting materials in the 1960s. More recently, with the advances in nanofabrication, characterization tools, and the drive to commercialize nanotechnology products, the size of the wires has reached the sub-10 nm level with corresponding interesting properties. There has been a significant increase in INW research as evidenced by the number of publications in peer-reviewed journals and focused sessions in international conferences. The future of INWs appears to be bright, with the field growing rapidly. This book attempts to provide a comprehensive and coherent account of INWs, as currently there are no books available apart from one or two edited volumes. Inorganic materials span across the entire periodic table, and therefore the coverage includes the growth, the characterization, the properties, and the applications of Group IV elements, compound semiconductors, oxides, nitrides, and other materials—all in the form of nanowires.

This book is organized as follows: Chapter 1 traces the evolution of nanotechnology and provides a classification of nanomaterials, thus providing a context to INWs. A summary that lists various nanowires and their potential applications is provided. Micron- and larger-sized wires of semiconducting and other inorganic materials have been grown since the 1960s and before. Chapter 2 covers these pioneering works, not only from a historical perspective, but also from a technical point of view to understand and appreciate what has come in the nano era, showing connectivity and continuity. Chapter 3 provides a detailed discussion of various growth techniques, including vapor–liquid–solid (VLS), laser ablation, sol-gel processing, template-assisted techniques, and direct spontaneous nucleation methods. Growth apparatuses—both laboratory scale and commercial, if any—are described along with source materials, growth conditions, substrate preparation, catalyst preparation, and post-growth processing. The bulk production of nanowires is also covered.

Chapter 4 critically evaluates and provides the most important aspects of classical thermodynamics that underline the nucleation and growth of nanowires. This chapter also reviews the current understanding of various kinetic processes involved with nanowire growth, that is, nanowire growth rates, nanowire faceting, surface diffusion, and adatom incorporation processes. Modeling and simulation provide an insight into growth mechanisms and an understanding of the process to enable the control of the dimensions and to improve the quality of the nanomaterials. There is very little of this in the literature at present. A review of the current status and the gaps therein is presented in Chapter 5. Nanowires exhibit differences in properties, for example, melting point, thermal conductivity, and other physical properties, from their bulky counterparts. Modeling results predicting the changes in properties and limited comparisons with the experiments are available. These results and other specific cases of characterization results are reserved for later chapters under each type of nanowire, as appropriate.

The growth of silicon, germanium, gallium arsenide, and other semiconducting nanowires is discussed in detail in Chapter 6, outlining growth recipes, precursors, and results. There is extensive work on the understanding of the physical and electronic properties of these nanowires, which is an important coverage here. Phase change materials have successfully penetrated CD and DVD applications. However, as the programming current required to read/write is very high in thin-film devices, their use in random access memory is facing difficulties. The programming current and power requirement are expected to reduce with the use of nanowires and hence the successful reporting of GeTe, GeSbTe, and In_2Se_3 nanowires to date. Growth, characterization, and property differences from the corresponding thin films are highlighted in Chapter 7.

Numerous metals have been grown in the form of nanowires. Chapter 8 covers boron, bismuth, silver, tungsten, gold, nickel, cobalt, copper, and other metallic nanowires. Numerous oxides, such as ZnO, InO, SnO, ITO, CuO, etc., have also been grown in the form of nanowires. Most of these are wide band gap semiconductors as well and have interesting applications in the ultraviolet (UV) regime for lasers, laser emitting diode (LEDs), and in field emission, as discussed in Chapter 9. Nitrides, especially group III nitrides, are popular with the nanowire community, and are discussed in Chapter 10. Chapter 11 deals with nanowire forms of all other materials such as antimonides, tellurides, phosphides, and silicides.

There is a tremendous interest in exploring Si, Ge, and Si_xGe_{1-x} nanowires for logic devices to allow the continuation of Moore's law scaling and go beyond, and for memory devices as well. These applications are covered in Chapter 12, with Chapter 13 focusing on optoelectronics. There are numerous nanowires in the UV (ZnO, GaN, SiC, etc.) and infrared (IR) (Insb, GaSb) regimes that are applied in detectors. Lasers in visible, UV, and IR regimes have been demonstrated using several nanowires. Chapter 13 focuses on the principles, development, and results of these applications. Chapter 14 considers applications in chemicals and biosensors, while Chapter 15 focuses on energy covering the use of nanowires in solar cells and batteries. Finally, all other applications, such as field emission devices and thermoelectric devices, are discussed in Chapter 16. First, the principles behind the device operation are covered, along with references to further sources. This is followed by a discussion of the need that the nanowires fill in to replace current candidates, device fabrication, performance results, and future prospects and challenges.

Any undertaking of this magnitude is only possible with the help of colleagues. The nanotechnology group at NASA Ames, particularly former and current members of the INW technology team, namely, Bin Yu, Xuhui Sun, H.T. Ng, Pho Nguyen, Aaron Mao, Sreeram Vaddiraju, Cun-Zheng Ning, Alan Chin, and Q. Ye, are acknowledged for educating one of the authors (MM) and for providing some of the figures used in this book. The works of several former students and associates: Shashank Sharma, Hari Chandrasekaran, Radhika C. Mani, Gopinath Bhimarasetti, Hongwei Li, Sreeram Vaddiraju, Ramakrishna Mulpuri, and Zhiqiang Chen at the University of Louisville; and colleagues: Miran Mozetic and Uros Cvelbar from Institute Jozef Stefan, Slovenia, have been of great help. Most importantly, the following students have provided considerable assistance to the writing of some of the chapters: Vivekanand Kumar (Chapter 3), Chandrashekhar Pendyala (Chapters 4 and 10), Praveen Meduri (Chapters 5 and 15), Jyothish Thangala (Chapter 9), Suresh Gubbala (Chapter 15), Boris Chernomordik (Chapter 15), and Santoshrupa Dumpala (Chapter 16). The hard work and contributions of all these students are gratefully acknowledged. Finally, this book would not have been possible without the assistance of Julianne Hildebrand, who diligently worked on the preparation of the manuscript, copyright permissions, and other logistical aspects.

M. Meyyappan
Moffett Field, California

Mahendra K. Sunkara
Louisville, Kentucky

Authors

M. Meyyappan is the chief scientist for exploration technology at the Center for Nanotechnology, NASA Ames Research Center in Moffett Field, California. Until June 2006, he served as the director of the Center for Nanotechnology as well as a senior scientist. He is a founding member of the Interagency Working Group on Nanotechnology (IWGN), established by the Office of Science and Technology Policy (OSTP). The IWGN is responsible for putting together the National Nanotechnology Initiative.

Dr. Meyyappan has authored or coauthored over 175 articles in peer-reviewed journals and has delivered more than 200 invited/keynote/plenary speeches on subjects related to nanotechnology around the world. His research interests include carbon nanotubes and various inorganic nanowires; their growth and characterization; and application development in chemical and bio sensors, instrumentation, electronics, and optoelectronics.

Dr. Meyyappan is a fellow of the Institute of Electrical and Electronics Engineers (IEEE), the Electrochemical Society (ECS), the AVS, the Materials Research Society, and the California Council of Science and Technology. In addition, he is a member of the American Society of Mechanical Engineers (ASME) and the American Institute of Chemical Engineers. He is the IEEE Nanotechnology Council Distinguished Lecturer on Nanotechnology, the IEEE Electron Devices Society Distinguished Lecturer, and ASME's Distinguished Lecturer on Nanotechnology (2004–2006). He served as the president of the IEEE's Nanotechnology Council in 2006–2007.

Dr. Meyyappan has received numerous awards including a Presidential Meritorious Award; NASA's Outstanding Leadership Medal; the Arthur Flemming Award given by the Arthur Flemming Foundation and the George Washington University; the 2008 IEEE Judith Resnick Award; the IEEE-USA Harry Diamond Award; and the AIChE Nanoscale Science and Engineering Forum Award for his contributions and leadership in nanotechnology. He was inducted into the Silicon Valley Engineering Council Hall of Fame in February 2009 for his sustained contributions to nanotechnology. He has received the Outstanding Recognition Award from the NASA Office of Education; the Engineer of the Year Award (2004) by the San Francisco Section of the American Institute of Aeronautics and Astronautics (AIAA); and the IEEE-EDS Education Award for his contributions to the field of nanotechnology education.

Mahendra K. Sunkara is currently a professor of Chemical Engineering and the founding director of the Institute for Advanced Materials and Renewable Energy (IAM-RE; web site: http://www.louisville.edu/iamre) at the University of Louisville (UofL). He received his BTech, MS, and PhD in

Chemical Engineering from Andhra University (Waltair, Andhra Pradesh, India) in 1986, Clarkson University (Potsdam, NY) in 1988, and Case Western Reserve University (Cleveland, OH) in 1993, respectively. He worked at Faraday Technology, Inc. in Dayton, OH from 1993 to 1996 as a Project Engineer and served as the technical leader/principal investigator on several SBIR research grants dealing with electrochemical technologies for environmental remediation and corrosion sensing and mitigation.

Dr. Sunkara joined UofL in 1996 as an assistant professor. He received external research contracts in excess of $10 million to support a research program and to establish an Institute for Advanced Materials and Renewable Energy at UofL. His research interests and projects include renewable energy technologies such as solar cells, Li Ion batteries, production of hydrogen from water and process development for growing large crystals of diamond, gallium nitride and bulk quantities of nanowires, processes for a set of novel carbon morphologies discovered within his group.

Dr. Sunkara has published over 100 articles in refereed journals and proceedings, four book chapters and was awarded seven U.S. patents along with several additional U.S. patent applications pending. Several national and international news articles appeared on his research work in the area of nanoscale materials and their applications in to Li Ion batteries and sensors, etc. In the last seven years, Dr. Sunkara delivered over 40 invited and keynote lectures in Germany, United States, Taiwan, Slovenia, and India. Three of his research articles appeared on the covers of prestigious journals, *Advanced Materials* and *Advanced Functional Materials*. He is the founding organizer of an annual statewide workshop on the theme of Materials Nanotechnology (KYNANOMAT) held since 2002. He was awarded the Ralph E. Powe Junior Faculty in Engineering award in 1999 and was the first recipient of the prestigious CAREER grant in Speed School from the National Science Foundation in 1999. In 2002, the *Louisville Magazine* placed him in the list of top 25 young guns in the city of Louisville. In 2009, he received the 2009 UofL President's distinguished faculty award for research.

1

Introduction

Nanotechnology has created a tremendous amount of excitement among the research community and industry in the last decade. There are various optimistic projections about the size of the economy nanotechnology will be directly responsible for or influence in another decade and beyond. Naturally then, there is a significant investment in research and development across the world to gain early advantages in creating intellectual property and commercial exploitation. There is also a strong realization about the need to educate the current and future generations of students and to provide continuing education to the practicing engineers and scientists about this emerging field. Nanotechnology has been pronounced, in fact, to be the technology of the twenty-first century and only time will tell how true this statement will turn out to be in the future.

Nanotechnology deals with the development of useful and functional materials, devices, and systems by manipulating matter at the nanometer scale, arbitrarily picked as 1–100 nm at least in one principal direction. The length scale is only a necessary condition but not a sufficient condition. If the length scale were to be the only defining criterion, then everything the computer industry has been doing in the last decade would qualify as nanotechnology; the critical dimension in a silicon CMOS reached the 100 nm level some time ago, and 32 nm feature devices will be going into production in the near future. But this is a routine, although extraordinarily challenging, miniaturization along Moore's law curve. In contrast, the vision in the National Nanotechnology Initiative (NNI) for nanotechnology calls for the ability to exploit the novel properties that arise because of going to the nanoscale [1]. Indeed properties of materials at the nanoscale differ from their bulk counterparts as will be seen shortly. It is important to realize that nanoscale is not a human scale. In that sense, the object or system to be created does not have to be nanoscale, rather it can be micro or macro or any useful size. This also points out the need to integrate hierarchically across the length scales from nano to micro to macro.

In the early days right after the unveiling of the U.S. NNI [2], it was not uncommon to hear proclamations such as "I have been doing nanotechnology for X number of years" where X was usually greater than or equal to 10. Of course, scientists have been dealing with small things for a long time. Photography and catalysts are old examples of using small particles. The Chinese pottery makers empirically figured out several hundred years ago that adding small gold particles—even though the color of bulk gold is yellow—enabled them to make ruby color vases. Damascus sabre steel used in swords from the seventeenth century was found to contain carbon nanotubes and cementite nanowires [3], which are attributed to the extraordinary mechanical properties and an exceptionally sharp cutting edge. In existing technologies using nanomaterials and processes, the role of nanoscale phenomena was not understood until the emergence of powerful tools recently. Going beyond serendipitous discoveries and gaining insight into mechanisms provide opportunities for improvement and the ability to design more complex systems in the future.

Atoms and molecules, which we study in chemistry, are generally less than a nanometer size. Condensed matter physics deals with solids characterized by an infinite array of bound atoms or molecules [4]. Nanoscience deals with the in-between mesoworld. Quantum chemistry does not apply and the systems are not large enough for applying classical laws of physics [3]. It is helpful to recall some definitions related to the nanoworld [4]. A cluster is a collection of units—atoms or reactive molecules—of up to about 50 units. A colloid is a stable liquid phase containing particles in the 1–100 nm range. A colloid particle is one such 1–100 nm particle. A nanoparticle is a solid particle in the 1–100 nm range that could be noncrystalline, an aggregate of crystallites, or a single crystallite. A nanocrystal is a solid particle, that is, a single crystal in the nanometer range.

Nanoscale materials exhibit a very high surface-to-volume ratio. A 30 nm iron particle has about 5% atoms on the surface, and a 10 nm particle has 20% of atoms on the surface. In contrast, a 3 nm particle has 50% of atoms on the surface [4]. This makes the surface or interface phenomena to dominate over bulk effects and strongly influence adsorption, solubility, reactivity, catalysis, etc. where surface effects are important. In addition, as mentioned earlier, the size-dependent properties make nanoscale materials attractive in various applications. As explained by Klabunde [4], the delocalization of valence electrons can be extensive in materials where strong chemical bonding is present. The extent of this delocalization can vary with the size of the system. The structure of materials also changes with size. These two effects together can lead to different properties, depending on size. Melting point of metals and semiconductors changes with size. Bulk gold melts at 1064°C whereas a 5 nm gold particle melts at a couple of hundred degrees sooner [5]. Specific heat is another property that changes with size. For example, a 6 nm Pd particle shows a 48% increase in specific heat at constant volume at 250 K [5]. The band gap changes with size for semiconductors and when the band gap lies in the visible spectrum, then a change in band gap also means a change in color. For magnetic materials such as Fe, Co, Ni, etc., the strength of internal magnetic field can be size-dependent and therefore, the coercive force—which is the force needed to reverse an internal magnetic field within the particle—is also size dependent [4].

In any application, the material selection is made based on one or more desirable properties of a material over its competition. When physical, chemical, optical, electrical, magnetic, and other properties change at the nanoscale, it is easy to understand the widespread impact of nanotechnology across all industrial sectors. Fundamental to this revolution is the development of nanostructured materials, and inorganic nanowires (INWs) form a class of nanomaterials that is expected to play a prominent role. In the last two to three decades, a variety of elemental and compound semiconducting, metallic, dielectric, and other materials have been grown successfully in the form of two-dimensional thin films using techniques such as chemical vapor deposition (CVD), metal organic chemical vapor deposition (MOCVD), and molecular beam epitaxy (MBE). These materials covered the wavelength range from ultraviolet (UV) to infrared (IR) and band gap range from 0.4 to 4.0 eV. Extraordinary dimensional control has been achieved with multiple quantum well layers of III–V material stacks such as GaAs/AlGaAs with layer thickness of 1 nm or less. The ability to grow thin epitaxial layers in a controlled manner has led to numerous successful commercial applications in logic and memory devices, lasers, detectors, microelectromechanical systems, sensors, photovoltaics, and others.

Attempts have been made in recent years to grow the above materials in the form of one-dimensional nanowires. The nanowires are single crystals with well-defined surface structural properties. When their diameter is less than the Bohr radius, the resulting quantum confinement is of great interest in studying the excitonic behavior of low-dimensional solids for its effect

TABLE 1.1

Material	Application
Silicon	Electronics, biosensors, solar cells
Germanium	Electronics, IR detectors
Tin oxide	Chemical sensors
Indium oxide	Chemical sensors, biosensors
Indium tin oxide	Transparent conductive film in display electrodes, solar cells, organic light-emitting diodes (LEDs)
Zinc oxide	UV lasers, photodetectors, UV LEDs, field emission device
Vanadium oxide	Chemical sensor
Gallium nitride	High-temperature electronics, UV detectors and lasers, automotive electronics, sensors
Boron nitride	Insulator
Gallium arsenide	Electronics
Indium phosphide	Electronics, optoelectronics, lasers
Zinc selenide	Photonics (Q-switch, blue green laser diode, photodetectors)
Indium selenide	Phase change memory device
Germanium telluride	Phase change memory device
Cadmium telluride	Solar cells
Copper, tungsten	Interconnects
Gold, silver	Biosensors

on electron transport, band structure, and optical properties. In electronics, their one dimensionality offers the lowest dimension transport channel for the best field effect transistor (FET) scalability. The interesting properties of various nanowires make them better candidates for a wide range of applications as seen in Table 1.1. It is also possible to conceive of systems wherein various nanowires can be employed for different functions as shown in Figure 1.1: silicon or germanium nanowires for processors, phase change nanowires for memory, oxide nanowires for sensing, thermoelectric nanowires for power generation, III–V nanowires for optoelectronic components, and some examples to build integrated systems.

Indeed, integration of various functional components with logic and memory has been the focus of "more than Moore" efforts proposed recently [6,7].

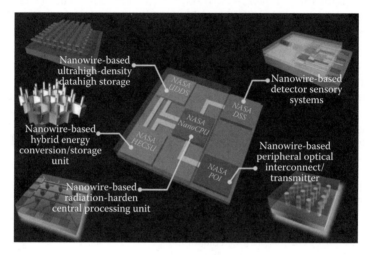

FIGURE 1.1
Future for inorganic nanowires. (Courtesy of H.T. Ng.)

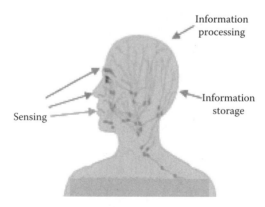

FIGURE 1.2
Integration of sensing with memory and logic. (Courtesy of B. Yu.)

The vision in Figure 1.1 is consistent with such integrated system development. We are already witnessing increasing incorporation of rich functions in electronic systems: miniature camera modules, global positioning systems (GPS), accelerometers, biometric identification, health monitoring systems, etc. One type of integration gaining attention is various types of sensors along with the brain provided by the microprocessor and memory subsystems [6] as depicted in the cartoon in Figure 1.2. As functional components in health, security, education, entertainment, power generation, and others find their way into integrated system, the INWs covered in this book not only provide advantages as nanoscale materials in terms of better performance but also rich diversity as in Figure 1.1.

References

1 M.C. Roco, R.S. Williams, and P. Alvisatos (eds.), *Nanotechnology Research Directions*, Kluwer Academic Publications, Dordrecht, the Netherlands (2000).

2 www.nano.gov

3 M. Reibold, P. Paufler, A.A. Levin, W. Kochmann, N. Patzke, and D.C. Meyer, *Nature*, 444, 286 (2006).

4 K.J. Klabunde (ed.), *Nanoscale Materials in Chemistry*, Wiley-Interscience, New York (2001), see Chapter 1.

5 O. Koper and S. Winecki, Chapter 8 in Ref. [4].

6 G.Q. Zhang, F. van Roosmalen, and M. Graef, The Paradigm of "More than Moore," *Proceedings of the ICEPT*, Shenzhen, China (2005).

7 G.Q. Zhang, Strategic Research Agenda of "More than Moore," Proceedings of the 7th International Conference on Thermal, Mechanical & Multi-Physics Simulation and Experiments in Micro-Electronics and Micro-Systems, ISBN: 1–4244–0275–1, Como, Italy, April 23–26, 2006, pp. 4–10.

2

Historical Perspective

The predecessor to the nanowires is the whiskers of metallic and nonmetallic single crystals, which received that name due to their appearance. One-dimensional anisotropic crystals or whiskers were found in natural ores in the 1500s. Growth of silver whiskers from rocks in the 1600s was thought to be similar to the growth of grass [1,2]. In the mid-twentieth century, intensive research work at the Bell Telephone Laboratories led to the laboratory preparation of whiskers of several materials. Early studies included the preparation and characterization of B, B_4C, C, Al_2O_3, SiC, and Si_3N_4; these whiskers were micron-sized (width or diameter) and exhibited high strengths of the order of 10^6 pounds per square inch (psi) [2]. The primary motivation to study them was their potential as a reinforcing additive for metals and plastics. Preparation of silicon whiskers and the vapor–liquid–solid (VLS) concept were first introduced by R.S. Wagner in a series of publications in the 1960s [3–6]. Since then, in the same decade, VLS technique was used to grow several other materials such as Al_2O_3, boron, GaAs, GaP, $GaAs_{1-x}P_x$, Ge, MgO, $NiBr_2$, NiO, selenium, and SiC as cataloged in Ref. [7].

In 1964, Wagner and Ellis [3] from the Bell Telephone Laboratories first proposed the VLS mechanism in a very brief seminal paper. Until then, one-dimensional whiskers crystal growth was explained by the so-called Frank theory [8], which suggested that the growth is aided by a structural defect. A screw dislocation present at the growth interface continuously adds new layers with growth occurring even at relatively low supersaturation levels and without the need for two-dimensional nucleation. Indeed, prior to Wagner's work, there were several studies on whisker growth of various materials, which used Frank's model to explain the observations [9–14]. The growth of mercury whiskers in a field emission tube by an evaporative technique was explained on the basis of the above growth model [11]. Johnson and Amick [12] reported the synthesis of silicon whiskers using diluted $SiCl_4$ with argon or H_2 [12], which also used screw dislocation theory to support the data. Carbon filaments were grown using thermal dissociation of hydrocarbon gases over Fe or Ni catalyst particles [14]. Analyzing a series of growth data from preceding years and new data on silicon growth, Wagner and Ellis [3] stated "...three important facts emerged: (a) silicon whiskers do not contain an axial screw dislocation; (b) an impurity is essential for whisker growth; (c) a small globule is present at the tip of the whisker during the growth." What they called "impurity" above is commonly known as catalyst in the nanowire literature, although a discussion is given later in Chapter 6.4,

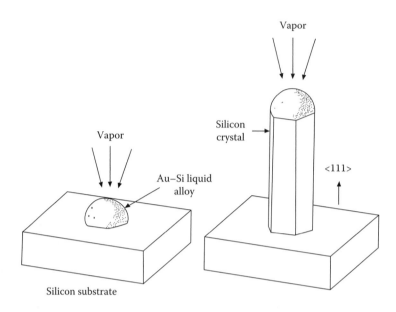

FIGURE 2.1
A schematic illustrating VLS mechanism as proposed by R.S. Wagner. (Reproduced from Levitt, A.P. (Ed.), VLS mechanism of crystal growth, in *Whisker Technology*, John Wiley & Sons, New York, 1970, 47–119. With permission.)

suggesting that "catalyst" is a misleading term in this context. Item (c) above is the logic commonly invoked to prove VLS mechanism in the nanowire literature, as will be seen throughout this book.

In the VLS mechanism, Wagner and Ellis suggested that the impurity forms a liquid alloy droplet and becomes a preferred site for deposition from the wafer (see Figure 2.1). Growth proceeds in two steps according to this mechanism. The first step involves the vapor–liquid system wherein material transfer from the vapor phase occurs directly into the liquid droplet. This causes the liquid alloy droplet to become supersaturated with silicon. The second step occurs in the liquid–solid system through the precipitation of material from the alloy droplet at the liquid–solid interface. It is clear from the preceding discussion that the catalyst (or impurity as Wagner calls it) plays a pivotal role in the VLS mechanism and a guidance to making the choice was presented by Wagner [7] in a list of eight criteria. (1) A good catalyst must form an alloy or liquid solution with the intended crystalline material to be grown. (2) The distribution coefficient of the catalyst given by $K_d = C_s/C_l$ must be less than 1, where C_s and C_l are solubilities in the solid and liquid respectively. (3) The equilibrium vapor pressure of the catalyst over the liquid alloy must be small since evaporation would otherwise change the droplet volume, leading to a change in whisker size midcourse. (4) The catalyst must be inert to the products of the chemical reactions generating the vapor source. (5) The interfacial energies of vapor–solid, vapor–liquid, and liquid–solid systems under consideration and the resulting wetting charac-teristics define the shape of the growing crystal. A large contact angle is nec-essary for the formation of whisker. (6) In growing whiskers of compound materials, the catalyst can be the excess of one of the constituent elements. For example, Ga can serve as a catalyst in growing GaAs. (7) The choice of the catalyst and growth temperature must be made to avoid the possibility of the catalyst forming a solid intermediate phase with one of the constituents of the vapor. (8) The solid–liquid interface must have well-defined crystal-lographic features in order to obtain unidirectional VLS growth.

The formation of liquid catalyst beads controls the diameter and the morphology of the whiskers. Wagner suggests [7] that there is a minimum whisker diameter as determined by the stability of a liquid droplet in its own vapor:

$$r_{min} = \frac{2\sigma_{LV}V_L}{RT \, l_n\sigma} \tag{2.1}$$

where

r_{min} is the minimum critical whisker radius
σ_{LV} is the liquid–vapor interfacial energy
V_L is the liquid molar volume
T is the growth temperature
σ is the degree of saturation given by

$$\sigma = \left(P - P_{eq}\right)/P_{eq} \tag{2.2}$$

where

P is the vapor pressure
P_{eq} is the equilibrium vapor pressure

In their early works, Wagner and coworkers [3–6] considered the growth of silicon whiskers using $SiCl_4 + H_2$. A schematic of their apparatus from Ref. [7] is reproduced in Figure 2.2. It is interesting to note that all the laboratory reactors for VLS growth in the nanowire literature since then are remarkably similar to Wagner's setup. A mixture of $SiCl_4$ and H_2 was introduced into the reaction tube heated by the furnace. The hydrogen flow was $1000\,cm^3/min$ and $SiCl_4/H_2$ ratio was 2%. A small particle of gold was placed on the {111} surface of a silicon wafer and heated to 950°C. The grown samples had a wide range of cross sections from 100 nm whiskers to 0.2 mm needles. The growth direction was <111> and the side faces of the whisker were usually {211} but occasionally {211} and {110}. They also used Pt, Ag, Pd, Cu, and Ni as catalysts instead of gold and obtained similar results. In contrast, the use of Zn, C,

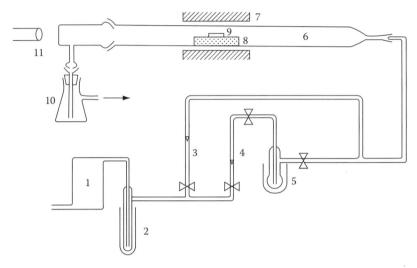

FIGURE 2.2
Growth apparatus used by Wagner in his pioneering work on the production of silicon whiskers by VLS technique. (Reproduced from Levitt, A.P. (Ed.), VLS mechanism of crystal growth, in *Whisker Technology*, John Wiley & Sons, New York, 1970, 47–119. With permission.)

Mn, Sn, and Ge did not produce whiskers, instead yielded polycrystalline nodules or films. Figure 2.3 shows growth versus temperature characteristics for gold and Pt catalysts along with growth on a bare silicon wafer. The growth on the bare silicon wafer is much smaller than that on catalysts; for example at 950°C, gold catalyst yields a growth rate 40 times higher than that obtained without the catalyst.

Though Wagner [3] originally advocated VLS growth as a suitable means to create P–N junctions and heterojunctions, he also proposed a technique for the mass production of whiskers [7]. The apparatus in Figure 2.4 has a feeder holding small particles of gold or platinum, which are introduced at the top of the reaction tube. The gas mixture of $SiCl_4/H_2$ is also introduced near the top of the tube. An extended section of the reaction tube is maintained at the growth temperature and the whiskers are collected at the bottom. Wagner suggested that growth rate can be adjusted to be several cm/s for catalyst particles smaller than 100 nm.

Since the pioneering work of Wagner, the VLS mechanism was used to explain observations on the moon [15] and in nature [16]. Carter [15] postulated that VLS-type growth probably occurred on the lunar surface, based on the analysis of lunar rock, which showed metallic iron stalks (0.015–0.15 μm diameter) with bulbous tips consisting of a mixture of iron and sulfur. Finkelman et al. [16] also invoked a VLS mechanism when they observed ~100 μm long rods of germanium sulfide capped by bulbs depleted in germanium in condensates from gases released by burning coal in culm banks.

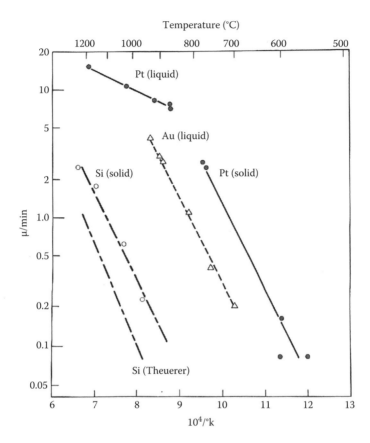

FIGURE 2.3
Deposition rate versus temperature for bare silicon wafer and catalyzed growth with Au and Pt. (Reproduced from Levitt, A.P. (Ed.), VLS mechanism of crystal growth, in *Whisker Technology*, John Wiley & Sons, New York, 1970, 47–119. With permission.)

In the early 1970s, E.I. Givargizov and his coworkers from the Academy of Sciences in the former Soviet Union published several papers [17–20] on the VLS-grown silicon whiskers, particularly discussing kinetic coefficients and rate-determining steps. Givargizov and Sheftal [17] conducted silicon whisker growth experiments with 50–100 nm thick films of Au, Ga, or In as catalysts. The whisker growth was in the [111] direction as in the case of Wagner's work. Ga and In yielded conical or pyramidal form of whiskers whereas gold produced cylindrical or prismatic form of whiskers. They postulated that the conical form in the case of Ga is due to the large distribution coefficient of gallium in silicon and also due to the removal of gallium from the liquid drop by the gaseous reaction products. The typical growth rates in their work were 1–3 μm/min, resulting in 10–60 μm whiskers over 10–20 min. In all of Givergizov's studies, most whiskers (~95%) grew perpendicular to the (111)

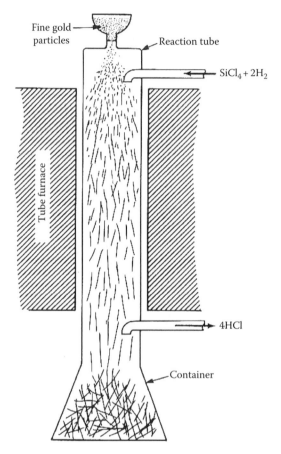

FIGURE 2.4
Apparatus proposed by Wagner for the mass production of silicon whiskers. (Reproduced from Levitt, A.P. (Ed.), VLS mechanism of crystal growth, in *Whisker Technology*, John Wiley & Sons, New York, 1970, 47–119. With permission.)

silicon substrate. They found a strong correlation between the uniformity in diameter and in height: the thicker whiskers grew more rapidly than the thin ones [18,20].

Givargizov et al. described the VLS growth as a sequence of four steps (see Figure 2.5):

1. Mass transfer in the gas phase
2. Chemical reaction on the vapor–liquid interface producing silicon atoms
3. Diffusion of silicon atoms in the liquid phase across the droplet
4. Incorporation of silicon into the crystal lattice

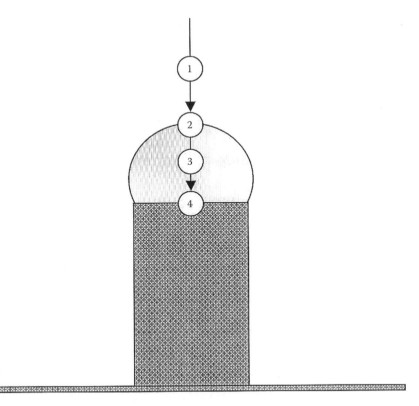

FIGURE 2.5
VLS process steps as described by Givargizov. (From Givargizov, E.I., *J. Cryst. Growth*, 31, 20, 1975.)

Step 3 can be immediately dismissed as rate-determining step since it would mean slower growth for thicker wires, which have thicker catalyst beads. The observed silicon whisker growth showed a strong temperature dependence of the kinetic coefficient with an activation energy of 48 kcal/mol [20]. This eliminates mass transport in step 1 as the critical step. Givargizov argues that the ratio of the liquid–vapor interfacial area to the liquid–solid interfacial area is nearly the same for whiskers of all diameters in his micrographs. Diffusion through the droplet in step 3 has already been eliminated. Then, if step 2 were the rate limiting, all whiskers—regardless of diameter—should have the same height, which is not the case. Thus, step 2 is also eliminated, and he concludes step 4 as rate determining with additional evidence that the kinetic coefficients for the polar faces of GaAs are quite different. Also, for GaAs whiskers on GaAs (111) substrate, most grew perpendicular to the substrate with some inclined to the substrate. The thin inclined whiskers, which are oriented in the <211> direction, grew faster than the thick perpendicular ones [20].

Both Wagner and Givargizov gave ample evidence that the growth temperature determines the growth rate, diameter, and stability. The whisker growth rate in their experiment increased with temperature. Note that direct vapor–solid deposition also increases with temperature. Givargizov [20] fitted his data on growth rate versus temperature to

$$V = b\frac{(\Delta\mu)^2}{kT} \tag{2.3}$$

where
V is the growth rate
$\Delta\mu$ is the effective chemical potential
T is the temperature
b is a constant

The strong decrease in growth rate (proportional to $1/T^2$) is attributed to the increase in both solubility and vapor pressure with temperature. The temperature also affects the stability of whiskers [7,20,21] in addition to the growth rate. Instabilities lead to whiskers with defects such as kinks and branches and can be attributed to high temperature gradients and unstable temperatures. These issues are also amplified by high concentrations of the source vapor.

In summary, the above studies were the pioneering contributions on whiskers that led to the nanowire research activities of the last decade. Even during the early days, it was understood that thinner films yielded more uniform catalyst bead diameters [18]. Givargizov used approximately 100 nm gold films to obtain the silicon whiskers. Much thinner films yield nanowires as we know today. Westwater et al. [22] appear to be the first to discuss nanoscale whiskers in 1995. They used a 5 nm Au film and added another innovation, changing the source gas to silane instead of silicon tetrachloride. The thinnest whiskers in their work were 10 nm wide. A few years later, this group used a 0.6 nm Au thin film and studied the stability of silicon nanowires [23] (see discussion related to Figure 6.7 in Section 6.2.2) as a function of silane concentration, temperature, and nanowire diameter. Ozaki et al. [24] also used a very thin film of gold around the same time and reported nanowhiskers of diameters 6–30 nm using silane as source gas. This is the period that the transition from large whiskers to truly nanoscale wires started to occur, leading to an explosion of activities in the growth of not only silicon and germanium nanowires but a whole variety of inorganic materials covered in this book.

References

1. A.P. Levitt (ed.), *Whisker Technology*, John Wiley & Sons, New York (1970).
2. A.P. Levitt, see Chapter 1 in Ref. [1], pp. 1–13.
3. R.S. Wagner and W.C. Ellis, *Appl. Phys. Lett.* 4, 89 (1964).
4. R.S. Wagner, W.C. Ellis, K.A. Jackson, and S.M. Arnold, *J. Appl. Phys.* 35, 2993 (1964).
5. R.S. Wagner and W.C. Ellis, *Trans. Met. Soc. AIME* 233, 1057 (1965).
6. R.S. Wagner, *J. Appl. Phys.* 38, 1554 (1967).
7. R.S. Wagner, "VLS mechanism of crystal growth," in Ref. [1], pp. 47–119.
8. F.C. Frank, *Discussions Faraday Soc.* 5, 48 (1949).
9. W.W. Webb and W.D. Forgeng, *J. Appl. Phys.* 28, 1449 (1957).
10. J.M. Blakely and K.A. Jackson, *J. Chem. Phys.* 37, 428 (1962).
11. R. Gomer, *J. Chem. Phys.* 28, 457 (1958).
12. E.R. Johnson and J.A. Amick, *J. Appl. Phys.* 25, 1204 (1954).
13. E.S. Greiner, J.A. Gutowski, and W.C. Ellis, *J. Appl. Phys.* 32, 2489 (1961).
14. R. Bacon, *J. Appl. Phys.* 31, 283 (1960).
15. J.L. Carter, *Science* 181, 841 (1973).
16. R.B. Finkelman, R.R. Larson, and E.J. Dowrnik, *J. Cryst. Growth* 22, 159 (1974).
17. E.I. Givargizov and N.N. Sheftal, *J. Cryst. Growth* 9, 326 (1971).
18. E.I. Givargizov and Y.G. Kostyuk, *Rost Krystallov (Crystal Growth)* 9, 276 (1972).
19. E.I. Givargizov, *J. Cryst. Growth* 20, 217 (1973).
20. E.I. Givargizov, *J. Cryst. Growth* 31, 20 (1975).
21. A. Mao, Masters thesis, San Jose State University (2004).
22. J. Westwater, D.P. Gosain, K. Yamauchi, and S. Usui, *Mater. Lett.* 24, 109 (1995).
23. J. Westwater, D.P. Gosain, S. Tomiya, S. Usui, and H. Ruda, *J. Vac. Sci. Technol. B* 15, 554 (1997).
24. N. Ozaki, Y. Ohno, and S. Takeda, *Appl. Phys. Lett.* 73, 3700 (1998).

3

Growth Techniques

3.1 Introduction

Nanowires are structures with at least one of their dimensions in the 1–100 nm range; typically they are several microns in length with diameter under 100 nm. Growth processes must be able to preferentially support growth in one dimension. The key requirements for growing nanowires are that there must be a reversible pathway or condition near equilibrium between a fluid phase such as solution, melt, or vapor and a solid phase, and also the adatoms in solid phase should have high surface or bulk mobility [1]. Nanowires have been synthesized using a variety of techniques and the present chapter aims at explaining the approaches in detail.

The nanowire synthesis methods can be divided into two broad categories: liquid- or solution-based techniques and vapor-phase techniques. The vapor-phase techniques include methods such as chemical vapor deposition (CVD) using catalyst metals, chemical vapor transport, reactive vapor transport, laser ablation, carbothermal reduction, chemical beam epitaxy (CBE), thermal evaporation and thermal decomposition, and plasma- and current-induced methods. Liquid- or solution-based techniques include approaches such as sol-gel synthesis, hydrothermal processes, and electrodeposition. Many liquid-phase techniques utilize templates for producing one-dimensional (1-D) materials though; techniques without the use of templates are being developed rapidly. This chapter provides a description of growth apparatus for these techniques along with source material, growth conditions, substrate preparation, catalyst preparation, and postgrowth processing. Bulk production approaches for nanowires are also covered.

3.2 Liquid-Phase Techniques

There have been a number of approaches for nanowire synthesis using liquid phase. These can be divided broadly into two categories: template-based and template-free approaches.

3.2.1 Template-Based Methods

Templates provide 1-D channels for guiding the growth or deposition of materials in 1-D form. There are three primary types of templates: positive template, negative template, and surface step templates. The negative template method is the most commonly used among the three. There are also three main deposition methods using these templates: electrodeposition, sol-gel, and CVD. Other less popular techniques involving templates include chemical precipitation or polymerization reactions and electroless deposition.

3.2.1.1 Template Preparation

Positive template approach: In a positive template approach, 1-D nanostructures such as carbon nanotubes (CNTs), DNA, and polymer are used as templates to deposit the material for nanowire production. This is schematically shown in Figure 3.1a. CNTs are commonly synthesized by CVD over a heated substrate coated with a metal catalyst at high temperatures (~900°C) using a carbon source gas such as methane. CNTs provide an ideal confined platform for chemical reactions. In principle, it is possible to obtain nanowires of any material by simply depositing it (using sputter coating, electrodeposition,

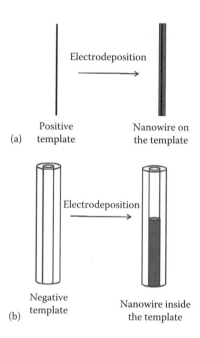

FIGURE 3.1
A simple schematic showing nanowire electrodeposition techniques based on (a) positive and (b) negative templates.

or thermal treatment) onto CNT. However, this is not possible for all types of materials because of weak metal–carbon interaction. Physical vapor deposition results in discontinuous structures on the CNT after deposition. However, titanium has strong interactions with the CNT and hence it can be used as a buffer layer to increase the adhesion for obtaining nanowires of other metals such as Au, Pd, Fe, Al, Pb, etc. [2].

DNA is also a preferable choice for positive template synthesis because of its small diameter (2 nm, which is the width of linear polynucleotide chain) and high specificity for molecular binding. DNA consists of double polymer in a helical structure with each polymer made up of repeating units called nucleotides, with a backbone structure made of sugar and phosphate groups. The aromatic bases attached to the sugar group in the backbone encode the DNA. Both phosphate and aromatic bases can serve as binding sites for different compounds. Metallic cations or cationic nanoparticles can bind to phosphate anionic site by electrostatic interaction whereas various transition metals such as Pt(II), Pd(II) form coordination complexes with the nitrogen atom of the aromatic bases. Its length can be suitably controlled from nanometers to microns by established molecular science techniques such as polymerase chain reaction (PCR), enzymatic digestion, DNA ligation, etc. It can also be modified with different functional groups, which allow specific surface attachment. DNA molecules, which are usually coiled and randomly structured at room temperature, must be stretched to serve as nanowire template. This is usually done by techniques such as molecular combing, electrophoretic stretching, and hydrodynamic stretching. The common strategy employed is to tether to a surface (DNA end binds to vinyl group on the surface) at one end and to stretch the molecule by an external force such as surface tension (as in molecular combing), electric field (as in electrophoretic stretching), and shear force generated by momentum gradient (as in hydrodynamic stretching). However, the stretching force must be smaller than the force needed to break the covalent bonds. A variety of nanowires have been assembled on to DNA strands using both types of binding [3].

Finally polymer chains such as poly-2-vinylpyridine (P2VP) have been used as a template for the nanowire formation. Since these polymers are thinner than DNA, it is possible to fabricate thinner nanowires.

Negative template approach: In a negative template approach, a prefabricated hollow cylindrical structure is used to deposit the material of interest inside the hollow pores in the 1-D form. This is shown schematically in Figure 3.1b. Potential templates include nuclear track-etched membranes (TEMs), anodic or anodized alumina membranes (AAMs), diblock copolymer, nanopore arrays in glass, mesoporous silica, and other porous or hollow structures.

A TEM is prepared by passing a heavy charged particle through membranes (~10–20 μm thick) of mica, glass, or plastics such as polycarbonate, polyester, polyethylene terephthalate (PET) as shown in Figure 3.2a (as suggested in Ref. [4]), which leaves a damaged hollow path (track with

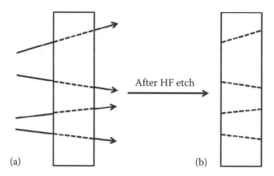

FIGURE 3.2
A nuclear track-etched membrane: (a) heavy nuclei passed through mica; and (b) track developed after HF etching.

minimum width of 2.5 nm) in the membrane. The track is further etched chemically. TEMs thus have pores that traverse the membrane as cylindrical channels. Plastic membranes are preferred over mica and glass because they are easy to dissolve and can be tailored to varied applications. Ion beam accelerators are used to generate the high-energy particles for preparing the TEMs. The track etch occurs by a phenomenon called coulomb explosion. When the material, placed in a scatter chamber and attached to a rotatable clamp, is irradiated with heavy charged particles such as Ni, U at high energy (about 30 MeV), the binding outer electrons in the target atoms of the membranes in the beam path are knocked off. The resulting charged constituents in the target atoms separate due to columbic repulsion to produce the etchable damaged track. Due to innovative techniques available now to create high-energy particles such as cyclotron (100 MeV), it is possible to process thicker materials (~100 μm thick) and also produce uniform-sized tracks.

The etching process involves passing the tracked membrane through HF or NaOH as shown in Figure 3.2b. HF is used in mica whereas an alkaline solvent like NaOH is used for etching plastics. During etching, the etchant reacts with the broken polymer chain to create cylindrical channels. The pore diameter can be controlled by controlling the etch time, temperature, and etchant concentration. An increase in the etch time increases the pore diameter whereas pore density is directly related to the irradiation dosage. However, the pore size and density are independent of each other. The typical pore sizes range from 20 nm to 15 μm, and pore density from 10^4 to 10^9 pores/cm^2.

AAM is another commonly used negative template, produced by well-known anodization process. Anodization involves placing an Al foil in a chemical acid bath (electrolytic solution) under a direct current which develops a self-assembled pore structure as shown in Figure 3.3a. The Al sheet becomes the anode whereas acidic solution becomes the cathode. During anodization an oxide film forms at the anode,

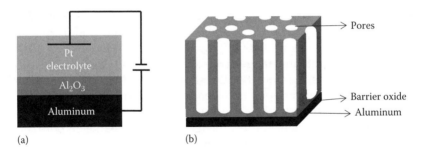

FIGURE 3.3
(See color insert following page 240.) Simplified schematics illustrating (a) an anodization process and (b) a nanoporous alumina channel.

$$2Al(s) + 3H_2O(l) = Al_2O_3(s) + 6H^+ + 6e^- \tag{3.1}$$

while hydrogen evolves at the cathode (Pt)

$$6H^+ + 6e^- = 3H_2(g) \tag{3.2}$$

The electrode potential at the anode is given by

$$E = -1.55 - 0.059\,pH \tag{3.3}$$

Thus, the reaction at the anode is determined by the pH, which in turn depends on the electrolyte and the temperature. As the anodization progresses, the growing oxide layer is also dissolved by the acid (reverse reaction of Reaction 3.1), leading to formation of pores as shown in Figure 3.3b. Thus the acid action is balanced by oxidation rate to form microscopic pores, 10–150 nm in diameter and lengths up to several microns. The bottom of each pore also consists of thin (10–100 nm thick) oxide barrier layer over the metallic Al surface. Also pores are formed only when Al anodization is done under acidic conditions (pH < 5) using acids such as H_2SO_4, $H_2C_2O_4$, and H_3PO_4. A flat nonporous barrier oxide forms under neutral or basic conditions (pH > 5). Usually porous oxide is produced by concentrated solutions (pH < 5) at higher temperatures (60°C–75°C) with lower voltages and currents, however, opposite conditions favor flat nonporous barrier oxides.

The creation of pores can be explained as follows. At the start of anodization, a thin oxide barrier film is formed first. However, due to imperfections in the native oxide surface, the electric field is higher at regions of thinner oxide (since oxide is nonconducting) and vice versa. The pore nucleation starts where the oxide is thinner and propagates due to the combined effect of enhanced electric field and local acid-enhanced partial oxide dissolution at these locations. It is possible to control the pore size and spacing by controlling anodization voltage and choosing the electrolyte. The pore diameter and pore spacing are linearly proportional to the applied potential with proportionality constants of 1.3 and 2.5 nm/V respectively, regardless of the electrolyte used [5]. The applied potential is determined by the type of electrolyte and its pH. A low potential range of 5–40 V for H_2SO_4, a medium potential range of 30–120 V for H_3PO_4, and a high potential range of 80–200 V for $H_2C_2O_4$ are used. Also lower the pH, the lower is the potential threshold for field-enhanced dissolution at the pore tip. So the variables involved in anodization process are applied voltage, anodizing time, current density, electrolyte and its pH, and solution temperature. These variables affect the resulting pore diameter, pore separation or pore density, pore length, or film thickness.

Most of the anodization is done in a one-step process, which gives a high pore density (10^{11} pores/cm²). The use of high-purity (99.999%) Al films produce well-ordered pores. The Al films are preannealed (500°C, 3 h), electropolished for 1–4 min to reduce the surface roughness and obtain a mirror smooth finish and then placed in a bath that is agitated to remove H_2 bubbles and local heating on the surface. Annealing removes any mechanical stress and enhances the grain size. However, one-step anodization does not produce long-range, highly dense pores, and therefore, a two-step anodization has been developed recently. As illustrated in Figure 3.4a and b, the first step involves a lengthy (1–2 days) anodization process after the pretreatment steps mentioned above are performed. The alumina pores formed are subsequently dissolved in chromic acid (H_2CrO_4) to obtain patterned hexagonal arrays of concave structure at the interface between the aluminum substrate and the porous alumina layer grown in the first step as shown in Figure 3.4c. These concave structures are used as initial sites to obtain highly ordered nanopore array in the second anodization process (Figure 3.4d). Unique structural properties of AAM, such as uniform pore size distribution, controllable pore diameter, high pore density (~10^{11} pores/cm²), and cylindrical shape of pores make them ideally suited for deposition of materials in 1-D format.

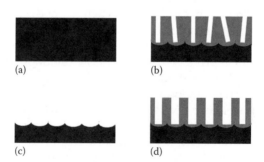

(a)   (b)

(c)   (d)

FIGURE 3.4

Schematic of stages in a two-step anodization process: (a) a fresh aluminum foil after pretreatment steps; (b) the aluminum foil with pores obtained after first anodization; (c) the aluminum foil with dissolved pores in chromic acid (H_2CrO_4) showing patterned hexagonal arrays of concave structure; and (d) the aluminum foil with highly ordered nanopore array using the above concave structures in (c) as initial sites in the second anodization process.

A diblock copolymer has two long chemically different polymer chains joined covalently in the middle. Each part of this polymer is composed of a monomer or block, which is incompatible with the block forming the other half of the polymer chain. Because of the inherent incompatibility of the blocks, they self-assemble in microphases with nanometer-scale domain sizes. The chain length and volume fraction of the block determine the ordered morphologies at equilibrium, such as sphere, cylinder, and lamellae. Of these, cylindrical morphology is preferred for producing nanowires along with the electrodeposition method.

Of the three most common negative templates described above, diblock copolymer and AAM are better in terms of packing density and size uniformity of nanowires. TEMs have lower pore density (10^9 pores/cm^2) and pores are randomly positioned and oriented, thus resulting in low yield. The lateral nanowire density obtained by diblock copolymer and AAM are comparable. However, the AAM template preparation requires strong acid etching and is also hampered by an insulating oxide barrier at the bottom of nanopores during electrodeposition.

Surface step template: The third type of template is the open planar surface, composed of stacked planes, which is highly oriented. Graphite, mica, molybdenum disulfide, and others are few examples of such materials that have stronger forces between the lateral planes than between the planes, thus giving rise to characteristic cleaving property. For example, highly oriented pyrolytic graphite (HOPG) is used as a surface template because it can be easily cleaved to expose a fresh conductive, smooth surface. It is easy to create steps between the two planes of a HOPG, and the atomic scale step edges can be used as a template to grow nanowires.

3.2.1.2 Deposition Methods

There are various methods for deposition of materials inside, outside, or on the templates for creating materials of choice in 1-D form. These methods include electrodeposition, sol-gel, CVD, chemical precipitation, or polymerization reactions, electroless deposition, etc. Each deposition method works better with a type of template for a certain type of material system.

Electrochemical methods: In electrochemical template synthesis approaches, the working electrode (cathode, where deposition is done) consists of a prefabricated template or an electroactive support for growing nanowires and the electrodeposition is carried out in a standard three-electrode electrochemical cell with a counterelectrode and a reference electrode. The cathode or working electrode is connected to the negative terminal of an external direct current power supply and anode to the positive terminal. Both these electrodes are immersed in the electrolyte that contains the metal ions to be deposited on the cathode during the experiment. The deposition starts from the base of the electrode resulting in well-defined structures through the templates as nanowires. The variables controlling the rate of electrodeposition are applied potential, current density, anions and cations in the solution, electrolyte concentration, temperature, and the nature of the electrode property (material, geometry, surface area, and surface condition).

Nanowires can be formed in a single/direct electrodeposition step or multiple/indirect steps depending on the material system. In the direct approach, the desired material is electrodeposited whereas postprocessing of electrodeposited layers is used in the indirect approach to obtain the desired materials inside the template. For example, ZnO nanowire arrays can be electrodeposited in a single step on ordered AAMs [6]. In contrast, SnO_2 nanowires are produced using a multistep method after thermally annealing the as-deposited Sn metal in AAM [7].

Electrodeposition can be performed on almost all types of templates described above. Electrodeposition on CNTs (diameter 20–40 nm, length 0.5–500 μm) has been performed recently to obtain cobalt hexacyanoferrate (CoHCF) nanowires [8]. First, the multiwalled carbon nanotubes (MWNTs) were shortened and functionalized by ultrasonication in a mixture of H_2SO_4 and HNO_3 (3:1 by volume) for 8 h followed by washing with water and drying at 60°C for 12 h. This results in attachment of carboxylic groups on the surface of the MWNTs. The indium tin oxide (ITO) substrates were also functionalized by dipping in a 4% polydiallyldimethylammonium chloride (PDDA) solution for 30 min. The carboxylic group–functionalized MWNTs can attach to PDDA functional groups forming a monolayer covalently attached on ITO substrates. The monolayer was used as a template for the electrodeposition to obtain CoHCF nanowires. The electrolyte was a mixed solution containing 2 mL 0.5 M KCl, 20 μL 0.1 M $CoCl_2$, and 10 μL 0.1 M $K_3Fe(CN)_6$. The applied potential was scanned from 0 to 1.1 V at a rate of 100 mV/s for 15 cycles. The CoHCF nanowires from this process were 70–100 nm in diameter and larger than the diameter of the MWNTs.

Electrodeposition on DNA is not practical; so electroless deposition is typically implemented. Electroless deposition on DNA is described toward the end of this section. It is also possible to implement electrodeposition onto DNA if it is attached to an ITO substrate.

The most widely used electrodeposition methods involve the negative templates such as TEMs, AAMs, and diblock copolymers. First set of experiments using the mica-etched track (by nuclear radiation) were done to produce 40 nm diameter nanowires of Sn, Zn, and In [9]. The mica hole density was $10^4/cm^2$ and mica was etched in 20% HF at 20°C for 1100 s to obtain a 100 nm diameter hole. Then, mica was mounted with epoxy on a glass and its bottom side was covered with an evaporated film of the same metal to be plated. The electrolyte for Sn deposition was a mixture of $SnSO_4$, H_2SO_4, tartaric acid, gelatin, and β-naphthol. The Sn electrodeposition was carried out by applying 0.7 V through the electrolyte, which also contained mica as cathode and tin as anode. This approach produced 100 nm diameter Sn nanowires. $In_2(SO_4)_3$ and $ZnCl_2$ were used as electrolyte to grow 70 nm diameter In and Zn nanowires under almost similar conditions.

SnO$_2$ nanowires were also deposited in a track-etched polycarbonate membrane with a thickness of 10 µm, 100 nm pore radius, and a pore density 6 × 10^8 pores/cm^2 [10]. Gold contact was made to the membrane and electrolyte was tin chloride, sodium nitrate, and nitric acid and the potential was varied from −0.3 to −0.5 V for a deposition time of 1 h. After deposition, gold was removed and membrane was dissolved in dichloromethane.

Electrodeposition is widely used to fill AAM templates to obtain nanowires. For example, cerium oxyhydroxide nanowires were prepared using AAM template [11]. Gold was sputtered on one side of the AAM to make it conducting and electrodeposition was carried out in ethyl alcohol 0.3 M CeCl$_3$·7H$_2$O solution using AAM/Au as working electrode, graphite as counterelectrode, and saturated calomel electrode (SCE) as reference electrode. The applied potential was varied from −3 V/SCE to −9 V/SCE. After the pores are filled with substrate, the aluminum film and the barrier layer can be removed to obtain free-standing porous alumina membrane. Al is removed by saturated HgCl$_2$ and barrier oxide by saturated solution of KOH in ethylene glycol. Similarly ZnO nanowires were obtained using AAM template in a 0.1 M zinc nitrate solution [6].

Diblock copolymer was also used to synthesize Co and Ni nanowires. Thurn et al. used polystyrene (PS) (0.71 volume fraction) and polymethylmethacrylate (PMMA) to self-assemble 14 nm diameter PMMA cylinders (Figure 3.5) hexagonally packed in a PS matrix with a lattice constant of

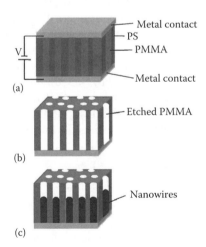

FIGURE 3.5
(See color insert following page 240.) Schematic showing the steps in nanowire electrodeposition using a template made using a diblock copolymer: (a) formation of vertically oriented hexagonal array of cylinders of diblock copolymer under the applied electric field and heating at temperatures above their glass transition temperatures; (b) removal of PMMA by UV exposure followed by rinsing in acetic acid to form a nanoporous structure; and (c) the resulting PS matrix with nanopores is filled by electrodeposition to form nanowires in a polymer matrix. (Recreated from Thurn-Albrecht, T. et al., *Science*, 290, 2126, 2000. With permission.)

24 nm in bulk [12]. PS matrix with nanopores can now be used for electrode-position to form nanowires in a polymer matrix. The polymer was spin-cast from toluene solutions onto a gold-coated silicon and annealed at 165°C for 14 h under an applied electric field to align the polymer parallel to the field lines (Figure 3.5a). The PMMA can be removed by UV exposure followed by rinsing in acetic acid to form a nanoporous (14 nm) structure (Figure 3.5b). The resulting PS matrix with nanopores is filled by electrodeposition to form nanowires in a polymer matrix. The electrolyte used was a Co salt with a standard three-electrode configuration and the template was made conductive by Au coating. The nanowires were grown from the base at the gold end at a rate of 1–10 nm/s when the current density was 30–300 A/m². The nanowire height was controlled by monitoring the total integrated current.

Apart from positive and negative templates, step edges on a highly oriented surface can also be used for electrodeposition. Electrodeposition on a single crystal surface often starts selectively at defect sites, such as the atomic step edges where the electric fields are high. These atomic scale step edges can be used as a template to grow nanowires. This method is also called as the step edge decoration. For example, molybdenum nanowires were obtained [13] by first electrodepositing MoOx on HOPG and then reducing it to form Mo nanowires in the presence of H_2 at 500°C for 1 h (Figure 3.6). The wire diameter reduced to about 30% but retained the shape. The nanowires were lifted off the HOPG surface by embedding in a polystyrene film. The electrolyte used was alkaline MoO_4^{2-} solution and a deposition potential of −0.7 to −0.9 V versus SCE was applied. It is evident from the above discussion on various types of templates that electrodeposition is one of the most widely used method to deposit conducting materials into, onto, or on the surface of templates. Using electrodeposition nanowires of various metals [13] (such as Au, Ag, Ni, Co, Pd, Pt, Cu), metal oxides [6,10] (such as SnO_2, ZnO), semiconductors [14] (such as CdS, Cu_xS), conducting polymers [15] (polypyrrole nanowires), metal alloys [16,17], and striped nanowires [18] (e.g., Au-Ag-Au) have been synthesized. Almost all kinds of conducting

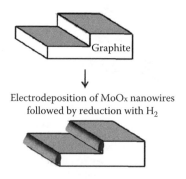

FIGURE 3.6

Schematic showing the step edge decoration method for the synthesis of nanowires by electrodeposition.

materials have been grown by electrodeposition to yield dense, continuous, and crystalline nanowires.

Semiconductor nanowires such as CdS were prepared by both direct and indirect electrodeposition methods. For example, direct deposition of CdS was done on AAM using an electrolyte containing Cd^{2+} and S in dimethyl sulfoxide [14]. The indirect approach involves first depositing Cd inside AAM channel and then sulfurizing the sample in a quartz tube furnace in Ar/H_2S (5%) atmosphere (pressure 0.1 MPa) at 500°C for 3 h [19]. Conducting polymer, such as polypyrrole (PPy), nanowires were deposited on AAM templates (pore diameter 200 nm) using 0.5 M solution of pyrrole and 1% sodium dodecyl sulfate (as a dopant) [15].

Metal alloy nanowires can also be fabricated by electrodeposition with ease. Ohgai et al. [16] deposited $Cd_{0.5}Te_{0.5}$ and $Ni_{0.8}Fe_{0.2}$ nanowires inside AAM by controlling the cathode potential and electrolyte composition during the electrodeposition. The CdTe bath contained both Cd^{2+} (0.1 M) and $HTeO_2^+$ (0.01 M) ions whereas NiFe bath contained Ni^{2+} (0.457 M) and Fe^{2+} (0.0215 M) ions. They performed a scan of cathodic potentials in these baths to determine the optimum conditions for $Cd_{0.5}Te_{0.5}$ and $Ni_{0.8}Fe_{0.2}$ deposition. The cathodic current was measured by varying the applied potential. Also the equilibrium potentials for Cd, Te, Fe, Ni were calculated based on the standard reduction potential values and the actual ion concentrations in the solution. A knowledge of the values of individual reduction potentials and from the cathodic polarization curve graph (current density versus potential) helps in determining the optimal ranges of applied potential. The optimal potential depends on the relative ionic concentrations in the solution, pH, and solution temperature. Once the potential range was known (−0.3 to −0.5 V for $Cd_{0.5}Te_{0.5}$ and −1.1 to −1.2 V for $Ni_{0.8}Fe_{0.2}$), a composition analysis of as-deposited nanowires using an energy dispersive x-ray (EDX) analyzer at different applied potentials was used to determine the optimum potential for the desired alloy composition. So it is possible to vary the potential to achieve the desired stoichiometry of as-deposited material. One can also vary the current density and solution composition at a fixed potential and obtain similar kind of results.

Another advantage of electrodeposition is the ability to control the nanowire length and hence the aspect ratio (wire diameter is limited by the pore diameter) by controlling the amount of charge passed (integral of applied current over time) at either a constant potential or constant current density. This can be used to obtain striped nanowires—which are nanowires with multiple segments of different metals—where one can change the plating solution at intervals after depositing each segment. For example, Au–Ag–Au striped nanowires (~750 nm per stripe, 8 stripes) and other such stripes were obtained using this approach [20] by changing the plating solution and by rinsing the material with distilled water in between two deposition steps. Alternatively, one can vary the applied potential in a solution containing all the metal ions constituting the nanowires. Only one ion will get deposited at a fixed potential, thus allowing the production of striped nanowires. This approach does not require changing the plating solution after deposition of each metal segment. These striped nanowires are potentially useful in photoconductivity and analytical chemistry as identification tags. Probably no method other than electrodeposition can fabricate such structures with a controlled composition. One can also incorporate metal–alloy–metal segments by combining the procedure for the metal alloys and striped metals.

The electrodeposition often requires the support of a conducting layer (a metal film) at the back of the membrane to serve as working electrode. In the case of large pore sizes, this film should be thick enough to support the membrane and seal the pores. Al at the back of the membrane is sufficient in most normal cases. Also once the full channel length is filled, the deposition must be stopped; otherwise, the material deposits on the membrane surface, forming hemispherical caps and further coalescing into a continuous film.

Electrodeposition is difficult to perform in very small diameter pores (few nanometres) as they are difficult to fill by a slow diffusion-driven transport through high-aspect-ratio structures. Several other factors such as capillary action, pore clogging, etc. also become issues in filling small pores. Performing deposition in an ultrasound bath can facilitate better mass transport of ions into pores. It is difficult to fill all the pores with DC electrodeposition and it requires a high potential (for the tunneling of electrons through the alumina barrier) and high current density. However, one can grow compact metal nanowires with AC electrodeposition at low current densities. Wire length can be increased at higher current density and multistep AC electrodeposition can fabricate long metal nanowires [21]. A pulsed electrodeposition method was also employed successfully for uniform deposition in the pores [22]. After each potential pulse, a relatively long delay follows, allowing the ions to diffuse into the region where they are depleted and hence uniform deposition can occur.

Electrodeposition is a low-temperature, inexpensive process and does not require reactors. It allows fabricating vertical, dense nanowire arrays with controlled size and stoichiometry. Almost all conducting materials can be fabricated with predefined sizes. However, the underlying substrate surface must be electrically conductive, which limits the range of materials that can be used as substrates. The applicability and scalability of this method is limited by the template itself. Also, the electrodeposition method is highly useful for metallic nanowires and has limited applicability for producing single-crystal nanowires of elemental and compound semiconducting materials systems. There are a number of other resources available on nanowire electrodeposition for further reading [4,23,24].

Sol-gel synthesis: The sol-gel template-based nanowire synthesis method relies on the capillary action (or electrophoresis) of the pores in the template to fill the pores with sol particles. In a simple process, as shown in Figure 3.7, the template is dipped directly into the relevant sol (containing the precursor

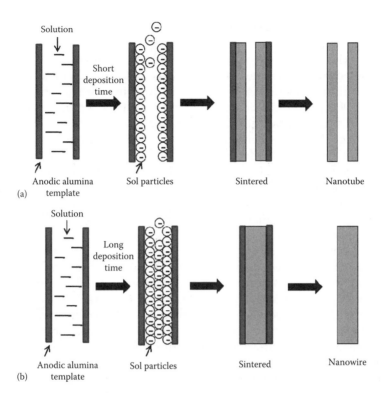

FIGURE 3.7
Schematic illustrating various stages involved in nanowire/nanotube formation using sol–gel synthesis within templates: (a) deposition over shorter time results in tubular structures; and (b) deposition over longer duration results in nanowire structures.

material) and heated for some time. Sols are the colloidal suspension of solid particles in a liquid dispersion. Sols are formed when the starting metal salt or precursor is processed through a series of reactions (e.g., hydrolysis) to form a colloidal suspension (the sol). Gels are the integrated networks (semisolid and liquid) comprising the sols along with the other liquid-phase solution ingredients formed by condensation reaction. Sol–gel process involves the transition of a system from a liquid sol to solid wet gel phase. The method is based on the hydrolysis and condensation reactions of precursors. The commonly used template is porous AAM. The pores of the template are filled slowly by compact stacking of the sol nanoparticles. If the deposition time is short, the particles deposit first on the walls, leaving voids in the center eventually making nanotubes (Figure 3.7a), whereas longer deposition time results in nanowires (Figure 3.7b). The final product is obtained after a thermal treatment to remove the gel. Alternatively sol-gel electrophoresis can also be used where the template pores are filled by movement of sols under an applied electric field.

For example, sol–gel method was employed to produce Eu_2O_3 nanotubes by filling the pore walls of AAM [25]. Initially, the AAM templates were dipped in the mixed solution containing europium nitrate and urea, which filled the pores of template. Upon heating at 80°C for 72 h, urea hydrolyses to form NH_4^+ and OH^- ions. The OH^- ions combine with europium ions to form $Eu[(OH)_x](H_2O)_y$ sol. Then the template was taken out from the sol and sintered in a tube furnace at 150°C for 1 h, changing the sol within the AAM pores into gel by the condensation reactions and gel was turned into Eu_2O_3 nanocrystals upon further sintering at 700°C for 10 h. The AAM template can then be removed to form Eu_2O_3 nanotubes. Another study demonstrated the fabrication of Co^{2+}-doped ZnO nanowire arrays using AAM templates [26].

Sol–gel method has been used to synthesize a variety of materials including compounds of transition and rare earth metals. Examples include metal oxides (e.g., ZnO, TiO_2, CeO_2, Cr_2O_3, WO_3, V_2O_5, ZrO_2, Eu_2O_3, etc.), complex metal oxides (e.g., $LiCo_{0.5}Mn_{0.5}O_2$, $CoFe_2O_4$, etc.), and semiconductors (e.g., CdS). Some of these compounds are very difficult to synthesize using other methods, especially compounds and complexes of rare earth metals (e.g., barium strontium titanate, $Ba_{0.8}Sr_{0.2}TiO_3$). However, sol–gel approach is viable as the corresponding salts or precursors could undergo hydrolysis and condensation reactions to form nanowires on the channels of the templates. Sol–gel method is a low-temperature solution-based process, which can be carried out without any major equipment. However, the method is relatively slow because it has a weak driving force due to slow diffusion of sols through the pores, but can be improved by heating the sol–gel template and also by enhancing the movement of sols by applying an electric field. The AAM is the most widely used template for sol–gel synthesis.

Template-assisted CVD: The nanochannels of a template such as AAM could also be filled by CVD. CVD inside nanochannel-Al_2O_3 (NCA) or AAM templates can be performed using a quartz tube horizontal furnace as shown in Figure 3.10. Deposition of silicon into the pores, for example, has been attempted using silane diluted in Ar at 900°C for 30 min [27]. The effort was unsuccessful in creating good quality nanowires due to pore blockage with deposition. This difficulty is easily understood by the fact that CVD readily occurs on any surface unless confined to a catalyst surface. Therefore, the use of catalysts (e.g., gold clusters) allows selective deposition inside the templates [28]. A gold layer was electrodeposited inside the channels closer to the top by electrodepositing Ag first and then Au inside the channel. Later, the Ag was etched to obtain a stripe of Au (250 nm) closer to the top of the AAM pore. The deposition was selective and was directed through the Au tip closer to the top of the membrane and growth occurred beyond the channels of the AAM template to obtain uniform diameter (50 nm) Si nanowires. This is not a popular technique as the use of catalysts defeats the purpose of using templates. In addition, low-pressure gas-phase deposition inside pores is not efficient. The diffusing molecules have no preference to start depositing from the pore bottom and due to large mean free path, they often stick at the side wall and pores are blocked. CVD with the aid of a catalyst has been more popular on flat substrates as will be discussed later.

CVD can also be done on step edges as they are exposed directly. This was demonstrated in preparing Pt nanowires using the corresponding organometallic vapor precursor on the step edges of a HOPG substrate [29].

Other filling methods: The pores of a template such as AAM membrane can be filled in several other ways as well. The pores of AAM can be filled by precipitates from a chemical reaction [30] or by polymers from a polymerization reaction [31] using a monomer and a polymerization reagent or various process combinations as in electroless deposition [32].

Chemical reaction–assisted deposition method involves the precipitate depositing from a liquid-phase reaction inside the pores of AAM template when it is immersed in the liquid bath. Iron oxide nanowires were obtained by carrying out iron oxide precipitation inside AAM pores [30]. Similarly polypyrrole (PPy) and poly-(3-methylthiophene) (PMT) were deposited inside the pores of AAM by carrying out polymerization reactions with their respective monomers and Fe^{3+} as a chemical oxidant for 2 h [31]. Chemical reactions were also carried out on CNT templates. Specifically, the deposition of gold nanocrystals followed by annealing was used to obtain polycrystalline gold nanowires [33]. The fabrication is based on the electroless deposition on a positive template such as DNA.

Electroless deposition is another method in which spontaneous reduction of a metal occurs from a solution of its salt in the presence of a reducing agent at a surface. First, the surfaces are activated where metal ions bind to the site on the template to form an activated complex. Then these complexes are treated by a reducing agent, resulting in 1-D metallic wires. The clusters act as nucleation centers to further catalyze the metallic growth of continuous nanowires. The binding affinity of DNA to different materials such as metallic cations such as Cu(II) at its phosphate anionic site by electrostatic attraction and various transition metals such as Pt(II), Pd(II) at N of aromatic bases to form coordination complexes were used to obtain a variety of nanowires [3]. For example, dimethylamine borane, a common reducing agent, was used to reduce Pd(II)–DNA complex and a second treatment of DNA with Pd bath on the preformed Pd nuclei resulted in the Pd nanowires on DNA. Electroless deposition unlike electrodeposition does not require any rectifiers, batteries, or anode. Unlike electrodeposition, which proceeds from bottom up in pores, electroless deposition occurs at all available surfaces on the template.

Comparison of deposition methods based on the templates: In a positive template method, nanowire diameter can be controlled by controlling the amount of material deposited, but the length is not controlled. In negative template method, nanowire length can be controlled, but not its diameter, which is predetermined by the pore diameter. Similarly, in the step edge decoration method, the diameter of nanowires can be controlled using the deposition time and applied potential but not the length. The diameter of the hemispherical nanowire in a step edge decoration technique during electrodeposition is given by [13]:

$$r(t) = \left(2i_{dep}t_{dep} \frac{V_m}{\pi n F l} \right)^{1/2}$$ (3.4)

where
  $r(t)$ is the radius of nanowire
  i_{dep} is the deposition current
  t_{dep} is the deposition time
  V_m is the molar volume of deposited material
  n is the number of electrons transferred per metal atom
  F is the Faraday constant
  l is the length of the nanowire

Most of the template-based methods require mild processing conditions and low costs. Template-based methods can control the nanowire length and diameters. However, these methods suffer from weak driving force of material deposition, which makes them time-consuming. Among the template-based methods, electrodeposition allows better control of nanowire size and length compared to others, but suffers from the requirement of conducting surface and other parameters, which must be controlled precisely. Electroless and other nonelectrodeposition template-based methods are relatively simpler, but the material fabricated is of poor quality.

3.2.2 Template-Free Methods

In the template-based methods, 1-D growth of nanowires is guided by the template and the growth depends primarily on the boundary between the template and bulk liquid, which occupies a small fraction of total system volume. In contrast, the nanowire synthesis using template-free methods from solutions often require stringent growth conditions because favorable conditions should be created to promote 1-D growth from the entire bulk liquid. These favorable conditions may be provided using methods such as hydrothermal, ultrasound, stirring, motion, and surfactant or catalyst seeds.

3.2.2.1 Hydrothermal Method

The term "hydrothermal" implies a regime of high temperatures and water pressure. Hydrothermal synthesis involves the crystallization and growth of a material from its high-temperature aqueous solution of soluble metal or metal–organic salt at high pressures. The solution is placed in an autoclave at temperatures between 100°C and 300°C and at pressures higher than atmospheric for 4–24 h. The pressure is kept high to prevent solution evaporation at temperatures employed in the autoclave. An autoclave is usually a cylindrical steel pressure vessel capable of withstanding high temperatures and pressure for a prolonged time and also inert to the solution. The inner lining is made up of materials such as Teflon, quartz, and glass depending on the requirement. The solution is placed in the autoclave through a cap at one of its end and closed tightly. A temperature gradient is maintained at two ends of the growth chamber so that the hotter end dissolves the solution (also called nutrient zone) and the cooler end (growth zone) causes seeds to take growth. The growth occurs by the precipitation of materials from the supersaturated solution. The growth conditions are often stringent compared to template-assisted methods. The nuclei are generated using a combination of harsh conditions, e.g., ultrasonication, horn sonication, magnetic stirrer, and mechanical stirring.

For example, ZnO nanorods [34] were produced using hydrothermal approach by treating a mixture of zinc nitrate, NaOH, and ethylenediamine in deionized water to form a solution. The resulting solution was treated in ultrasonic water bath followed by heating in an autoclave for 20 h at 180°C to obtain nanorods. Ethylenediamine was used as an oxidizing agent. V_2O_5 nanowires were synthesized by treating an aqueous solution of $VOSO_4 \cdot xH_2O$ and $KBrO_3$ in an autoclave at 160°C for 24 h [35]. Polyol synthesis has also been employed in which a metal precursor is heated along with ethylene glycol. Ethylene glycol can form linear complexes with metal cations and also suppress hydrolysis of metal precursor. TiO_2, SnO_2, In_2O_3, and PbO nanowires [36] were synthesized using a polyol-mediated process. Bi nanowires [37] were also synthesized using the same approach.

For hydrothermal synthesis of nanowires, the starting material composition and its homogeneity, purity, and quality must be carefully controlled. The variables affecting the product are reaction temperature, pH of the solution, and the type of oxidants used. The disadvantage with this approach is the requirement of high temperatures and pressures and also the expensive reactor chamber. The growth mechanism responsible for 1-D growth is not fully understood; however, theories relate the reactions on the starting material (i.e., nanowire precursor) occurring during the seed formation in the solution and underlying structural changes carried out by oxidizing agents to growing nanowires. The mechanism can be termed as solution-solid type with no underlying principle for guiding 1-D growth and is not well understood. During crystallization, the conditions that favor the growth of nanowires versus nanoparticles or any other shape have not been properly studied and remain a subject of continued research.

Hydrothermal synthesis using substrates: The rapid heating or cooling conditions in the solution during hydrothermal synthesis result in the nucleation of seed crystals. With the use of a substrate or a substrate with seeds, relatively milder conditions were found to help with the nanowire growth. Also, the quality of resulting nanowires tends to be much better when using a substrate or a seed. There are two steps involved here: (1) the formation of seeds onto the substrates and (2) the growth of nanowires using these seeds on the substrates. For example, ZnO nanowires were synthesized on Zn substrate by hydrothermal method [38]. Here, a zinc foil was placed in an aqueous solution containing NaOH and heated in an autoclave at 80°C for 12 h. The use of alkaline conditions helped the formation of Zn seeds on Zn foil. Further growth on the seeds resulted in ZnO nanowires on the Zn foil.

It is also possible to use a different substrate other than its own metal substrate and grow the required seeds first followed by nanowire growth. For example, ZnO seeds on any substrate such as Si wafer, glass, or Zn foil can be formed by spin-coating the ZnO precursor (zinc acetate dehydrate solution in methanol) and annealing the film for decomposing zinc acetate to form ZnO seeds. Immersing the seeded substrate into an aqueous solution containing zinc nitrate hexahydrate and hexamethylenetetramine (HMTA) at 90°C led to ZnO nanowire arrays on the substrate [39,40].

3.2.2.2 Sonochemical Method

Sonochemical, like hydrothermal, is based on seed formation in bulk solution but by using ultrasonic waves to generate seeds, which leads to the growth of nanowires. Note that, in hydrothermal process, the temperature gradient inside the autoclave is responsible for seeding. Sonochemical is based on a phenomenon called cavitation, where vapor bubbles form when the pressure of a liquid falls below its vapor pressure. The cavitation in solutions (bubble formation, collapse, and energy release) can be induced by applying ultrasonic waves (usually 15–400 kHz). Compression waves produced in the liquid rupture the liquid and create numerous microscopic bubbles due to local pressure falling below the vapor pressure. There is an enormous release of localized heat with temperatures of about 5000 K and pressure 100 atm, and rapid cooling rate (10^{10} K/s) when these bubbles with very high kinetic energies collapse. The high local temperatures and pressures coupled with rapid cooling provide a unique way of driving chemical reactions at extreme conditions. The reaction should be done at inert conditions and also the starting solution should be degassed, so that only the required bubbles are created. There are three regions around an imploding vapor bubble, as shown in Figure 3.8: the inner most region where the conditions are most extreme and often the reactions driving radical species (most often H$^{•}$ and OH$^{•}$) are generated, the interfacial region between the vapor bubble and bulk solution where growth reactions (mostly metal cations reduction with the help of generated reducing radicals) occur and then the bulk solution at ambient condition.

For example, selenium nanowires have been synthesized using sonochemical method [41]. First spherical Se colloids were prepared by reducing

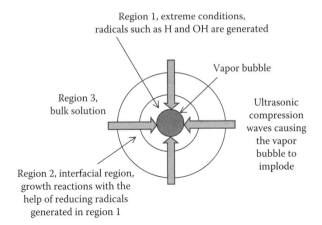

Region 1, extreme conditions, radicals such as H and OH are generated

Vapor bubble

Region 3, bulk solution

Ultrasonic compression waves causing the vapor bubble to implode

Region 2, interfacial region, growth reactions with the help of reducing radicals generated in region 1

FIGURE 3.8
A schematic showing three regions around an imploding vapor bubble during sonochemical method for nanowire synthesis.

selenious acid with hydrazine in an aqueous solution and then colloids were filtered by a membrane. After drying, they were dispersed in an alcoholic solution and sonicated for about 30 s (42 kHz, 0.15 W/m²). These colloids aggregated into irregular shapes during acoustic cavitation and seeds of Se formed on the colloids (self seeding by Se colloids) due to localized heating and consequent solubility variation. These seeds served as the source for Se nanowire growth by solid–solution–solid mechanism. The mechanism is based on the anisotropic extended spiral chain of selenium seeds, which serve as a template and direct the 1-D growth. Kumar et al. synthesized magnetite nanorods [42] by ultrasound irradiation of aqueous iron acetate in the presence of β-cyclodextrin, which acts as a size stabilizer. Zhang et al. synthesized gold nanobelts [43] using aqueous solution of $HAuCl_4$ and α-D-glucose, which acts as a size stabilizer. Sonochemical method is rapid and nonhazardous but the control of shape and size distribution of the products is not easy.

3.2.2.3 Surfactant-Assisted Growth: Soft Directing Agents

Surfactants are soft directing agents and they fall in the middle of classification between hard template methods such as AAM and completely template-free methods such as sonothermal or hydrothermal. They are included here under template-free methods. Surfactants are organic compounds containing a hydrophilic head and hydrophobic long alkyl chain tail (Figure 3.9a). The alkyl chains self-assemble to form a variety of shapes when added to water. Microemulsions are clear, stable, and isotropic mixtures of oil, water, and surfactant. Two types of emulsions are oil dispersed in water (direct) and water dispersed in oil (reversed). Surfactants assemble to form micelles

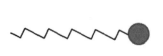

(a) A surfactant

(b) Oil in water, normal micelle

(c) Water in oil, reverse micelle

FIGURE 3.9
Schematics showing (a) surfactant shape, (b) its self-assembly to form normal micelle, and (c) reverse micelle.

(Figure 3.9b) where the long alkyl chain forms the inner core and the head groups form the outer part. Reverse micelles are formed in water-in-oil droplet cases as shown in Figure 3.9c and their size can be controlled by the amount of water added. The size of reverse micelles increases linearly with the amount of water added. This relation is given by R (nm)$=Kw$, where K is a constant and w is the water content, $w=[H_2O]/[S]$ where [S] is surfactant concentration [44]. The reverse micelles can be used as nanoreactors due to their variable droplet size and exchange of aqueous content. Surfactants along with hydrothermal treatment can further enhance nanowire formation. Surfactant-assisted nanowire synthesis involves trial and error in the selection of appropriate nanowire precursor, surfactant, temperature, pH, and concentration of the reactants. Nanorods of CeO_2 [45], ZnO [46], Au [47], and nanocrystals of PbS [48,49] and others have been synthesized using this approach. Usually in these approaches, the nanowire precursor is decomposed thermally and unique properties of surfactant enhance the nanowire growth. Oxalate-based metal precursors have low thermal stability and presence of cavities, which can be used in seed formation. Use of NaCl decreases the viscosity of the melt and increases the mobility of components, thus assisting the nanowire formation.

3.2.2.4 Catalyst-Assisted Solution-Based Approaches

Catalysts can be used in a fluid environment to direct the growth of nanowires under extreme conditions. Korgel and coworkers [50] used supercritical hexane to disperse both the solute and gold catalyst. They used diphenyl silane dissolved in supercritical hexane at 500°C and 200–270 bar pressure for Si nanowire synthesis. Diphenyl silane decomposes at these temperatures and dissolves into the gold clusters to yield silicon nanowires. The apparatus and corresponding results using this method are discussed in more detail later in the section on bulk production methods.

In summary, template-based methods provide good control over the dimensions of fabricated nanowires with high yield and also ordered arrays of nanowires. However, in order to obtain free-standing nanowires, the templates must be removed, which may not be easy in some cases. Template removal processes may also adversely affect the surface quality of the nanowires where harsh conditions are used. One of the key advantages of nanowires grown bottom-up compared to nanowires produced by top-down etching of thin films is the extraordinary surface quality; however, the template approach compromises this unique feature when the need arises to remove the template during device fabrication. Template methods also require many preprocessing steps accompanied by multiple washing. Impurities are difficult to remove, which may pose problems in performing any surface-bound chemistry on nanowires. Template-free methods naturally do not suffer from template removal problems. However, they are ill-defined in the absence of surfactants and thus lead to poor yield, irregular morphology, polycrystallinity, and low aspect ratio.

3.3 Vapor-Phase Techniques

Vapor-phase synthesis is the primary method for synthesizing nanowires in a number of material systems. Several vapor-phase methods have been developed to an advanced stage compared to liquid-phase methods. For this reason, the focus in the remaining chapters of the book will be mostly on the generic science and technology of vapor-phase methods for producing nanowires in a variety of materials systems. These methods can be broadly divided in to two categories: substrate-based and substrate-free direct methods.

Substrate-based methods: Figure 3.10a represents a typical substrate-based approach where the substrate is placed downstream to the source. The reactor

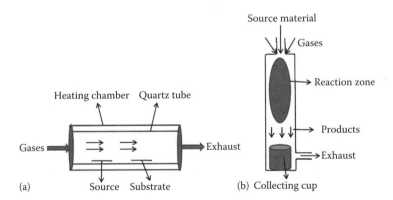

FIGURE 3.10
Schematics of two general methods used for vapor-phase synthesis of nanowires: (a) substrate-based method using a horizontal chemical vapor deposition reactor; and (b) direct gas-phase nanowire synthesis method.

operating under vacuum is heated and the source vapors react to produce the responsible growth species, which then deposit onto the catalyst supported by the substrate. The substrate has been found to be essential for nanowire synthesis in the vapor-phase methods for the following reasons: (1) the substrate supports the catalyst clusters and allows control of cluster temperature and (2) induces preferential precipitation at substrate–catalyst cluster interface. So the use of a substrate is essential for maintaining 1-D growth in vapor-phase methods involving catalyst clusters. Direct precipitation from vapor-phase precursors typically lead to spherical particles. Some common substrate-based methods are: CVD, thermal evaporation, laser ablation, arc discharge, plasma oxidation, CBE, and metal organic chemical vapor deposition (MOCVD).

Substrate-free methods: There have been very few direct gas-phase (substrate-free) methods developed to synthesize nanowires. Figure 3.10b shows a simplified schematic of a vertical reactor where a microwave plasma discharge (reaction zone) generated by passing reactive gases is confined inside a vertical quartz tube [51]. The source material is allowed to fall under gravity and it is collected in a cup. The key advantage of this approach over a substrate-based method is bulk production. Since the use of a substrate limits the amount of nanowire which can be produced, this method can theoretically yield bulk quantities of nanowires. A detailed discussion about this method is presented later under the section on bulk production of nanowires.

3.3.1 One-Dimensional Growth Concepts

The fundamentals of growth mechanisms underlying various substrate-based methods are briefly discussed in the following. The mechanistic details and the corresponding thermodynamics and kinetic considerations will be discussed in greater detail in Chapter 4. The known mechanisms are vapor–liquid–solid (VLS), low-melting metal mediated, self-catalytic, and vapor–solid–solid (VSS).

3.3.1.1 Vapor–Liquid–Solid Schemes Using Foreign Metal Clusters

As discussed in Chapter 2, in a traditional VLS method, the solute-containing vapor-phase species react selectively on the catalyst clusters supported on a substrate. The use of catalyst metal clusters enhances the selectivity of vapor-phase reactions on their surfaces compared to that on bare substrates. At the constant synthesis temperature, continuous dissolution from the gas phase leads to the supersaturation of the solute in the molten alloy cluster, eventually leading to precipitation of solute at the droplet–substrate surface interface. For the typical catalyst-assisted VLS growth, the resulting alloy supports a tip-led growth in which the catalyst cluster sits at the tip of the nanowire and leads the 1-D growth. This is schematically shown in Figure 3.11a.

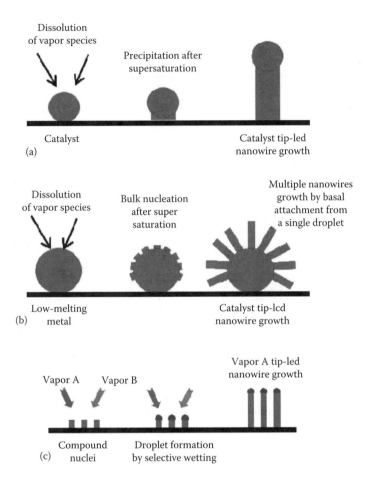

FIGURE 3.11
(See color insert following page 240.) Schematics illustrating the growth mechanism of nanowires: (a) classical VLS growth mechanism; (b) low-melting metal mediated growth; and (c) reactive vapor transport mechanism.

The wire diameter is controlled by the droplet size and one catalyst cluster leads to the growth of one nanowire.

3.3.1.2 Vapor–Liquid–Solid Schemes Using Low-Melting Metal Clusters

The low-melting metal clusters can also catalyze the 1-D growth of their compound materials as nanowires in a mechanism similar to that described above for traditional VLS mechanism. The main underlying concept is that the corresponding metal must be molten at the synthesis temperature and the gas phase must be sufficiently reactive so that selectivity of reaction on the metal clusters could be accomplished compared to bare substrates. As there are no foreign metals, these techniques fall under a generic scheme of self-catalysis. Such techniques using low-melting metals have been implemented for oxides, nitrides, phosphides, and sulfides among others. In these techniques, the crystals of compounds nucleate first and then support molten metal clusters. The continuous supply of metal and other reactive species ensures that the droplets establish liquid-phase epitaxy with the underlying crystals. Such precipitation supports 1-D growth through tip-led growth (Figure 3.11c) similar to that of catalyst-assisted growth.

3.3.1.3 Vapor–Liquid–Solid Schemes Using Large Size, Molten Metal Clusters

The catalysts used in traditional VLS methods are mostly Au, Ni, Pt, etc. which have high solubilities for many inorganic nanowires of interest. The precipitations from such high molten metal catalysts do not yield nanometer-scale nucleation and thus is confined by the droplet size. Several other metals such as tin, zinc, gallium, tin, etc. exhibit low solubility for several solutes of interest and thus can support nanometer scale nuclei from micron or larger-sized clusters. Such phenomena lead to bulk nucleation and growth of a high density of nanowires from a single droplet. This is schematically shown in Figure 3.11b. The 1-D growth of nuclei occurs due to preferential growth via basal attachment. This phenomenon can apply to a variety of materials systems including self-catalysis schemes. A good example is the growth of high-density oxide nanowires resulting from their own molten metals.

3.3.1.4 Vapor–Solid–Solid Scheme

Vapor–solid–solid is not a well-known phenomenon but has been observed quite frequently. In some cases, this is very similar to that of VLS in that a solid cluster catalyzes 1-D growth. In many other cases, the underlying mechanism is not clear as no clusters at the tip are observed.

3.3.1.5 Oxygen-Assisted Growth (OAG) Scheme

Oxide-assisted growth (OAG) of nanowires is similar to VSS scheme. The oxide sheath around the metal cluster (first nucleated on the substrate) allows enhanced and selective growth at its tip. The steps are as follows: (1) metal clusters are formed on the substrate; (2) thin solid oxide sheath is formed around the growing nanowire to prevent lateral growth; (3) the tip is covered by semiliquid suboxide species, selectively enhancing the absorption at the tips followed by the 1-D growth [52].

3.3.2 Source Generation and Reactors for Vapor-Phase Synthesis of Nanowires

There are a variety of reactor setups used for nanowire synthesis depending on how the source is generated and hence differ primarily with source

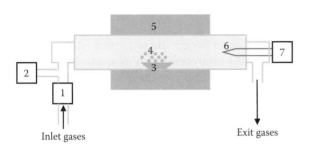

1: Mass flow controller
2: Pressure gauge
3: Alumina boat containing the
   nanowire source materials,
   placed near the center of the
   tube furnace
4: Evaporated species
5: Tube furnace
6: Cold finger
7: Temperature controller for
   the cold finger

FIGURE 3.12
Schematic of a typical reactor setup used for thermal evaporation.

material, process conditions, substrates, and the catalyst choice. Each of the
reactor setups is discussed in detail in the following.

3.3.2.1 Thermal Evaporation

A typical reactor setup used for thermal evaporation contains a horizontal
tube furnace with provisions for gas flow, and pressure gauge with or without control as shown in a simplified schematic in Figure 3.12. The furnace
may have a single, two, or three heating zones. More than one zone allows a
hotter source generation zone with a relatively cooler growth zone. Note that
vertical reactors with similar setup can also be used. The experiments are
typically carried out at low pressures (achieved using a mechanical pump)
so that the reactions over catalysts can be selective. Low-pressure operation
can also help to reduce oxygen contamination whenever it is critical. The
typical operating pressure is about 1–100 Torr. The desired source material to
be heated is placed in an alumina crucible. The crucible material should have
high melting point, good chemical stability, and hardness so that it can withstand high temperatures and chemical corrosion. The tube furnace heats up
the source through conduction, gas convection, and radiation. Alternatively,
the source material can also be heated by resistive heating of the crucible.

It is important to bake and purge the reactor before each experiment to
remove adsorbed gases from reactor walls and reduce contamination and
oxygen partial pressure. The source gases diluted in an inert gas such as
argon are used. The pure source vapors produced from thermal evaporation (e.g., ZnO vapor) or source vapor mixed with other feed gases (e.g., in
vapor mixed with NH_3 for InN growth) react on the substrate kept at a distance from the source boat. Note that the source materials do need to be at
very high temperatures close to their melting, decomposition, or sublimation
points to produce enough vapors. In downstream, there can be a provision
for cold finger or a substrate where temperature is lower. This is the reason dual zone furnaces are preferred. Due to low pressures used, the nanowire growth occurs over a prolonged deposition time (few hours) if long

nanowires are desired. Typically, the nanowire products are collected near the downstream end of the tube furnace.

Several different types of growth can be conducted using this reactor, for example, catalyst-assisted and self-catalyzed. Also, one can easily produce vertical arrays, epitaxial arrays on single crystal substrates, as well as randomly oriented nanowire films. In the case of catalyst-assisted growth, substrates coated with catalyst layers are placed downstream. In the case of self-catalyzed schemes, the low-melting metal clusters are generated in situ on substrates during evaporation of solid oxide sources. A wide variety of nanowires such as metal oxides, metal sulfides, and semiconductor materials have been synthesized using thermal evaporation. This approach is most widely used to prepare metal oxides nanowires. Almost all kinds of metal oxides nanowires (both low-melting and high-melting metals) can be prepared by passing O_2 over the metals in a horizontal tubular reactor. For example, SnO_2 nanowires were produced from SnO powder using self-catalysis technique [53]. SnO powder was heated at 680°C for 8 h in an alumina crucible, keeping the total pressure at 200 Torr and passing argon and oxygen gases. ZnO nanowires were produced by heating Zn powder at 850°C–900°C for 30 min and using Au-coated sapphire substrate. The growth of nanowires followed traditional foreign metal cluster–catalyzed VLS mechanism with molten Au at their tips [54]. Similarly, the synthesis of GaO [55], MgO [56], WO_3 [57], and many other metal oxide nanowires were also reported in the literature and will be further discussed in Chapter 9. ZnS nanowires were synthesized using ZnS powder source and Au catalyst–coated substrate [58]. Cu_2S nanowires were prepared by placing ZnS powder on Cu sheets in a thermal reactor and using argon as carrier gas [59]. CdS nanowires were obtained by placing CdS nanopowders as source and Si wafer as substrate and passing Ar as carrier gas in a horizontal tubular reactor [60]. Si nanowires were obtained by simply placing a Si substrate at 1100°C, 4000 He pressure [61].

Sometimes graphite is mixed with the source material to create vapors at lower temperatures in a process called carbothermal reduction. For example, graphite when heated with ZnO reduces ZnO to Zn vapors as shown in the following:

$$ZnO(s) + C(s) \rightarrow Zn(g) + CO(s)$$

$$CO(g) + ZnO(s) \rightarrow Zn(g) + CO_2(g)$$

The argon carrier gas used along with reacting gas such as oxygen carries the product CO and CO_2 gases to the exit. Nanowires of tin oxide [62], indium oxide [63], and others have been synthesized using carbothermal reduction.

3.3.2.2 Laser Ablation

Laser ablation is another low-pressure, high-energetic, nonequilibrium, rapid condensation process for producing nanowires. The schematic of

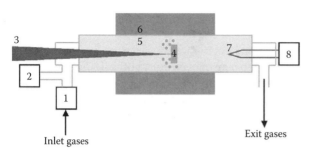

1: Mass flow controller
2: Pressure gauge
3: Laser source
4: Ablation target composed
of the nanowire source
materials
5: Quartz tube
6: Tube furnace
7: Cold finger, used to collect
the nanowire products
8: Temperature controller for
the cold finger

FIGURE 3.13
Schematic of a typical reactor setup used for laser ablation.

a typical laser ablation setup is shown in Figure 3.13. The setup is almost similar to that for thermal evaporation, except there is a port for introducing the laser beam and a target containing the source in the middle of the quartz tube. The laser, when it hits the target, generates a plume containing vapor fragments of the target. The target is made by pressing the source materials and the catalyst metal together into solid cylindrical pellets. The chamber temperature is maintained using the tube furnace at a level needed to sustain the catalyst-molten alloy or for the nanowire growth (typically between 800°C and 1200°C). The variables in this method include laser power, target composition, gas-phase composition, temperature, pressure, residence time of the vapor-phase species, and cold finger conditions. Laser ablation in general is not a scalable process and therefore is suitable for laboratory investigation.

Laser ablation method for nanowire synthesis has been extensively developed by Lieber et al. [64]. A large range of semiconductor compounds such as binary III–V nanowires (GaAs, GaP, InP), binary II–VI nanowires (ZnS, ZnSe, CdSe), ternary III–V nanowires (GaAsP, InAsP), alloys (SiGe) have been synthesized by them [65]. Similarly, metal oxides nanowires [66] of Fe, Zn, Sn, Ga, etc., metal nanowires [67] (e.g., B, Si, Ge etc.) have been obtained. Si nanowires have been produced using Si–Fe targets with a neodymium:yttrium-aluminum-garnet (Nd:YAG) laser irradiation (2.0 W average power, 10 Hz pulse rate, λ 532 nm) in 50 sccm of Ar flow [64]. The $Si_{0.9}Fe_{0.1}$ target was ablated to create the source vapors and the nanowires were deposited on the cold finger while the tube furnace was maintained at 1200°C. The resulting silicon nanowires were 10 nm thick and several microns long and consisted of Fe-rich clusters at their tips, indicating the underlying VLS mechanism.

In a different variation, the laser ablation apparatus was used without a catalyst in the target for producing Fe_2O_3 nanowires, nanobelts, and nanorods. In these experiments, magnetite powder source was placed at the center of the quartz tube and an alumina wafer substrate coated with Au film was placed below the target [68]. A pulsed laser energy source generated by a Compendex series excimer 102 laser (20 Hz, 30 kV, 300 mJ per pulse) was used to ablate a spot of 1 mm × 5 mm on the target for 1 h at ambient temperature and pressure.

The laser–material interaction has been studied extensively and the laser power density or irradiance (W/cm²) employed for synthesizing nanowires is usually high enough to generate vapors through absorption of laser energy by the material. The material can undergo vaporization or sublimation upon optical energy absorbance. The important factors that determine the amount of materials evaporated are laser irradiance (power density, W/cm²), optical properties (absorption and reflection), and thermal conductivity of the material. However, laser irradiance is significant only for generation of the vapors, which need to be transported to the catalyst site and therefore synthesis of nanowires depend on additional factors such as substrate temperature and gas-phase composition.

The laser ablation differs from thermal evaporation in the heating method of the target. In thermal evaporation, the vapor-phase composition changes during the course of the experiment since different materials in the source generate different amount of vapors based on varying melting points and vapor pressures of the source constituents. However, in laser ablation, the vapor-phase composition is same as the source composition due to complete vaporization of the targeted spot irrespective of the difference in the evaporation point of the constituent elements. The complete vaporization is achieved by the extremely high heating rate of the target (order of 100 K/s) confined to a small spot on the target. Thus laser ablation allows one to retain the target stoichiometry in the deposited structure. Also, laser ablation requires a lower substrate temperature than thermal evaporation because of high heating rate of the target. Unlike thermal evaporation where the whole material absorbs heat and increases its temperature, it is possible to selectively heat a small portion of the target in laser ablation, thereby increasing the efficiency and saving energy. However, because of small amounts of vapors generated, only small amount of nanowires can be produced.

3.3.2.3 Metal Organic Chemical Vapor Deposition

Metal organic chemical vapor deposition (MOCVD) is a special case of CVD method, where one or more of the gas-phase precursor is a metal-organic (MO) compound. As shown in the schematic in Figure 3.12, a typical MOCVD setup consists of a reactor with provision for inlets for reacting gases and MO precursors and connection to a vacuum pump. Inside the reactor, a substrate is placed directly facing the gas flow with provision for accurate temperature control using a heater.

Typically, methyl- or ethyl-based precursors are used, which exhibit lower decomposition temperatures, thus allowing growth reactions at lower temperatures compared to other CVD methods. MO precursors are transported to the reactor along with a carrier gas, which is passed through a bubbler, as shown in Figure 3.14. The vapor pressure of the MO precursor is a critical

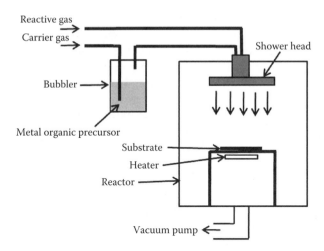

FIGURE 3.14
Schematic of a typical reactor setup used for MOCVD.

parameter that must be controlled, as this determines the concentration of the precursor in the vapor phase and subsequently the deposition or growth rate of the nanostructure. So, the bubbling cylinder is kept at a constant low temperature ($\sim 0°C$). The low temperature in the bubbler allows only small and controlled amount of precursor vapor in the chamber, which helps to prevent the reaction inside the reaction chamber before reaching the substrate. The carrier gases are usually H_2, Ar, or N_2.

Also, the gas connecting lines should be heated to avoid MO precursor condensation onto the walls of the connecting lines. Gas-phase reactions inside the reaction chamber before reaching the substrate need to be avoided. Some of the precautions include the following:

1. Flow only small amounts of MO precursor by keeping bubbler at low temperatures and pressure.
2. The ambient temperature inside the reaction chamber should be near room temperature so that no significant reaction occurs.
3. The lateral convection of gas flow can also be avoided by introducing a nonreacting gas, such as N_2, through the shower head, which acts as a pushing gas, forming an annular curtain of gas flow.
4. The distance between the shower head and substrate can be varied.

The pressure inside the reactor is typically kept between 10 and 100 Torr and hence the gas flow is laminar and the chemicals reach the substrate surface by diffusion. The growth on the substrate surface occurs by the reaction between the decomposing MO precursor and reactive gas (e.g., ammonia for nitrides). Substrate rotation is necessary to obtain uniform growth over the entire wafer.

The parameters influencing nanowire growth by this approach are MO precursor flow rate (mol/min), pressure inside the reactor, substrate temperature, and substrate material. The MO precursor flow rate from the bubbler depends on the vapor pressure of metal organic precursor (function of temperature), flow rate of the carrying gas (sccm), and reactor chamber pressure. The relationship between the saturated vapor pressure of the MO precursor and its temperature (constant temperature bath) is given by the well-known Antoine equation:

$$\log(P_{MO}, \text{Torr}) = A - \frac{B}{T(K)} \quad (3.5)$$

where A and B are material-specific constants and their values for commonly used MO precursors are given in Table 3.1. A quick observation from Table 3.1 is that for the same metal, methyl-based precursor has higher vapor pressure than ethyl-based precursor at a given temperature. This is because of stronger molecular forces of attraction (due to van der Walls forces with the increasing aromatic groups added) in ethyl-based MO precursor than the methyl-based one. Thus less gas molecules are present in TEAl (due to less escape from liquid phase to gas) and hence it has lower vapor pressure compared to TMAl. The partial pressure of MO precursor can be regulated by maintaining the temperature of the bath containing the precursor. The molar flow rate of MO precursor inside the reactor based on ideal gas law is given by

$$F_{MO}(\text{mol/min}) = \frac{P_{MO}}{P_{tot}} \times F_{tot}(\text{mol/min}) \quad (3.6)$$

where
F_{tot} is the total flow rate
P_{tot} is the total pressure inside the reactor chamber

TABLE 3.1

Antoine Equation Constants for Commonly Used Metal Organic Precursors

Formula	Name	P (Torr at 298 K)	A	B
$Al(CH_3)_3$	TMAl	14.2	10.48	2780
$Al(C_2H_5)_3$	TEAl	0.041	10.78	3625
$Ga(CH_3)_3$	TMGa	238	8.50	1825
$Ga(C_2H_5)_3$	TEGa	4.79	9.19	2530
$In(CH_3)_3$	TMIn	1.75	9.74	2830
$In(C_2H_5)_3$	TEin	0.31	8.94	2815
$Zn(C_2H_5)_2$	DEZn	8.53	8.28	2190
$Mg(C_2H_5)_2$	DEMg	0.05	10.56	3556

This can be further simplified as

$$F_{MO}(mol/min) = \frac{P_{MO}}{P_{tot}} \times V_{carrier-gas}(sccm) \times \frac{(1\,mol)}{(22,400\,cm^3)} \qquad (3.7)$$

the standard cubic centimeter per minute (sccm) is obtained at standard conditions (1 atm, 298 K) and 1 mol of a gas occupies 22.4 L or 22,400 cm³ of volume at standard conditions.

The nanowire growth can occur with or without the aid of a catalyst depending on the process conditions. Also, thin film growth can be achieved at higher substrate temperatures due to increased precursor decomposition and random surface diffusion. Traditionally MOCVD has been used for the growth of III–V and II–VI based semiconductor compound thin films. Metal organic chemical vapor epitaxy (MOCVE) is a special case of MOCVD where the epitaxial growth on the substrate is achieved using a highly oriented single crystalline substrate. MOCVD has been used to grow various nano-wires including III–V semiconductors, nitrides, and oxides. ZnO nanowires were grown on GaAs(002) substrate using diethylzinc (DEZn) maintained at 0°C [69] with the pressure in the reaction chamber maintained at 15 Torr and a substrate temperature of 600°C. The Ar and O_2 gases were flown at 120 and 85 sccm. Similarly, GaO nanowires were synthesized using trimethylgallium (TMGa) precursor on Si substrate at 550°C [70].

3.3.2.4 Chemical and Molecular Beam Epitaxy

Molecular beam epitaxy (MBE) is a technique for epitaxial growth of materials under extreme vacuum conditions on the substrate surface. The growth occurs by the interaction of molecular beams from two or more beam sources at the substrate. As shown in Figure 3.15, a typical MBE setup consists of three vacuum chambers: the load chamber, transfer chamber, and the growth

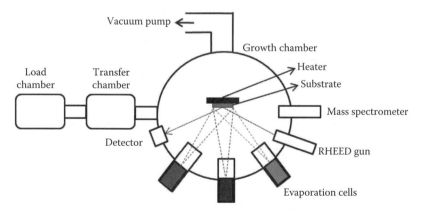

FIGURE 3.15
(See color insert following page 240.) Schematic of a typical reactor setup used for MBE.

chamber. The substrate is inserted into the load chamber and subsequently placed inside the growth chamber. These chambers operate at decreasing pressures ranges. The load, transfer, and growth chambers have pressures in the ranges of 10^{-4}, 10^{-8}, and 10^{-9} Torr respectively. The growth chamber is an ultrahigh vacuum (UHV) chamber and the pressure is kept at about 10^{-9} Torr by using ion pump, turbomolecular pump, and titanium sublimation pump together. The mean free path of the gas molecules at these low pressures is larger than the dimensions of the growth chamber. This implies that the gas molecules emitted from the evaporation cells do not collide and react with other gaseous species until they strike the substrate. Also for a controlled monolayer growth, the pressure should be kept as low as possible.

The slower deposition rate in MBE results in highly controlled growth. The MBE chamber has provision for liquid nitrogen cooling. The chamber wall at low temperatures acts as a sink for impurities and does not allow molecules to escape because of high adsorption at lower temperatures. For the same reason, a proper bake out of the chamber at high temperatures (~200°C) for 20 h is required during maintenance to desorb the gas molecules from the chamber surface.

As shown in Figure 3.15, the growth chamber consists of evaporation cells, reflection high energy electron diffraction (RHEED) gun and detector, mass spectrometer, and substrate. The evaporation cells can be either based on thermal evaporation or electron-beam evaporation. In MBE, highly pure substances cast in solid form are heated in evaporation cells until they start to sublimate. The columnar shape of these cells causes the beam to travel straight to the substrate. There may be a shutter in front of the mouth of these molecular beams to control the amount and duration of material reaching the substrate. So one or more of the gaseous elements condense and react at the substrate under controlled conditions to form single crystal materials. A mass spectrometer is inserted into the growth chamber to detect the molecules present inside at any time. RHEED is used to monitor the growth of the crystal layers. It works by detecting the diffraction pattern of the signal using a detector after reflection from the growing crystal. The oscillation of RHEED diffraction signals gives the time required to grow a monolayer. The substrate is heated and also rotated to obtain uniform growth. It is possible to obtain either homo or hetero structures by carefully controlling the shutter opening and closing times of different beams.

MBE has been used for the growth of III–V and II–VI semiconductor compound nanowires. For example, GaAs nanowire arrays were obtained on a GaAs (111) substrate coated with Au catalyst at temperature 450°C–550°C for 20–30 min [71]. The gold spots on the substrate surface were patterned using a nanochannel alumina (NCA) template by lithography. The growth occurred by Au tip-led VLS mechanism. MBE also allows one to obtain segmented nanowires (or hetero structures) similar to electrodeposition by controlling the opening and closing of different beam shutters inside the UHV chamber. As an example, GaP–GaAsP–GaAs heterostructures were catalytically grown from Au particles on GaAs (111) substrate [72]. First, the substrate was coated with 4 nm thick film of Au in an e-beam evaporation chamber and then annealed later at 550°C for 10 min to form Au nanoparticles on surface. The beam sources of As_2 and P_2 were from a dual filament AsH_3 and PH_3 gas cracker operating at 950°C while Ga was supplied from an evaporation cell. The three segments in GaP–GaAsP–GaP were grown by maintaining the supply of every required beam, except the one not required in that segment, for 3 min (GaP segment) and 24 min (GaAsP segment). For example, the GaP segment was grown by opening the beam shutters for Ga and PH_3 and closing the AsH_3 shutter. In each case, Ga was used to control the exact time while keeping other beams flowing.

CBE is an intermediate between MOCVD and MBE. In CBE, MO precursors are used and they are transported as beams to the substrate surface and decomposed to form required products. InAs nanowires were synthesized using CBE on InAs (111) substrate, which were lithographically patterned with Au catalyst [73]. Trimethylindium (TMIn) and tertiarybutyl-arsine (TBAs) were used as precursors.

3.3.2.5 Plasma Arc Discharge–Based Techniques

The dielectric breakdown of gases to generate a plasma under the influence of an electric field is an effective method to produce nanowires. Direct current (DC) arc discharge is an important method under this category. The arc discharge occurs due to the electrical breakdown of the gas, which produces a plasma discharge, resulting in a flow of current through normally nonconductive air. Plasma is a partially ionized gas containing neutrals, free, and charged species, and electrons making it electrically conductive. As shown in Figure 3.16, a typical DC arc discharge schematic can be of two types: current-carrying

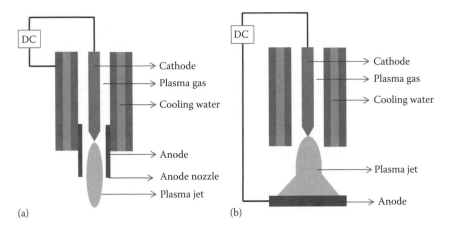

(a) (b)

FIGURE 3.16
Schematic of a typical setup used with DC arc discharge. (a) A current carrying arc and (b) a transferred arc.

arc and transferred arc. Both types consist of a cathode, which emits electrons, a plasma gas, cooling water, and a nozzle to confine the plasma. In a current-carrying arc type, the anode can be any material whereas in the transferred arc case, the treated material is the anode. The arc is ignited and plasma is formed with the application of required voltage (higher than the ionization potential of the plasma gas) between the anode and cathode. Note that the whole apparatus is maintained at low pressures ($\sim10^{-2}$ Torr) so that the plasma gas ionization is effectively achieved. The typical voltage applied is 30 V and the arc current is $\sim$100 A and the plasma temperature is about 10,000 K. In both cases, the growth of nanostructures occurs by the direct exposure of reactive radicals in the plasma with the substrate.

Ga_2O_3 nanowires have been obtained using a current-carrying arc type discharge [74]. GaN with 5 wt% Ni/Co metals in Ar/O_2 ambient atmosphere was used. Direct current of 55–65 A with a voltage of 13–15 V and total pressure of 500 Torr ($P_{Ar}/P_{O_2}=4{:}1$) was applied for a few seconds. Similarly, transferred arc type discharge was used to obtain Ta_2N nanocrystals using a Ta anode and N_2–NH_3 as the plasma gas [75].

3.4 Bulk Production Methods

Nanowires are beginning to find applications in several fields such as optoelectronics, solar cells, light-weight nanocomposites, catalysis, lithium ion batteries, gas sensing, and others. In many of these applications, nanowires are needed in large quantities. Thus far, very few synthesis techniques are able to produce nanowires in bulk quantities. In this section, we discuss some of the methods that can produce nanowires in bulk quantities including hot filament CVD (HFCVD), supercritical fluid approach, plasma foil or substrate oxidation, and direct gas-phase reaction in a plasma jet discharge. Out of these four methods, HFCVD and foil oxidation make use of a substrate, which limits the ability to produce large quantities of nanowire powders. On other hand, those two reactors can be used for producing nanowire arrays of select set of materials. The preferred method for high-throughput nanowire production is gas-phase synthesis, wherein the reacted species are swept away from the reaction zone quickly. In this context, direct gas-phase synthesis is the best approach. Supercritical approach is closer to a gas-phase medium but is expensive and involves extreme conditions.

3.4.1 Hot Filament CVD Method

HFCVD is used to synthesize metal and metal oxide nanowires of high-melting metals such as tungsten, molybdenum, tantalum, nickel, etc., which can be difficult to synthesize by other methods. As shown in Figure 3.17, the

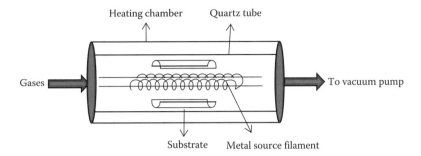

FIGURE 3.17
Schematic of a HFCVD reactor.

reactor consists of a quartz tube with the metal filaments wound on top of two ceramic tubes, which are connected to an electrical feed through. The filament temperature is controlled by controlling the applied power to the filament using a variac power supply. The substrates in the form of boats made up of quartz are heated up to required temperatures either using an external furnace or radiatively using the hot filaments. The hot filament setup differs slightly from that of thermal evaporation in the source material. Here, the source material is metal wire filament wound on ceramic tubes. In HFCVD, the chemical vapor transport of metal oxide vapors onto substrates kept at temperatures lower than the decomposition temperature is used for producing metal oxide nanowires, and onto substrates kept at temperatures higher than the decomposition temperature is used for producing the respective metal nanowires. Examples include the synthesis of tungsten oxide nanowire arrays using tungsten filaments and oxygen flow at low partial pressures in the range of 0.05–0.3 Torr at a filament temperature around 1600°C and at a substrate temperature around 400°C–800°C [76]. Similarly, the experiments using substrates at temperatures closer to or higher than 1450°C result in tungsten nanowires [77].

Nanowires of other transition metal oxides (tantalum oxide, nickel oxide, molybdenum oxide) can also be synthesized by changing the filament wire to their respective metal sources. In all of these cases with other metals, the conditions such as filament temperature, substrate temperature, and oxygen partial pressure play an important role in determining the density of nucleation and the resulting metal oxide characteristics. HFCVD is more effective than thermal evaporation in generating vapors from high-melting metals. In hot filament case, the metals are in wire shape having higher surface area than in bulk powder shape as in thermal evaporation. This leads to effective heating of wire filament to generate sufficient quantities of metal vapors, which would not be possible in thermal evaporation of high-melting metals. In this reactor, the nanowire growth on large area substrates both in the thin film and array fashion can categorize this method under bulk production of nanowires. The easy scalability of the process can

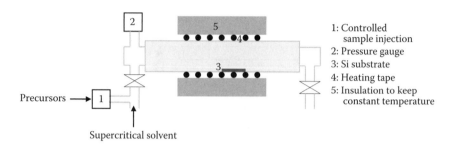

1: Controlled
   sample injection
2: Pressure gauge
3: Si substrate
4: Heating tape
5: Insulation to keep
   constant temperature

Precursors

Supercritical solvent

FIGURE 3.18
Schematic of a supercritical fluid-based reactor. (Courtesy of J.D. Holmes.)

be demonstrated by varying the lengths of the filaments and increasing the area of the substrates onto which these nanowires are grown.

3.4.2 Supercritical Fluid Approach

A supercritical fluid method developed by Korgel et al. [50] uses supercritical hexane containing the dispersion of uniform-sized, alkanethiol-stabilized gold catalyst nanoclusters, and the dissolved solute material of interest for growing nanowires. The schematic of the high-pressure reactor setup used in this work is shown in Figure 3.18. For silicon nanowire synthesis, diphenylsilane dissolved into the supercritical hexane is used at 500°C and 200–270 bar pressure for 1 h [50]. Diphenylsilane decomposes selectively on gold clusters at these temperatures, resulting in silicon nanowire growth. Similar experiments were also reported on the synthesis of Ge nanowires using a similar experimental setup at temperatures of 300°C–500°C [78].

The growth mechanism is similar to that described for traditional VLS techniques but using floating catalyst. The dissolution into molten metal cluster happens from solution phase rather than in the vapor phase, thus giving the name, the solution–liquid–solid (SolLS) method. The supercritical fluid environment offers rapid mass transport of the solute, thus giving much more control on the solution chemistry for the dissolution kinetics and the nanowire growth kinetics. Also relatively size-monodisperse Au nanocrysals can be maintained in the supercritical fluid environment to grow uniform-size nanowires.

3.4.3 Direct Oxidation Schemes Using Plasma

Direct oxidation approach to obtain bulk quantities of metal oxide nanowires has been developed primarily by Sunkara and coworkers [79–81]. There is no transport of chemical vapors from source to substrate in this approach and it involves exposing metal foils/other substrates or low-melting metals such as Ga, Sn, Zn, etc. to low-pressure, weakly ionized, fully dissociated cold oxygen plasma. A simplified schematic of a direct oxidation scheme is shown

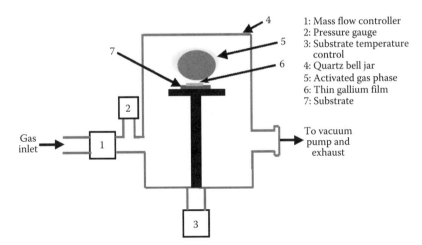

1: Mass flow controller
2: Pressure gauge
3: Substrate temperature control
4: Quartz bell jar
5: Activated gas phase
6: Thin gallium film
7: Substrate

Gas inlet

To vacuum pump and exhaust

FIGURE 3.19
Schematic of a reactor setup for substrate or foil exposure in a plasma.

in Figure 3.19. The reactor setup basically contains a source to generate a plasma upon introduction of O_2 and other gases in the chamber under low pressures. Also the metal foil or low-melting metal pool can be placed on the substrate and exposed to the plasma. The chamber is maintained typically at few torr pressure and the plasma exposure time can vary depending on the requirement. For example, Nb_2O_5 nanowires were obtained by exposing Nb foil to weakly dissociated O_2 plasma for 90 s [80]. The applied RF power was 1.5 kW and chamber pressure was about 0.1 Torr. Sharma et al. [79] synthesized gallium oxide nanowires and tubes by exposing a substrate covered with molten gallium in a microwave plasma reactor. The reaction conditions included a substrate temperature of 550°C, power of 700 W, total pressure of 40 Torr, and H_2 and O_2 flow rates of 100 and 8 sccm, respectively.

Nanowire growth by direct plasma oxidation follows bulk nucleation and basal growth as explained earlier in Section 3.3.1.3. The Nb foil exposure follows the VSS mechanism of bulk nucleation and basal growth of Nb_2O_5 nanowires, whereas gallium oxide nanowire growth in a molten gallium pool involves VLS mechanism of bulk nucleation and basal growth. The large number density of resulting nanowires makes this technique suitable and interesting for large-scale production.

3.4.4 Direct Gas-Phase Reactions Using Plasma Discharges

Direct gas-phase reaction of metal powders to produce bulk quantities of metal oxide nanowires is a recent technique developed by Sunkara and coworkers [51]. This is a follow-up to the earlier work on the direct oxidation of foils or low-melting metals supported on a substrate. As this method does not depend upon the use of substrates, it can potentially yield

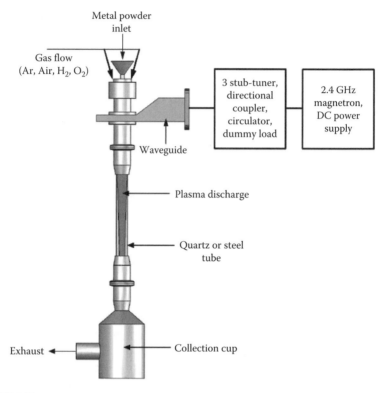

FIGURE 3.20

Schematic of a reactor setup for direct gas-phase microwave plasma oxidation. (Reproduced from Kumar, V. et al., *J. Phys. Chem. C*, 112, 17750, 2008. With permission.)

nanowire production at a rate of kilograms per day. A simplified schematic in Figure 3.20 shows a vertical open-ended microwave plasma reactor. A highly dense microwave plasma discharge, confined by a quartz tube, is generated using a microwave power source operating at 1.5 kW and 2.45 GHz. The reactor can operate at pressures ranging from a few torr to atmospheric pressure and at powers ranging from 300 W to 3 kW. The gases such as O_2, air, and H_2 are supplied together at the top of the reactor and metal powders are poured directly into the plasma cavity zone and allowed to flow down by gravity with gas flow. As the metal powder falls through the plasma flame region, it heats up, melts, and reacts with the reactive species present in the plasma and solidifies quickly as it goes past the flame region. The product is collected in a cup at the bottom of the reactor in the form of powder.

Kumar et al. [51] demonstrated oxidation of different metal powders such as Sn, Zn, and Al under almost similar processing conditions to obtain respective metal oxide nanowires. The typical processing conditions involved microwave plasma power of 1.5 kW, micron size metal powders, 10 slpm of air, 100 sccm of H_2, and 500 sccm of O_2. The plasma flame can process about

5 g/min of metal powder, which translates to a production capacity of 5 kg of metal oxide nanowires per day when operated continuously. The reactor can be operated continuously with a recycle stream for unreacted metal particles and a continuous collection system for metal oxide nanowires using filter bags. Experiments performed at higher plasma powers, yielded spherical, unagglomerated metal oxide nanoparticles. The results obtained using various metal oxides suggest that the mechanism of nanowire nucleation and growth in the gas phase is similar to that observed in experiments with metal particles supported on substrates.

Direct reaction of metal with the gases in the gas phase is the best of all the methods discussed for the bulk production of nanowires. HFCVD and direct substrate/foil oxidation using plasma are limited by the use of substrate and supercritical scheme is relatively expensive and operates at extreme process conditions. Also the direct gas-phase reaction method using microwave plasma is a very fast, low-cost method, and does not require any catalyst, substrate, template, or vacuum components.

3.5 Future Developments

Various growth techniques discussed in this chapter are summarized in a schematic shown in Figure 3.21. As described earlier, the advantage of liquid-phase techniques over others are lower temperatures, low cost, less hazardous, simpler equipments, high yield, and more uniform nanowires. They can be flexible and tunable compared to vapor-phase methods. But the liquid-phase techniques suffer from a number of disadvantages such as template removal, surfactant costs, applicability to only certain types of materials systems depending upon the technique, crystallinity of resulting nanowires, low mobility of adatoms, more time needed to form compared to gas phase, and template removal problem if used.

All the gas- and liquid-phase methods mentioned so far are bottom-up approaches where the nanowire is formed from adatom additions, however, there are some other methods as well involving the top-down approaches where nanowires are carved out of a bigger substance, for example, electrospinning/lithography, electrophoresis, and etching from thin films. The general disadvantage of the top-down techniques is that relatively poor quality of the nanowire surface compared to the grown/synthesized wires.

At the time of this writing, commercial reactors for large wafer growth of nanowires are not common. Taking VLS growth by CVD as an example, conventional commercial CVD reactors from the microelectronics industry may be used. But a major constraint of these cold wall reactors with a heated substrate platform is the inability to separate the source production

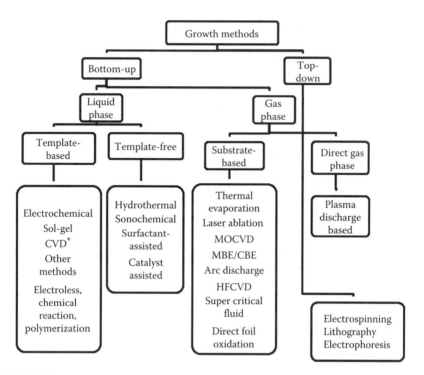

FIGURE 3.21

Summary tree of the various growth techniques discussed in this chapter. *Note*: Chemical vapor deposition is usually performed on flat substrates and is a gas-phase technique. Here it is included under liquid phase, because it can also be performed on templates in some cases, which is mainly associated with liquid-phase methods.

and growth aspects. These two proceed at two distinct temperatures in most cases, preferably growth at temperatures lower than that required for source production by judicious selection of catalyst seeds. Regardless of such clever selections, conventional commercial CVD reactors are "single zone" reactors. Future equipment development must address this need to accommodate low-temperature growth on processed wafers with valuable features already in them. Also, a truly bulk production method has not been commercialized yet and hence at the time of this writing, it is impossible to obtain kilogram (or even gram quantity) quantities of nanowires commercially. However, several methods described in this chapter (direct gas-phase reaction using plasma as an example) have potential to be commercialized.

As will be seen in Chapters 12 through 16, the use of nanowire covers areas well known for the corresponding materials in their film form. One-dimensional nanowires have the potential to provide improved performance in chosen applications, and to exploit this potential, large wafer, and bulk processing capabilities and equipment development will play a key role.

References

1. T. J. Trentler, K. M. Hickman, S. C. Goel, A. M. Viano, P. C. Gibbons, and W. E. Buhro, *Science* 270, 1791 (1995).
2. Y. Zhang and H. Dai, *Appl. Phys. Lett.* 77, 3015 (2000).
3. Q. Gu, C. Cheng, R. Gonela, S. Suryanarayanan, S. Anabathula, K. Dai, and D. T. Haynie, *Nanotechnology* 17, R14 (2006).
4. H. He and N. J. Tao, *Encycl. Nanosci. Nanotech.* 2, 755 (2003).
5. D. Crouse, Y.H. Lo, A. E. Miller, and M. Crouse, *Appl. Phys. Lett.* 76, 49 (2000).
6. M. J. Zheng, L. D. Zhang, G. H. Li, and W. Z. Shen, *Chem. Phys. Lett.* 363, 123 (2002).
7. Y. H. Chen, X. T. Zhang, Z. H. Xue, Z. L. Du, and T. J. Li, *J. Inorg. Mater.* 20, 59 (2005).
8. L. Qian and X. Yang, *Talanta* 69, 957 (2006).
9. G. E. Possin, *Rev. Sci. Instrum.* 41, 772 (1970).
10. M. Lai, J. A. G. Martinez, M. Gratzel, and D. J. Riley, *J. Mater. Chem.* 16, 2843 (2006).
11. P. Bocchetta, M. Santamaria, and F. D. Quarto, *Electrochem. Solid State Lett.* 11, K93 (2008).
12. T. Thurn-Albrecht, J. Schotter, G. A. Kastle, N. Emley, T. Shibauchi, L. Krusin-Elbaum, K. Guarini, C. T. Black, M. T. Tuominen, and T. P. Russell, *Science* 290, 2126 (2000).
13. M. P. Zach, K. H. Ng, and R. M. Penner, *Science* 290, 2120 (2000).
14. D. Routkevitch, T. Bigioni, M. Moskovits, and J. M. Xu, *J. Phys. Chem.* 100, 14037 (1996).
15. C. M. Hangarter, M. Bangar, S. C. Hernandez, W. Chen, M. A. Deshusses, A. Mulchandani, and N. V. Myung, *Appl. Phys. Lett.* 92, 073104 (2008).
16. T. Ohgai, L. Gravier, X. Hoffer, and J. P. Ansermet, *J. Appl. Electrochem.* 35, 479 (2005).
17. Y. H. Huang, H. Okumura, G. C. Hadjipanayis, and D. Weller, *J. Appl. Phys.* 91, 6869 (2002).
18. C. D. Keating and M. J. Natan, *Adv. Mater.* 15, 451 (2003).
19. S. Gavrilov, L. Nosova, I. Sieber, A. Belaidi, L. Dloczik, and T. Dittrich, *Phys. Status Solidi A* 202, 1497 (2005).
20. B. D. Reiss, R. G. Freeman, I. D. Walton, S. M. Norton, P. C. Smith, W. G. Stonas, C. D. Keating, and M. J. Natan, *J. Electroanal. Chem.* 522, 95 (2002).
21. P. Wang, L. Gao, Z. Qiu, X. Song, L. Wang, S. Yang, and R. I. Murakami, *J. Appl. Phys.* 104, 064304 (2008).
22. S. Banerjee, A. Dan, and D. Chakravorty, *J. Mater. Sci.* 37, 4261 (2002).
23. Y. Xia, P. Yang, Y. Sun, Y. Wu, B. Mayers, B. Gates, Y. Yin, F. Kim, and H. Yan, *Adv. Mater.* 15, 356 (2003).
24. K. Nielsch, F. Müller, A. P. Li, and U. Gösele, *Adv. Mater.* 12, 582 (2000).
25. G. Wu, L. Zhang, B. Cheng, T. Xie, and X. Yuan, *J. Am. Chem. Soc.* 126, 5976 (2004).
26. M. Wang, X. Cao, H. Wang, G. Hua, Y. Lin, and L. Zhang, *Chem. Lett.* 35, 1234 (2006).
27. M. Lu, M. K. Li, L. B. Kong, X. Y. Guo, and H. L. Li, *Chem. Phys. Lett.* 374, 542 (2003).

28. T. E. Bogart, S. Dey, K. K. Lew, S. E. Mohney, and J. M. Redwing, *Adv. Mater.* 17, 114 (2005).
29. M. Aktary, C. E. Lee, Y. Xing, S. H. Bergens, and M. T. McDermott, *Langmuir* 16, 5837 (2000).
30. L. Suber, P. Imperatori, G. Ausanio, F. Fabbri, and H. Hofmeister, *J. Phys. Chem. B* 109, 7103 (2005).
31. Z. Cai, J. Lei, W. Liang, V. Menon, and C. R. Martin, *Chem. Mater.* 3, 960 (1991).
32. Z. Shi, S. Wu, and J. A. Szpunar, *Nanotechnology* 17, 2161 (2006).
33. D. C. S. Fullam, H. Rensmo, and D. Fitzmaurice, *Adv. Mater.* 12, 1430 (2000).
34. B. Liu and H. C. Zeng, *J. Am. Chem. Soc.* 125, 4430 (2003).
35. F. Zhou, X. Zhao, C. Yuan, and L. Li, *Cryst. Growth Des.* 8, 723 (2008).
36. X. Jiang, Y. Wang, T. Herricks, and Y. Xia, *J. Mater. Chem.* 14, 695 (2004).
37. Y. Wang and K. S. Kim, *Nanotechnol.* 19, 6 (2008).
38. Z. Li, X. Huang, J. Liu, and H. Ai, *Mater. Lett.* 62, 2507 (2008).
39. L. Sunyoung, M. H. Jin, M. Sunglyul, and K. Sang-Hyeob, *Mater. Res. Soc. Symp. Proc.* 1018, 5 (2007).
40. L. E. Greene, B. D. Yuhas, M. Law, D. Zitoun, and P. Yang, *Inorg. Chem.* 45, 7535 (2006).
41. B. Gates, B. Mayers, A. Grossman, and Y. Xia, *Adv. Mater.* 14, 1749 (2002).
42. R. V. Kumar, Y. Koltypin, X. N. Xu, Y. Yeshurun, A. Gedanken, and I. Felner, *J. Appl. Phys.* 89, 6324 (2001).
43. J. Zhang, J. Du, B. Han, Z. Liu, T. Jiang, and Z. Zhang, *Angew. Chem. Int. Ed.* 45, 1116 (2006).
44. M. P. Pileni, *Nat. Mater.* 2, 145 (2003).
45. A. Vantomme, Z. Y. Yuan, G. Du, and B. L. Su, *Langmuir* 21, 1132 (2005).
46. C. Xu, G. Xu, Y. Liu, and G. Wang, *Solid State Comm.* 122, 175 (2002).
47. N. R. Jana, L. Gearheart and C. J. Murphy, *J. Phys. Chem. B* 105, 4065 (2001).
48. J. Xiang, S. H. Yu, B. Liu, Y. Xu, X. Gen, and L. Ren, *Inorg. Chem. Comm.* 7, 572 (2004).
49. B. Zhang, G. Li, J. Zhang, Y. Zhang, and L. Zhang, *Nanotechnology* 14, 443 (2003).
50. J. D. Holmes, K. P. Johnston, R. C. Doty, and B. A. Korgel, *Science* 287, 1471 (2000).
51. V. Kumar, J. H. Kim, C. Pendyala, B. Chernomordik, and M. K. Sunkara, *J. Phys. Chem. C* 112, 17750 (2008).
52. R. Q. Zhang, Y. Lifshitz, and S. T. Lee, *Adv. Mater.* 15, 635 (2003).
53. Y. Chen, X. Cui, K. Zhang, D. Pan, S. Zhang, B. Wang, and J. G. Hou, *Chem. Phys. Lett.* 369, 16 (2003).
54. Y. Dai, Y. Zhang, Y. Q. Bai, and Z. L. Wang, *Chem. Phys. Lett.* 375, 96 (2003).
55. J. Zhang and F. Jiang, *Chem. Phys.* 289, 243 (2003).
56. H. Y. Dang, J. Wang, and S. S. Fan, *Nanotechnology* 14, 738 (2003).
57. K. Hong, M. Xie, R. Hu, and H. Wu, *Nanotechnology* 19, 085604 (2008).
58. Q. Li and C. Wang, *Appl. Phys. Lett.* 83, 359 (2003).
59. J. Niu, J. Sha, Z. Liu, Z. Su, and D. Yang, *Mater. Lett.* 59, 2094 (2005).
60. C. Ye, G. Meng, Y. Wang, Z. Jiang, and L. Zhang, *J. Phys. Chem. B* 106, 10338 (2002).
61. H. Pan, S. Lim, C. Poh, H. Sun, X. Wu, Y. Feng, and J. Lin, *Nanotechnology* 16, 417 (2005).

62. S. H. Sun, G. W. Meng, M. G. Zhang, X. H. An, G. S. Wu, and L. D. Zhang, *J. Phys. D* 37, 409 (2004).
63. X. C. Wu, J. M. Hong, Z. J. Han, and Y. R. Tao, *Chem. Phys. Lett.* 373, 28 (2003).
64. A. M. Morales and C. M. Lieber, *Science* 279, 208 (1998).
65. X. Duan and C. M. Lieber, *Adv. Mater.* 12, 298 (2000).
66. Y. Sun, G. M. Fuge, and M. N. R. Ashfold, *Chem. Phys. Lett.* 396, 21 (2004).
67. X. M. Meng, J. Q. Hu, Y. Jiang, C. S. Lee, and S. T. Lee, *Chem. Phys. Lett.* 370, 825 (2003).
68. J. R. Morber, Y. Ding, M. S. Haluska, Y. Li, J. P. Liu, Z. L. Wang, and R. L. Snyder, *J. Phys. Chem. B* 110, 21672 (2006).
69. W. Lee, M. C. Jeong, and J. M. Myoung, *Nanotechnology* 15, 254 (2004).
70. H. W. Kim and N. H. Kim, *Appl. Surf. Sci.* 233, 294 (2004).
71. Z. H. Wu, X. Y. Mei, D. Kim, M. Blumin, and H. E. Ruda, *Appl. Phys. Lett.* 81, 5177 (2002).
72. C. Chen, M. C. Plante, C. Fradin, and R. R. LaPierre, *J. Mater. Res.* 21, 2801 (2006).
73. L. E. Jensen, M. T. Bjork, S. Jeppesen, A. I. Persson, B. J. Ohlsson, and L. Samuelson, *Nano Lett.* 4, 1961 (2004).
74. Y. C. Choi, W. S. Kim, Y. S. Park, S. M. Lee, D. J. Bae, Y. H. Lee, G. S. Park, W. B. Choi, N. S. Lee, and J. M. Kim, *Adv. Mater.* 12, 746 (2000).
75. W. Lei, D. Liu, J. Zhang, L. Shen, X. Li, Q. Cui, and G. Zou, *J. Alloy Comp.* 459, 298 (2008).
76. J. Thangala, S. Vaddiraju, R. Bogale, R. Thurman, T. Powers, B. Deb, and M. K. Sunkara, *Small* 3, 890 (2007).
77. S. Vaddiraju, H. Chandrasekaran, and M. K. Sunkara, *J. Am. Chem. Soc.* 125, 10792 (2003).
78. T. Hanrath and B. A. Korgel, *J. Am. Chem. Soc.* 124, 1424 (2002).
79. S. Sharma and M. K. Sunkara, *J. Am. Chem. Soc.* 124, 12288 (2002).
80. M. Mozetic, U. Cvelbar, M. K. Sunkara, and S. Vaddiraju, *Adv. Mater.* 17, 2138 (2005).
81. M. K. Sunkara, S. Sharma, R. Miranda, G. Lian, and E. C. Dickey, *Appl. Phys. Lett.* 79, 1546 (2001).

4

Thermodynamic and Kinetic Aspects of Nanowire Growth

4.1 Introduction

Vapor–liquid–solid (VLS) growth of one-dimensional (1-D) materials involves three sequential steps: (1) selective and rapid dissolution of solutes of interest from the gas phase into molten metal droplet; (2) diffusional processes of solute through and on the surface of the droplet; and (3) precipitation of solute from molten metal droplet [1]. The original concept was conceived through accidental discovery using chemical vapor deposition of Si onto Au metal clusters as discussed in Chapter 2. It was then extended to a number of catalytic metals such as Fe, Co, and Ni. As the 1-D growth is enhanced with the catalytic metal at the tip, the technique is termed as "catalyst-assisted vapor–liquid–solid" growth. However, there are several questions to be asked: (a) What is the catalytic action of the metal cluster in the growth of a nanowire? (b) Can one use any metal cluster for growing nanowires? and (c) What factors inhibit nucleation from the molten metal cluster and lead to tip-led growth of a single nanowire per metal cluster?

The questions can be answered by looking at the general schematic as seen in Figure 4.1, showing various processes that occur during a typical VLS growth of a nanowire with a molten metal droplet at its tip. These processes include (1) material dissociation and adsorption at the metal surface; (2) bulk diffusion in the catalyst and surface diffusion on the droplet; (3) adatom diffusion from the substrate onto the nanowire surface and the metal droplet; (4) the adsorption and desorption processes therein; (5) precipitation processes at the catalyst–nanowire interface; (6) material competition for the gas-phase species via direct impingement and adatom flux. All these processes can be divided into three broad categories: (a) dissolution at vapor–liquid interface; (b) diffusional processes for enhancing adatom flux to liquid–solid interface; (c) precipitation at liquid–solid interface.

The precipitation from the supersaturated melt droplet should follow simple rules of nucleation and growth. The primary question is, how do certain metals support tip-led growth by avoiding multiple nucleation events from individual droplets? It is necessary to understand the underlying thermodynamics of nucleation from molten metal clusters and factors controlling the diameter and growth direction of the resulting nanowires under such

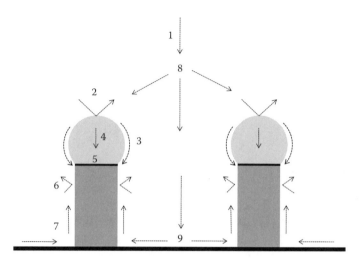

FIGURE 4.1
Schematic of the various processes contributing in VLS nanowire growth: (1) supply of species in the vapor phase; (2) catalytic adsorption and desorption of the species at the metal droplet surface; (3) surface diffusion of the adsorbed species toward the metal–solid interface; (4) bulk diffusion of the adsorbed species in the metal droplet toward the interface; (5) material incorporation at the droplet–solid interface; (6) adatom adsorption and desorption processes on the nanowire surface; (7) adatom diffusion from the substrate to the nanowire; (8) material distribution in vapor phase (direct impingement on droplets) due to proximity of other competing growth processes; and (9) distribution of adatom flux due to proximity of other competing growth processes.

conditions. In many practical applications, it is essential to promote tip-led growth of one nanowire using a metal cluster. In these applications, it is often important to choose the metal and obtain a control on the growth direction, surface faceting, and growth kinetics. The nucleation from supersaturated metal droplets needs to be understood in terms of which factors affect the nucleation and how one can choose a metal for promoting the tip-led growth of nanowires for a number of technologically important materials.

Irrespective of the type of nucleation and growth mechanisms involved, the growth kinetics and morphology of resulting nanowires depend directly on the dissolution kinetics at the vapor–liquid interface and diffusional processes for adatoms. The use of certain precious metals and heterogeneous catalyst metals can enhance the dissociative adsorption on their surfaces. Under molten state, such enhanced dissociative adsorption of gas-phase species involving solutes of interest (for example, dichlorosilane or silyl radical for Si) can enhance the selective dissolution of solutes. The true catalytic action of certain metals—such as Au, Fe, Co, Ni, Pt, Pd, etc.—is selective enhancement of dissolution kinetics. Although there is very little known about the molecular mechanisms involved with the dissociative adsorption and subsequent dissolution of solutes from various gas-phase precursors, one can use simple thermodynamic and kinetic arguments to understand the factors that control the growth kinetics. Thus, the kinetic modeling of such processes is of fundamental interest for controlled growth of nanowires in a number of important, practical applications.

In this chapter, the thermodynamic and kinetic aspects of various processes involved in nanowire growth using a generic VLS scheme are discussed. These concepts can be applied to a variety of other schemes such as vapor–solid–solid, etc.

4.2 Thermodynamic Considerations for Vapor–Liquid–Solid Growth

In a typical VLS scheme, the molten metal droplets are created and exposed to gas phase containing solutes. In the initial stages, the dissolution into molten metal droplet can setup supersaturation, which drives the precipitation. Such supersaturation can often lead to nucleation throughout its surface or can just lead to precipitation at the substrate–droplet interface followed by homoepitaxy. Thermodynamic formulations related to nucleation phenomena from supersaturated molten metal melts are discussed in the following. The discussion includes a rational basis for choosing a metal for avoiding nucleation and promoting tip-led growth. Following this, several models for the molten metal droplet–solid interface thermodynamics are discussed to understand which conditions can lead to different morphologies: straight, tapered, and helical nanowires.

4.2.1 Thermodynamic Considerations of Nucleation from Molten Metal Droplets

4.2.1.1 Gibbs–Thompson Relationship

The formation of a molten metal droplet can be explained by the minimization of its free energy. The difference in chemical potential (μ) between a curved surface (radius, r) and planar surface (radius, $r=\infty$) can be obtained by Gibbs–Thompson equation.

$$\mu_r - \mu_\infty = \frac{2\gamma\Omega}{r} \tag{4.1}$$

where

  γ is the surface energy of the droplet
  Ω is the molar volume of the species within the
    nuclei

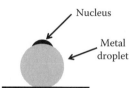

The above equation can be used to describe critical
diameter of pure phase nuclei and the vapor–liquid
equilibrium differences for spherical droplet and the
planar surface represented in Figure 4.2. Assuming
the same reference state for gas-phase species, one
can express chemical potential of gas-phase species
in terms of partial pressures. At equilibrium, the
chemical potentials are given by

FIGURE 4.2
Schematic of the formation
of a solid crystal nucleating
out of a metal droplet due
to the supersaturation in
the metal droplet.

$$\text{For the droplet} \qquad \mu_l = \mu_g \qquad (4.2)$$

$$\text{For the planar surface} \qquad \mu_l^\infty = \mu_g^\infty \qquad (4.3)$$

Subtracting Equation 4.3 from Equation 4.2 results in

$$\mu_g = kT \ln\left[\frac{p}{p^\infty}\right] \qquad (4.4)$$

where

  p/p^∞ represents vapor-phase supersaturation within the gas phase
  k is the Boltzmann constant
  T is the temperature

Using Equations 4.1 through 4.4, the concentration (partial pressure, p) of the
gas phase in equilibrium with a liquid droplet of known size is given as

$$kT \ln\left[\frac{p}{p^\infty}\right] = \frac{2\gamma\Omega}{r} \qquad (4.5)$$

It can be observed that the partial pressure of the species has to be greater in
equilibrium with a droplet than for a planar surface, indicating the need for
supersaturation. One can write the partial pressure of gas-phase solutes that
will be in equilibrium with a droplet of radius, r, P_r as following:

$$P_r = P^\infty \exp(2\gamma\Omega / rkT) \qquad (4.6)$$

where P^∞ is the partial pressure of the gas-phase solutes in equilibrium with
a planar surface. One can also interpret the above equation for estimating the

critical nucleation for pure phase condensation (say, molten metal droplets from metal vapor).

$$r_c = \frac{2\gamma\Omega}{RT \ln\left[\dfrac{p}{p^\infty}\right]} \tag{4.7}$$

Here r_c is the critical size of nuclei. Supersaturation, in this case, can be varied by changing the partial pressure, p.

4.2.1.2 Nucleation from Molten Metal Alloy Droplet

Nucleation of a solid-phase growth species out of a spherical metal droplet is shown in Figure 4.2. Thermodynamically, Gibbs free energy minimization is the criterion to satisfy. Gibbs free energy change for the formation of the nuclei is expressed as

$$\Delta G_T = \Delta G_V \times \tfrac{4}{3}\pi r^3 + \Delta G_S \times 4\pi r^2 \tag{4.8}$$

where the first term on the right is the free energy change due to change in volume and the second term is the effect of the curvature of the nucleus determined by the surface area created. For the growth of nuclei to be thermodynamically favorable, $\Delta G_T < 0$. The critical nuclei size can be estimated from the condition:

$$\frac{\partial \Delta G_T}{\partial r} = 0 \tag{4.9}$$

This leads to the expression for critical radius r^* as

$$r^* = \frac{-2\sigma}{\Delta G_V} \tag{4.10}$$

where
    r^* is the critical nuclei radius
    σ is the interfacial energy

ΔG_V is the volume free energy given by the expression:

$$\Delta G_V = \frac{RT}{\Omega} \ln\left(\frac{C}{C^*}\right) \tag{4.11}$$

where

    C and C^* represent solute concentration within the liquid alloy and equilibrium concentration, respectively

    C/C^* then becomes the driving force for the nucleation process

The critical nucleus diameter d_c is given by

$$d_c = \frac{4\sigma\Omega}{RT \ln\left(\dfrac{C}{C^*}\right)} \qquad (4.12)$$

C^* is a function of temperature and can be determined as the equilibrium composition from the liquidus line of the binary phase diagram. On the other hand, the supersaturation, C, that exists within the droplet in the initial stages for nucleation is difficult to determine experimentally and also can depend upon the process conditions. Gosele and coworkers [2] suggested that the composition in the liquid droplet can be correlated with the partial pressure of solute in the gas phase under steady stages of growth. Assuming that this condition holds in the early stages, Equation 4.10 can be rewritten as

$$r_c^{min} = \frac{2\Omega^s \sigma^s}{k_B T \ln\left(\dfrac{P_{Si}}{P_{Si}}\right) + k_B T \ln\left(\dfrac{P_{Si}}{P_{Si}^{eq}}\right)} \qquad (4.13)$$

where

    P is the pressure in the reference state

    Ω^s is the molar volume of silicon

    σ^s is the surface energy of the droplet

The driving force in Equation 4.13 (denominator on the right), comprises of two supersaturations; one in the liquid droplet that causes nucleation and another in the vapor phase that maintains the spherical shape of the droplet. Again, Equation 4.13 is difficult to use due to the unknown values for the reference state (P_{Si}) and the equilibrium partial pressure P_{Si}^{eq}. Equation 4.13 may be more appropriate during steady-state growth under precipitation kinetics limited conditions.

4.2.1.3 Nucleation from Various Molten Metal Droplets

Equation 4.13 can be used to estimate the critical size of nuclei for different metal–solute systems. Let us consider the Au–Si system in which the equilibrium solubility of Si at the experimental temperatures is always in excess of

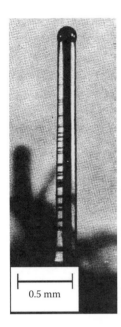

FIGURE 4.3

SEM micrograph of a nickel droplet tip-led growth of a silicon whisker demonstrating the VLS growth mechanism. (Reproduced from Wagner, R.S. and Ellis, W.C., *Appl. Phys. Lett.*, 4, 89, 1964. With permission.)

19.8 at% (the eutectic composition). Even for a dissolved concentration of 50 at%, the resulting critical size for nucleation can be as large as several microns. If one were to expose submicron size Au droplets to the gas phase, multiple nuclei will not be able to form as the nuclei size will be greater than the droplet size. Under these conditions, the precipitation at solid–liquid interface will lead to one nanowire with the droplet at its tip ("tip-led" growth mode) with no nucleation step. Experiments using "catalytic" transition metals such as Ni [1], Al [3], Cu [4], Fe [5], and Au [5] have led to the metal droplet tip-led growth of individual nanowires as shown in Figure 4.3. If the droplet is very large when compared to the critical nucleus size, multiple nucleation out of the droplet is possible. This has indeed been observed in the case of silicon whisker growth with Ni catalyst [6], which has a phase diagram similar to Au–Si with a high solubility eutectic with silicon. A bunch of silicon nanowires have been observed to grow out of a very large ca. 0.5 mm size single droplet as shown in Figure 4.4.

The difference between low-melting metals and catalytic metals is the considerably lower eutectic solubility of growth species in the low-melting metals. A side-by-side comparison of the binary phase diagrams of Au–Si and Ga–Si systems [7] is presented in Figure 4.5. Gallium forms a eutectic with silicon at about 29°C with an extremely low silicon fraction

FIGURE 4.4

SEM micrograph of multiple silicon whiskers growing out of one large nickel droplet. (Reproduced from Wagner, R.S. et al., *J. Appl. Phys.*, 35, 2993, 1964. With permission.)

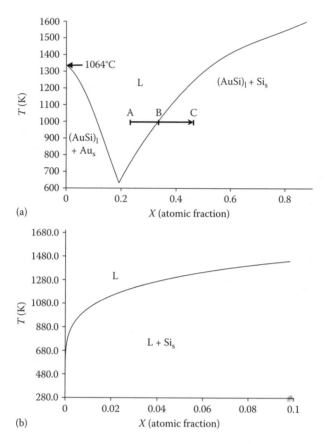

(a)

(b)

FIGURE 4.5

Comparison of eutectic compositions in case of (a) Au–Si system with ~18% solubility of Si in Au and (b) Ga–Si system with ~10^{-5}% Si solubility in Ga. (Reproduced from Sharma, S. et al., *Encyclopedia of Nanoscience and Nanotechnology*, Nalwa, H.S. (ed.), Vol. 10, The American Scientific Publishers, Los Angels, CA, 2003, 327. With permission.)

~10–5 at%. At 1000 K, with an arbitrary dissolved silicon concentration of less than 3 at% in gallium, Equation 4.12 yields the critical nucleus diameter to be 10 nm. The small nuclei size allows for multiple nucleation events on the droplet surface. Once nucleation occurs, further growth can occur via basal attachment or liquid-phase epitaxy. Figure 4.6 shows the variation of critical nuclei diameter as a function of dissolved silicon concentration for Au–Si and Ga–Si systems [7]. Less than 10 at% dissolved silicon at 1000 K will result in 5 nm nuclei while greater than 40 at% Si will be needed in an Au pool for the nuclei diameters to be less than 100 nm, which is not realizable within the limits of ideal solution approximation. On the contrary, multiple nanowires have been observed to grow out of a metal droplet in the case of a low-melting metal, such as Ga [8–10] and In [9,11], mediated

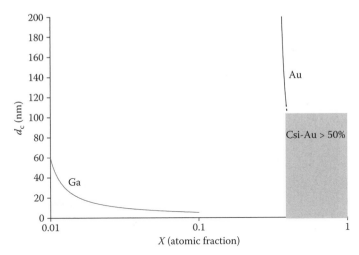

FIGURE 4.6
Critical nuclei diameter as a function of the dissolved silicon concentration (atomic %) in Ga–Si and Au–Si binary solutions ($T = 1000\,K$). (Reproduced from Sharma, S. et al., *Encyclopedia of Nanoscience and Nanotechnology*, Nalwa, H.S. (ed.), Vol. 10, The American Scientific Publishers, Los Angels, CA, 2003, 327. With permission.)

FIGURE 4.7
SEM micrograph of high density of bismuth nanowires growing out of a gallium droplet (low-melting metal mediated) due to spinodal decomposition and spontaneous nucleation. (From Bhimarasetti, G. and Sunkara, M.K., *J. Phys. Chem. B*, 109, 16219, 2005. With permission.)

growth as seen in Figure 4.7. As explained before, the extent of supersaturation within the molten metal alloy droplet can be dependent upon process conditions and under some process conditions, it is possible to achieve tip-led growth of a single silicon nanowire with a Ga droplet at the tip as shown in Figure 4.8.

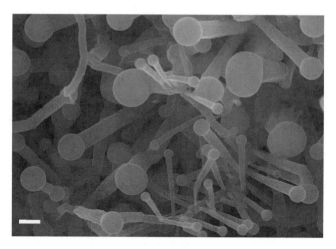

FIGURE 4.8
SEM micrograph of gallium tip-led silicon nanowires synthesized via exposing a gallium coated silicon wafer to hydrogen plasma. The scale bar is 200 nm. (Courtesy of P. Meduri.)

4.2.1.4 Thermodynamic Estimation of Supersaturation for Spontaneity of Nucleation

Most VLS processes with low-melting metals are typically carried out under isothermal conditions. Under isothermal conditions, the information about the maximum supersaturation (similar to the concept of under-cooling for pure metals), at which the melts become completely unstable, is important to determine the expected critical nuclei size. An attempt is made using thermodynamic stability analysis to estimate such supersaturation limit [12]. For a binary system, the equilibrium between the solution and the solid exists when the free energy of mixing is the lowest for any given temperature. This gives rise to the liquidus line in the binary phase diagram:

$$\left(\frac{\partial G_l}{\partial X}\right)_{T,P} = 0 \qquad (4.14)$$

where
  G_l is the Gibbs free energy
  X is the composition

The concentration that satisfies the above equation is the equilibrium solubility of the solute at a given temperature. Metastable regions in the phase diagram are known to exist where the solution is supersaturated with the solute species and do not segregate. Continuous nucleation and growth from such metastable melts are quite possible, as predicted by classical nucleation theory. Alternatively, it is also possible to destabilize the supersaturated melt such that the phase segregation occurs spontaneously. The critical limit (neutral stability) of the solution beyond which the solution spontaneously decomposes into two distinct phases is defined by its second-order derivative:

$$\left(\frac{\partial^2 G_l}{\partial X^2}\right)_{T,P} = 0 \tag{4.15}$$

The necessary condition for a stable supersaturated melt is that its second-order derivative is less than zero. In order to determine the point of neutral stability, let us consider the change in free energy of solution. From Gibbs–Duhem relationship, the change in free energy of the solution can be written as

$$\Delta G_l = X\,d\mu + (1-X)d\mu_S \tag{4.16}$$

where
  X is the mole fraction of solute
  μ and μ_S represent the chemical potentials of solute and metal droplet, respectively

After expanding the chemical potential terms based on regular solution model and defining an interaction parameter, ξ, the Gibbs energy change for solution as a function of T and X as the following:

$$\Delta G_l = RT(X\ln X + (1-X)\ln(1-X)) + \xi X(1-X) \tag{4.17}$$

The interaction parameter ξ is determined by the relation $\xi = a - bT$, where a and b are material-specific property constants. Assuming that the critical temperature is at $X=0.5$, the composition, X, at which the solution tends to spontaneously decompose into two phases (solid and liquid phases) at a constant temperature T is determined by solving Equations 4.12 and 4.16 simultaneously. In this manner, the loci of points in T and X at which the second differential is zero are determined to describe the spinodal composition line of solute (Si or Ge) in solvent (Ga or In) binary systems. The equilibrium phase diagram and the spinodal curve for Ge–Ga binary system are shown in Figure 4.9. The point of neutral stability or the spinodal limit is assumed to be the supersaturation at which spontaneous nucleation of the solute from within the low-melting melts occurs. The following equation based on the energy minimization of nuclei formation could be used for estimating the nucleus size (or the resulting nanowire diameter) similar to Equation 4.12:

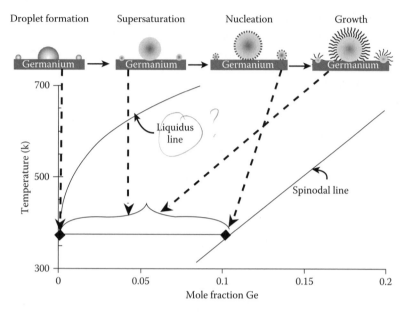

FIGURE 4.9

Plot showing the liquidus line and the spinodal line for Ge–Ga binary system obtained using the regular solution model using interaction parameter of 16.7 kJ/mol [5]. The proposed nanowire nucleation and growth mechanism involves: (a) formation of Ga droplets on Ge substrate; (b) supersaturation of the Ga droplets; (c) spontaneous nucleation of Ge on Ga at the point of instability; and (d) growth of the Ge nuclei in 1-D at composition between the liquidus and spinodal curves. (Reproduced from Chandrasekaran, H. et al., *J. Phys. Chem. B*, 110, 18351, 2006. With permission.)

$$d_c = \frac{4\Omega\sigma}{RT\ln\left(\dfrac{X}{X^1}\right)} \tag{4.18}$$

where

    d_c is the diameter of the resulting nanowire

    σ is the interfacial energy

    X/X^1 is the ratio of solute concentration at the point of instability to the corresponding equilibrium solubility at a given temperature, T

The schematic of the proposed nucleation and growth mechanism is also illustrated in Figure 4.9. As indicated in Figure 4.9, after the nucleation, further growth could occur at solute supersaturation levels below the spinodal limit at any given temperature. Subsequent dissolution into the metal droplet beyond nucleation stage sets up a solute concentration supersaturation level that is well below the instability limit for yielding the necessary steady growth rate based on the dissolution rate.

Interfacial energy of the droplet–crystal interface needs to be determined accurately for a reliable estimation of critical nuclei diameter. For a molten metal droplet in contact with a solid surface at an angle θ, the relation between the surface and interfacial energies for the droplet to be in equilibrium is governed by the relation

$$\gamma_{lv} \cos\theta = \gamma_{sv} - \gamma_{sl} \tag{4.19}$$

where
γ_{lv} is the liquid–vapor surface energy
γ_{sv} is the solid–vapor surface energy
γ_{sl} is the solid–liquid interfacial energy, which is used to calculate the critical nuclei size (Equation 4.18)

Surface and interfacial energy data are not readily available in the literature for all the material systems. The surface energy needs to be determined experimentally for systems where data are not available. In the case of Ga–Sb system, surface energies of Ga [13] and Sb [14] are available, which can also be estimated using optical measurement of droplet contact angle [13] or via surface tension measurements [14]. Thermodynamic data can also be used to calculate interfacial energy in the case of Ga–GaSb [15]. Ga in the liquid state is nonwetting on solid GaSb. The contact angle is assumed to be in the range of 150°–180°, which is used to calculate a range for the surface energy of GaSb. Now, using the surface energy for Sb and GaSb, and assuming nonwetting behavior, a range of interfacial energy of Sb–GaSb is estimated. For a known atomic fraction (concentration supersaturation) determined from the phase diagram at the synthesis temperature, molar volume, and an average value of the interfacial energy, the critical nuclei size for a GaSb crystal precipitating out of Sb droplet is estimated using Equation 4.18.

4.2.1.5 Rational Choice of Metal for Tip-Led Growth of Nanowires (Avoiding Nucleation)

In general, if one were to choose a metal for growing given nanowires using tip-led growth mode, it is reasonable to use the eutectic solubility or the liquidus curve. This can be illustrated with a model system, such as Ga–Sb, for which the phase diagram is shown in Figure 4.10. This system has two eutectic points: region I is in the excess Ga part of the phase diagram, with a low solubility of ~10–4 at% Sb in Ga (similar to most low-melting metal mediated growth systems); and region II is in the Sb rich part, with a high solubility of 10.6 at%, Ga in Sb.

Operating in region II of the Ga–Sb phase diagram should lead to the tip-led growth of individual GaSb nanowires with Sb at the tips. The large

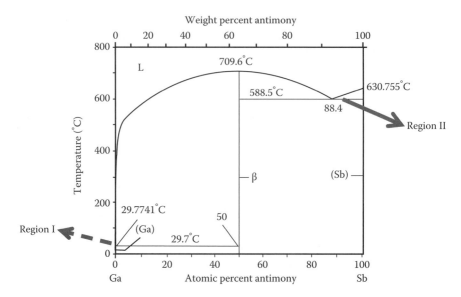

FIGURE 4.10
Binary phase diagram of the Ga–Sb system. The two eutectic points are marked out. Region I is the eutectic in the Ga-rich zone, with very low solubility of Sb in Ga and Region II is the eutectic in the Sb-rich zone, with a high solubility of Ga in Sb. (From Pendyala, C. et al., *Semicond. Sci. Tech.*, submitted, 2009.)

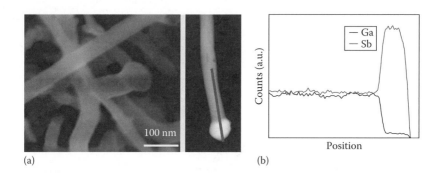

(a)   (b)

FIGURE 4.11
(a) SEM micrograph and TEM EDS line scan on GaSb nanowires. The SEM micrograph clearly shows GaSb nanowires with a distinct tip-led growth. (b) The line scan shows that the wire is GaSb and the tip comprises of excess antimony. (From Pendyala, C. et al., *Semicond. Sci. Tech.*, submitted, 2009.)

equilibrium solubility of Ga in Sb should not lead to spontaneous nucleation, but instead allow for the tip-led growth with Sb acting as the growth mediating metal. Figure 4.11a shows a SEM micrograph of GaSb nanowires with distinct tip-led growth. TEM EDX line scan along the length of the nanowire confirms the wire to be GaSb, with Sb (group V species) at the tip [16] (see

Figure 4.11b). A group V species droplet tip-led nanowire growth is unlike the generally observed group III metal droplet tip-led growth. The observed result can be explained based on the region II of Figure 4.10. In this region of high solubility of Ga in Sb, the critical nuclei size is comparable to the size of Sb droplets used. This does not allow the spontaneous nucleation step to occur, thus causing one Sb droplet to lead the tip-led growth of one GaSb nanowire. This result validates the mechanism that eutectic or equilibrium solubility can be used as a criterion to predict the mode of nucleation and nanowire growth for VLS schemes.

This nucleation and growth model is generic and can be extended to other material systems with suitable binary phase behavior. Figure 4.12 is the phase diagram of In–Sb system, which has a higher solubility at both the eutectic regions. At the In-rich region, the solubility of Sb in In is ~1 at% and at the other eutectic, the solubility of In in Sb is ~31 at%. Following the same mechanism, the high solubilities of both the eutectics in In–Sb system should lead to tip-led growth with In tips at the In-rich eutectic and Sb tips at the Sb-rich eutectic. Figure 4.13a is the SEM micrograph of the as-synthesized nanowires and the SEM EDX line scan (Figure 4.13b) confirms the wire to be InSb with a Sb-rich tip. The synthesis conditions moved the system to Sb-rich region II of the phase diagram, which led to the tip-led growth of InSb nanowires with Sb at the tips.

The examples shown here illustrate the significance of both eutectic and equilibrium solubility in the design of nanowire growth via VLS schemes. If tip-led growth is the requirement, metals with high eutectic

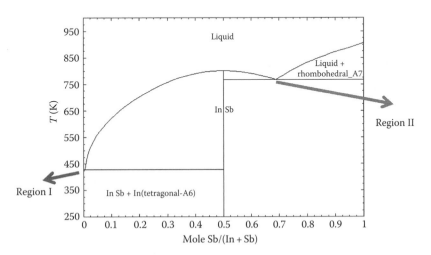

FIGURE 4.12
Binary phase diagram of the In-Sb system. The two eutectic points are marked. Region I is the eutectic in the In rich zone with ~1% solubility of Sb in In. Region II is the eutectic in the Sb-rich zone with ~31% solubility of In in Sb. (From Pendyala, C. et al., *Semicond. Sci. Tech.*, submitted, 2009.)

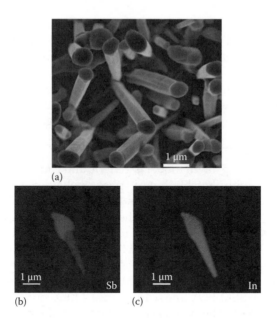

FIGURE 4.13

(a) SEM micrograph of as synthesized InSb nanowires showing a distinct tip-led growth. SEM EDX mapping of a InSb nanowire (b) Sb map and (c) In map showing the wire to be InSb with the tip being Sb rich. (From Pendyala, C. et al., *Semicond. Sci. Tech.*, submitted, 2009.)

solubility of the growth species should be chosen with the use of small droplet sizes.

4.2.1.6 Experimental Conditions for Promoting Tip-Led Growth Using Any Molten Metal

The discussions so far on the modes of nucleation and the use of eutectic solubility can be applied to design experiments to yield nanowires with desired growth mechanism. Tip-led growth of nanowires is very important for devices with individually addressable nanowires. The choice of material system determines the experimental conditions that can be employed to yield tip-led growth.

It is evident that for metal–solute systems with a reasonably high solubility, the critical nucleus size is expected to be larger than the metal droplet size and does not lead to multiple nucleation. For the metal–solute systems with very low solubility, if the dissolved solute concentration reaches a supersaturation limit, then these systems will undergo spontaneous nucleation. Again, if the dissolved concentration does not reach such supersaturation limit, then the nucleation can be avoided and a tip-led growth can be established. It has been experimentally found that such conditions can be established for systems involving low-melting metals for growing III–V compounds such

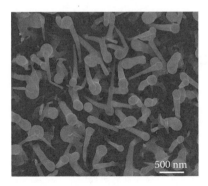

FIGURE 4.14
SEM micrograph of In tip-led InN nanowires grown via reactive vapor transport. (Courtesy of C. Pendyala.)

as oxides, nitrides, etc. In these cases, it is clearly observed that "tip-led" growth using low-melting metals can be established even though the solubility of nitrogen or nitride in metal melt is expected to be very low. In these experiments, both group III metal and group V species are supplied onto the substrate in a reactive vapor transport mode. Figure 4.14 provides an example involving indium droplet-led growth of InN nanowires. Growth of III–V semiconductor nanowires such as GaN [11], GaSb [9], InN [10], and InSb [9] has also been accomplished using similar schemes. Since the mediator metal is one of the growth species, this approach has been called the "self-catalytic scheme," which is gaining popularity to avoid contamination from the use of foreign metals.

4.2.2 Interfacial Energy and Tip-Led Growth

The spontaneous nucleation from larger droplets can provide high density of nanowires and their production in large quantities. Such nucleation and growth, however, do not allow for either epitaxial or vertical nanowire arrays, which are only possible with tip-led growth mode. For many practical applications that require nanowire arrays, it is preferable to understand the growth of individual nanowire with one droplet at its tip.

4.2.2.1 Role of Interfacial Energy in the Nanowire Growth Stability

4.2.2.1.1 Cylindrical/Tapered Morphologies

There are many reports, especially for silicon nanowires, of straight cylindrical as well as tapered morphologies. To put things in perspective, the growth of straight nanowires can be categorized as a special case of conical nanowires (conical angle=0). The metal droplet (characterized by the material and the contact angle with the growth front) controls the interfacial energy droplet–nanowire interface [17]. The stability of the droplet is

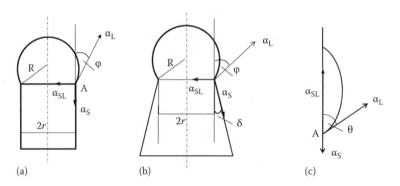

FIGURE 4.15
Schematic of whisker geometry during growth: (a) at constant radius ($\delta=0$); (b) at decreasing radius ($\delta>0$); and (c) three-phase equilibrium (when droplet wets the lateral facets). (Reproduced from Nebol'sin, V.A. and Shchetinin, A.A., *Inorg. Mater.*, 39, 899, 2003. With permission.)

controlled by the interplay between the various interfacial and bulk tensions. Under mechanical equilibrium, the droplet will always be in contact with the terminal nanowire growth face and will never contact the lateral facets (Figure 4.15).

Growth of a new facet always occurs with minimization in Gibbs free energy and the condition for growth to occur is the droplet wetting the surface of the newly formed step. Using this condition, the range of contact angles φ where whisker growth is thermodynamically possible is given by the inequality

$$\frac{\alpha_S}{\alpha_L} < \frac{\sin\varphi + \cos\varphi}{\cos\delta - \sin\delta} \tag{4.20}$$

where
 α_S and α_L are the solid–vapor and liquid–vapor interface tensions, respectively
 δ is the conical angle

The plot of the inequality against φ for silicon whiskers growing in Au-Si system, as shown in Figure 4.16, provides a measure of the available regions of growth where the conditions follow Equation 4.20. For the case of silicon nanowire growth, the region between y (α_S/α_L)=1.33 and curve 1 is the region for stable nanowire growth for a straight nanowire. In this region, the droplet shape is favorable for nanowire growth. In the regions that do not fit the criteria, the nucleated crystals can be dissolved back into the droplet or form negative whiskers (dissolution of the substrate), which makes them unfavorable for nanowire growth. The region suitable for conical nanowire growth is larger than that for straight nanowires covered between curve

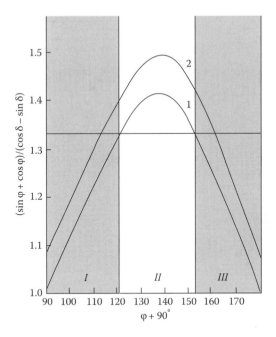

FIGURE 4.16

Plot of $\sin\varphi + \cos\varphi/\cos\delta - \sin\delta$ against contact angle of the metal droplet on the whisker tip: (I) for constant radius ($\delta=0$); (II) for conical whisker ($\delta>0$). (I and III are the regions of whisker etching and vaporization. Growth criterion not met in these regions. The part of region (II) above the line is the region of steady-state whisker growth. Here, the droplet shape favors whisker growth. (Reproduced from Nebol'sin, V.A. and Shchetinin, A.A., *Inorg. Mater.*, 39, 899, 2003. With permission.)

2 and $y=1.33$. This shows that the formation of conical nanowires is more favorable since the corresponding area is larger.

From Equation 4.20, it is clear that materials with large α_L will have a better chance to follow the condition for stable growth. Such a condition could be used for the selection of a specific metal system (liquid) for leading stable growth of a specific nanowire material (solid). In reality, the metal droplet tends to be a rich alloy with significant solute fraction (necessary for tip-led growth) whose properties vary widely from that of pure molten metal.

4.2.2.1.2 Spiral, Springlike Morphologies

Growth of amorphous spiral nanowires has been observed in some cases as seen in Figure 4.17. Here, a stable metal droplet theory cannot explain the observations of spiral growth. Instability of the catalyst droplet on the nanowire growth front has been proposed as the cause for the growth [18]. Also, the amorphous nature of the wires eliminates the role of any crystal imperfections in the growth process. Precise reasons for the lack of crystallinity of the structures are not known.

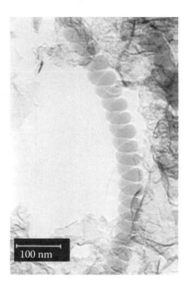

FIGURE 4.17
TEM image of amorphous boron carbide nanosprings. (Reproduced from McIlroy, D.N. et al., *J. Phys.: Condens. Matter*, R415, 2004. With permission.)

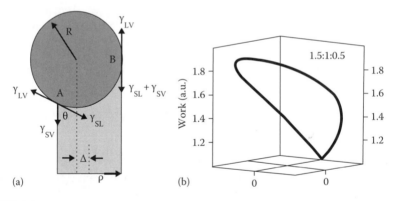

FIGURE 4.18
(a) Schematic of the catalyst droplet of radius R on a nanowire of radius ρ, showing the asymmetry that causes the imbalance of forces. (b) Plot of work of adhesion of the droplet to the nanowire showing the asymmetrical distribution which causes preferential material deposition at the cusp. (Reproduced from Nebol'sin, V.A. and Shchetinin, A.A., *Inorg. Mater.*, 39, 899, 2003. With permission.)

A symmetric droplet will lead to a cylindrical or conical nanowire, whereas periodic asymmetry is needed for helical growth. Contact angle asymmetry is the reason for the observed helical growth, which can be explained based on the interplay of the various interfacial energies acting nonuniformly on the metal droplet. In the case of an asymmetric droplet on the nanowire surface as shown in Figure 4.18a, the droplet is displaced from the center of the nanowire axis. This creates a nonzero torque on the droplet that can cause it to shear off the wire. The work required to shear the droplet from the nanowire surface can be expressed as

$$W_A = \gamma_{SV} + \gamma_{SL} - \gamma_{LV} = \gamma_{SV}(1 - \cos\theta) \tag{4.21}$$

where

W_A is the work of adhesion acting on the metal droplet
γ_{SV} is the solid–vapor interfacial surface tension
γ_{SL} is the solid–liquid interfacial surface tension
γ_{LV} is the liquid–vapor interfacial surface tension
θ is the angle between γ_{SV} and the plane containing γ_{SL} and γ_{LV}

Work of adhesion then becomes a function of the interfacial tensions and the contact angle. The asymmetry of the work of adhesion is very evident in Figure 4.18b. Material addition on the growth front from the droplet is maximum at the minimum point on the work of adhesion curve. So the material addition rate (growth rate) is asymmetric and maximum at the farthest point of the droplet displacement, which makes the droplet move back toward the nanowire. Because there is a cusp in the work of adhesion curve, any displacement will drive the system into a spiral trajectory, causing the growth of spiral nanowires.

4.2.2.2 Role of Interfacial Energy in Nanowire Faceting

The development of lateral facets on nanowire surfaces is a phenomenon that has been observed but has not been fully understood. Along with faceting, another prominently observed feature is the sawtooth morphology on the nanowire facets as shown in Figure 4.19. Ross et al. [19] proposed a mechanism based on the change in the contact angle between the droplet and the nanowire terminal facet for this periodic instability as illustrated in Figure 4.20. Silicon nanowires, expanding or contracting, preferentially grow in <111> direction as the surface energy is minimum for the close packed <111> singular surfaces. This leads to a hexagonal cross-section of the terminal facet where alternating sides are wider and narrower. The observation of sawtooth morphology is also prominent on the narrower facets and not so much on the wider facets.

Under the case of expanding nanowire, the droplet contact angle increases with increasing nanowire

FIGURE 4.19
SEM micrograph of silicon nanowire showing "SAWTOOTH" faceting on the lateral facets. (Reproduced from Ross, F.M. et al., *Phys. Rev. Lett.*, 95, 146104, 2005. With permission.)

100 μm

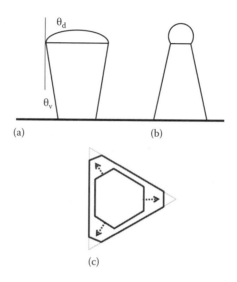

FIGURE 4.20
Schematic of the model used to explain sawtooth faceting. (a, b) Represent the expanding and contracting nanowire growth respectively. (c) Shows the forces on the facets during the oscillatory growth. (Reproduced Ross, F.M. et al., *Phys. Rev. Lett.*, 95, 146104, 2005. With permission.)

diameter. This causes an inward force, which leads to the formation of a narrowing inclined facet. This slowly leads to the formation of a triangular cross-section due to the faster disappearance of three narrower planes, which at one point becomes unfavorable due to higher surface energy. This leads to growth of outward growing facets, and the narrower facets start growing back, until energetically unfavorable condition is met and the reverse process starts occurring. The amplitude and period of such oscillations are larger for narrow facets than for wide facets. Amplitude for such periodic oscillations can be expressed [19] as

$$\lambda \sim L \frac{\Gamma_c}{w_d} \frac{1}{4\theta_w^2}$$

(4.22)

where
L is the length of the nanowire
Γ_c is the critical force
w_d is the work done against the surface tension of the droplet
θ_w is the widening angle of the facet

The observation of narrow facets of the hexagon having more pronounced sawtooth faceting is due to the larger amplitude and period of such oscillations for the narrower facets. Although the model explains the observed growth, it fails to recognize the processes that lead to the conical growth of nanowire. The observed faceting and periodic oscillatory sawtooth growth on silicon nanowires can also be explained following a different route [20]. In this case, the droplet does not wet the lateral surface and the condition for the droplet to be in mechanical equilibrium with the nanowire terminal facet is given by [20]:

$$\alpha_{SL} + \alpha_L \cos\theta = \alpha_S(111) \tag{4.23}$$

where

α_{SL} is the solid–liquid interfacial energy
α_L is the liquid surface energy
θ is the contact angle of the droplet with the silicon surface as indicated in Figure 4.21I

The formation of an inclined plane at an angle δ to the three-phase line (see Figure 4.21II) by the movement of the droplet over the surface, is possible only if

$$\alpha_{SL} + \alpha_L \cos(\varphi + \delta) > \alpha_S(111) \tag{4.24}$$

(or)

$$(\varphi + \delta) < \theta \tag{4.25}$$

From Equations 4.23 through 4.25, the condition for faceting of nanowire at constant radius can be given as [20]:

$$\delta + \frac{\arcsin\alpha_{SL}}{\alpha_L} > \frac{\arcsin(\alpha_L - \alpha_{SL})}{\alpha_L} \tag{4.26}$$

Initially, the angle between the lateral surface and growth direction changes due to the bending effects. At a favorable angle, the formation of a facet occurs and the liquid droplet slides on the facet. As the droplet rises on the facet, the condition for faceting (Equation 4.26) is not satisfied. This leads to the droplet moving on the facet at its equilibrium contact angle, creating an expanding isotropic surface. This process continues till the faceting condition is fulfilled again and this continuous process leads to the faceting of nanowires. Periodic local elevations (sawtooth morphology) occur when the system satisfies the condition [20]:

$$\delta + \varphi > \theta \tag{4.27}$$

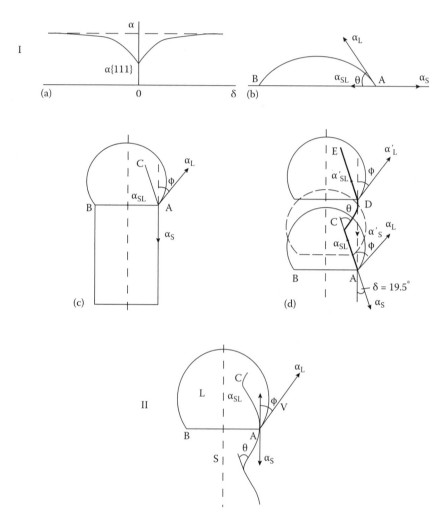

FIGURE 4.21
(I) Schematic illustrating lateral faceting of a silicon whisker during growth from a liquid droplet: (a) orientation dependence of the surface free energy near a singular facet; (b, c) equilibrium configurations of the liquid droplet on the singular facet and whisker tip, respectively; (d) lateral faceting of the whisker. (II) Schematic illustrating the formation of local elevations with periodic oscillations on the lateral facets. (Reproduced from Nebol'sin, V.A. et al., *Inorg. Mater.*, 39, 899, 2003. With permission.)

Following this condition, the growth occurs analogous to faceting and when the condition (Equation 4.27) becomes unfavorable, the droplet starts to move in the other direction causing a local elevation. This process leads to the observed sawtooth formations.

4.2.2.3 Role of Interfacial Energy on the Nanowire Growth Direction

The faceting of 3-D crystal has been explained by Wulff based on Gibbs theorem of minimization of free energy, given as

$$F = \sum \sigma_i \alpha_i = \min \tag{4.28}$$

where
 σ is the surface energy of a facet
 α is the area of the facet

Wulff theorem relates the surface energy of a facet σ to its distance from the center of the crystal h as

$$\frac{\sigma_a}{h_a} = \frac{\sigma_b}{h_b} = \frac{\sigma_c}{h_c} \tag{4.29}$$

In the case of diamond octahedral crystal formation studied via kinetic Monte Carlo simulation [21], the {111} facets are formed preferentially due to their lower surface energy. Incorporation of second-order interactions yielded a shift from only {111} faceted octahedrons to mixed {111} and {100} faceted cubo-octahedrons. The system shifted to predominantly {110} faceting along with {111} and {113} planes when third neighbor interactions were also included as seen in Figure 4.22.

Considering only the first neighbor interactions, the adatoms were well separated out and grew without any external influence, forming {111} facets. The introduction of second- and third-order interactions increases the external influence and rough {111} surfaces were created with kinks and steps, which increase the surface energy of the {111} facet. Now, the surface energies of the other facets become smaller than the rough {111} facets, leading to shift from {111} equilibrium faceting. This is in case of 3-D crystal formation, where the entire surface is a vapor–solid interface that follows Gibbs rule and Wulff theorem. The question then is: Can the same arguments be extended for the case of nanowire growth where the interface is liquid–solid and spans a few nanometers?

Gosele and coworkers [22] proposed a model to understand the observed, "apparent" diameter dependency of growth direction in the case of silicon whiskers. They observed that the preferred orientation of silicon nanowires changed from <110> to <111> at a crossover radius of 20 nm. Larger diameter silicon nanowires preferentially grew in <111>. Only the direction-dependent properties were considered for calculations since the energy difference of the two growth directions were needed. The two growth direction-dependent terms are the solid–liquid interfacial tension and the silicon surface tension

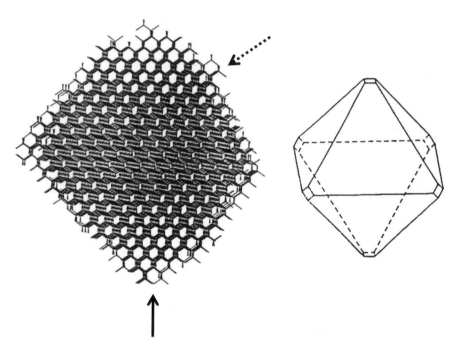

FIGURE 4.22

KMC-simulated diamond crystallite with 4000 atom size and the schematic of the octahedral crystal. The crystal ideally should be completely {111} faceted. The observations of small {100} facets shows that the free energy of {111} facet is increasing, which makes growth of {100} facets possible. The roughness of {111} surface is caused by the high density of coverage of growth species on the surface. The dashed arrow indicates the {111} surface with high coverage and the solid arrow indicates the small {100} facets. (Reproduced from Sunkara, M.K., PhD dissertation, Monte-Carlo simulation of diamond nucleation and growth, Case Western Reserve University, Cleveland, Ohio, 1993. With permission.)

contribution from the edge of the interface as shown in Figure 4.23. The free energy of an interface can be expressed as [22]:

$$F = \Delta z \sigma_s L + \sigma_{ls} A \qquad (4.30)$$

where

Δz is the interfacial thickness on the silicon surface
σ_s is the silicon surface energy
L is the circumference of the interface
σ_{ls} is the liquid–solid interface tension
A is the interfacial area

The first term on the right-hand side represents the line tension at the liquid–solid interface and the second term represents the liquid–solid surface energy on the nanowire growth front. This model proposes that diameter is

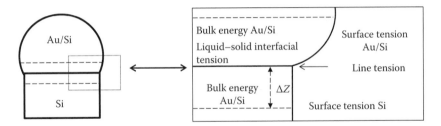

FIGURE 4.23
Schematic of the liquid–solid interfacial region and the associated interfacial energies. (Reproduced from Schmidt, V. et al., *Nano Lett.*, 5, 931, 2005. With permission.)

the factor that changes the free energy by varying the contributions from the liquid–solid interfacial energy and the solid surface energy. In fact, following the arguments for a 3-D crystal growth, one can argue that it is the rate of material incorporation or the adatom flux on the nanowires (a diameter dependent property), which controls the free energy. At smaller diameters, the interface does not remain atomically smooth and the free energy of a rough {111} plane can be higher than the energy of other smooth planes. Thus, the growth direction is not just a function of diameter alone.

4.3 Kinetic Considerations of Nanowire Growth under VLS Growth

4.3.1 Kinetics of Vapor–Liquid–Solid Equilibrium

A planar liquid surface is usually referred to as the reference state. The pressure of the vapor in equilibrium with the surface is the equilibrium reference vapor pressure. For species i, any vapor pressure above this pressure creates a supersaturation, which leads to the formation of a spherical droplet. This pressure is the equilibrium pressure of the system necessary to maintain the spherical droplet. The driving force (supersaturation) expressed in terms of chemical potentials can be related to the formation of a droplet of a diameter d as

$$\mu_{i,v}^{eq} = \mu_{i,v}^{\infty} + \frac{4\Omega\alpha}{d} \tag{4.31}$$

Equation 4.31 shows the increase in the chemical potential of the species i, from the planar interface to the spherical droplet, by the creation of droplet under the existing supersaturation. At operating pressures higher than the equilibrium pressures, a supersaturation is created, which causes the

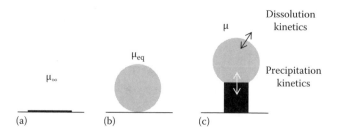

FIGURE 4.24
Schematic illustrating the relation of chemical potentials and various stages of droplet and nanowire formation: (a) when the vapor is in equilibrium with a reference state (planar liquid film); (b) when the vapor is at a supersaturation to support droplet formation (at equilibrium vapor pressure); and (c) when the vapor is at a pressure higher than equilibrium vapor pressure, which controls the dissolution kinetics.

vapor of species i to dissolve into the droplet as seen in Figure 4.24. The supersaturation can be expressed as

$$\left[\mu_{i,v} - \mu_{i,v}^{eq}\right] = \left[\mu_{i,v} - \mu_{i,v}^{\infty}\right] - \frac{4\Omega\alpha}{d} \tag{4.32}$$

or

$$\Delta\mu_{i,v} = \Delta\mu_{i,v}^{\infty} - \frac{4\Omega\alpha}{d} \tag{4.33}$$

Here

$\Delta\mu_{i,v}$ represents the change in chemical potential of species i in the vapor phase and the liquid droplet

$\Delta\mu_{i,v}^{\infty}$ represents the same difference at the reference state

Equation 4.32 suggests that the vapor-phase supersaturation above droplets depends upon the droplet diameter. Such vapor-phase supersaturation can directly influence the dissolution rate and the nanowire growth rate if it is considered to be dissolution kinetics–limited. Under such a case, the dissolution rate and subsequently the growth rate can be assumed to be proportional to the net impingement rate (R), which is determined by the partial pressure of species while correcting for equilibrium partial pressure expected for a spherical droplet of radius, r. Equation 4.6 determines the equilibrium partial pressure over a spherical droplet (p_r) compared to that over planar surface (p^{∞}). The net impingement flux is estimated using pressure difference ($p - p_r$). The expression for growth rate relating to the net impingement flux can be written as

$$R \propto \left\{\frac{p - p^{\infty}\exp\left(2\sigma_{vl}\Omega/rkT\right)}{(2\pi mkT)^{\frac{1}{2}}}\right\}^{n} \tag{4.34}$$

The *n*th power dependence in Equation 4.34 is included due to unknown, exact relationship between net impingement flux and the growth rate. Equation 4.34 also suggests that there may be a size below which the growth rate would go to zero as the supersaturation needed to drive the growth and ceases to exist after that point.

Givargizov [23] suggested a similar model as that shown in Equation 4.33 but suggested that the above chemical potential difference is for solute species in the vapor and solid phases. This analysis is not accurate, because the model correlates chemical potentials of solute species in the gas phase to that in the solid phase. This can be valid only under the case of precipitation-limited kinetics, though. In such a case, the chemical potential of solute in the gas-phase composition and the liquid phase can reach equilibrium and are related. Under these conditions, the solute concentration, *c*, can be related to the vapor pressure of the solute containing gas-phase precursors. In any case, the pre-cipitation kinetics is directly proportional to the concentration supersaturation that exists in the droplet. Assuming that the precipitation kinetics limits the growth of nanowire, the Givargizov model relates the growth rate (*V*) to the supersaturation (driving force for crystallization) in the following manner:

$$V^{1/n} = \frac{\Delta\mu_0}{kT}b^{1/n} - \frac{4\Omega\alpha}{kT}b^{1/n}\frac{1}{d}$$

(4.35)

where
   V is the growth rate
   b is a constant

A plot of $V^{1/n}$ versus $1/d$ can be used to determine the diameter dependence of growth rate. It can be observed that as diameter increases, the growth rate increases correlating with experimental observation of thicker whiskers growing faster than thinner ones.

4.3.2 Role of Direct Impingement in Growth Kinetics

In the case of nanowire growth illustrated in Figure 4.25, where the material addition occurs via impingement from the vapor phase, for small diameters, monocenter nucleation is assumed to occur out of the alloy droplet on top of the nanowire surface [24]. Here, the droplet and nanowire sizes are assumed to be the same for the sake of simplicity. The nucleation rate is determined using supersaturation in the liquid phase and relating the supersaturations in the gas phase to the liquid phase. The balance on number of molecules of species *i* in the droplet can be expressed with the change in concentration *C* in the droplet [24] as

$$\frac{2}{3}\pi R^3 \frac{dC}{dt} = \pi R^2 \chi_{vl}J - 2\pi R^2 \frac{r_i C}{\tau_i} - \pi R^2 \frac{V_{NW}}{\Omega_m}$$

(4.36)

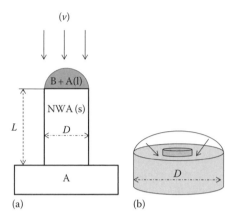

FIGURE 4.25
Schematic of (a) nanowire growth of nanowire of material A from an alloy of catalyst B on the top surface of the crystal; and (b) 2-D nucleation mechanism of crystal growth at the liquid solid interface. (Reproduced from Dubrovskii, V. et al., *Tech. Phys. Lett.*, 30, 682, 2004. With permission.)

where the first term on the right-hand side represents the number of molecules adsorbed, the second term represents the number of desorption events, and the third term represents the molecules lost due to crystallization. χ_{vl} is the coefficient of condensation of molecules from vapor onto the liquid droplet, J is the adatom flux, τ_i is the average lifetime of molecules on the liquid surface, V_{NW} is the nanowire growth rate, and Ω_m is the specific volume of the molecule. Accounting for the difference in supersaturation in the vapor and liquid phases, the whisker growth rate can be expressed as

$$V_{NW} = V_o(\Phi - \mu) \qquad (4.37)$$

where
    Φ is a measure of supersaturation in the gas phase determined by the balance of adsorption and desorption processes
    μ is the supersaturation in the liquid phase
    V_o is a coefficient dependent on the equilibrium concentration, molar volumes of the growth species in solid and liquid phases, and the average lifetime of the molecules on the droplet surface controlled by the desorption processes

In the case of growth controlled by processes at the vapor–liquid interface, the supersaturation in the gas phase is much larger than that in the liquid droplet. The solution of Equation 4.36 then becomes trivial and the result is a diameter-independent growth rate determined only by adsorption and desorption processes. However, this is not typically observed in most experimental studies. If the processes at the liquid–solid interface are assumed to be the limiting factors, the vapor–liquid interface is at equilibrium. In this case, the growth involves nucleation of new layers and the growth rate becomes a function of supersaturation within the liquid phase (or related to vapor phase through equilibrium). For large diameters, Equation 4.36 yields the following expression for growth rate of nanowires:

$$V_{NW} = \exp\left(\frac{-\alpha}{\ln(1+\Phi) - \dfrac{D_o}{d}}\right) \qquad (4.38)$$

This relation shows an exponential dependence on growth parameters. For small diameters, the growth rate can be expressed [24] as

$$V_{NW} = \frac{\chi_{vl} J h}{\pi}\left(\frac{k_B T}{\alpha_{ls}}\right)^2 \left[\frac{\ln\left\{\dfrac{\beta\alpha\dfrac{3}{2}}{\left[\left(\ln(1+\Phi) - \dfrac{D_o}{d}\right)^{\frac{3}{2}}\left(\dfrac{d}{D_o}\right)^2\right]}\right\}}{\left\{\ln(1+\Phi) - \dfrac{D_o}{d}\right\}^2 - \alpha\left\{\ln(1+\Phi) - \dfrac{D_o}{d}\right\}}\right] \qquad (4.39)$$

where
　χ_{vl} is the coefficient of condensation on the liquid surface
　d is diameter of nanowire
　J is the flux of molecules from the vapor phase on the droplet surface
　h is the monolayer step height
　β is a constant
　a is a dimensionless parameter related to the liquid–solid interfacial
　　energy
　D_o is determined by the difference in surface energies of the solid–vapor
　　and liquid–vapor interfaces

Equations 4.38 and 4.39 show that the processes at the liquid–solid interface actually play a limiting role in nanowire growth as they explain the observed increase in the nanowire growth rate with increasing diameter, which is a typical observation for direct impingement-controlled growth.

4.3.3 Role of Surface Diffusion in Growth Kinetics

In the case of nanowire growth where adatom flux on the nanowire from the substrate plays an important role [25], the growth rate can be calculated based on the adatom diffusion on substrate and nanowire surfaces only. The processes occurring at the metal–semiconductor interface are not considered. The model primarily addresses issues in metal organic chemical vapor deposition (MOCVD) where the growth rates are high. Figure 4.26 is the schematic of the processes considered and the system coordinates for nanowire growth. The metal particle is assumed to be hemispherical and only the material diffusion to the wire surface and top is considered. Material balance on the substrate surface yields the number density of adatoms [25] as

$$D_S \nabla^2 n_S - \frac{n_S}{\tau_S} + R_S = \frac{\partial n_S}{\partial t} \tag{4.40}$$

where
  D_S is the surface diffusivity of adatoms on the substrate
  n_S is the number density of adatoms on the substrate surface
  τ_S is the average lifetime of diffusion of adatoms on the surface
  R_S is the effective impingement rate of adatoms on the substrate

The adatom number density on the nanowire surface is given [25] by the material balance

$$D_w \frac{\partial^2 n_w}{\partial z^2} - \frac{n_w}{\tau_w} + R_w = \frac{n_w}{\partial t} \tag{4.41}$$

where
  D_w is the surface diffusivity of adatoms on the nanowire
  n_w is the number density of adatoms on the nanowire surface

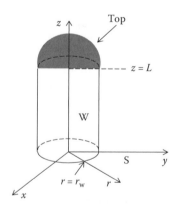

FIGURE 4.26
Schematic of the nanowire growth model with the coordinate system used in the calculations. The diffusion equations are solved at the domains W (on the wire) and S (on the substrate). (Reproduced from Johansson, J. et al., *J. Phys. Chem. B*, 109, 13567, 2005. With permission.)

τ_w is the average lifetime of diffusion of adatoms on the nanowire surface
R_w is the effective impingement rate of adatoms on the nanowire surface

The solution for Equation 4.41 uses the flux (solution from Equation 4.40, as the initial condition for inlet flux at the base of the nanowire). Using expressions for adatom flux, the nanowire growth rate is given by

$$\frac{dl}{dt} = \frac{2\Omega R_w \lambda_w}{r_w} \tanh\left(\frac{L}{\lambda_w}\right) - \frac{2\Omega J_{Sw}}{r_w \cosh\left(\dfrac{L}{\lambda_w}\right)} + 2\Omega R_{top} \qquad (4.42)$$

where
λ_w is the diffusion length of adatoms on the nanowire surface
J_{Sw} is the adatom flux from the substrate to the nanowire
R_{top} is the material addition on the catalyst particle

Here, the first term represents material diffused directly onto the nanowire sides, the second term stands for the adatom diffusion from the surface onto the nanowire, and the third term represents material deposited directly onto the metal particle.

The initial stages of growth are not modeled well as the droplet does not achieve steady-state supersaturation and the growth rates are not representative. In the later stages of growth when $L \gg \lambda_w$, the role of surface diffusion is not significant, and the growth rate can be expressed as

$$\frac{dl}{dt} = 2R\left(1 + \frac{\lambda_w}{r_w}\right) \qquad (4.43)$$

The diffusion lengths λ_w and λ_s are assumed to be the same for simplification of the model. For MOCVD-grown InAs (111), calculations point to a diffusion length of 650 nm. Diffusion length of species on the InAs {110} side facets in chemical beam epitaxy (CBE) was calculated to be 10 μm by Jensen and coworkers [26]. The difference lies in the fact that CBE is a high-vacuum technique and the species do not crack until they reach the droplet, thus leading to longer diffusion lengths. In MOCVD, the species are cracked more before they reach the incorporation site, leading to the observed smaller diffusion lengths. Compensating for the Gibbs–Thomson effect, the attachment rate R can be expressed as

$$R(r_w) = \frac{p - p^{\infty} \exp\left(-2\sigma\Omega_1 / rkT\right)}{\sqrt{2\pi m k_B T}} \qquad (4.44)$$

In the case of silicon nanowires using gold catalyst, there are conflicting reports on the diameter dependence of the nanowire growth rate. The results from Equation 4.43 and Schubert et al. [27] contradict Givargizov's [23] observations. In his analysis, Givargizov did not include the two pressures involved in Gibbs–Thomson effect. Calculations at the conditions used by Schubert et al. [27] show that p and p^∞ differ by 10 orders of magnitude. Hence, the Gibbs–Thomson effect is negligible and thinner wires grow faster than thicker wires. Similar calculations performed at the experimental conditions employed by Givargizov [23] had comparable values for p and p^∞, i.e., lower supersaturation conditions were prevalent. Here, Gibbs–Thomson effect becomes important and faster growth of thicker wires is observed.

4.3.4 Direct Impingement and Diffusion

In the case of material addition on the nanowire by impingement from the top onto the droplet and from the sides of the nanowires [28] as seen in Figure 4.27, assuming equal contribution rate of materials via vapor-phase impingement and surface adatom diffusion, the growth rate is given as

$$R_L = 2R\left(1 + \frac{\lambda}{r}\right) \tag{4.45}$$

Combining Equation 4.45 and Equation 4.34 for R,

$$R \propto \left\{\frac{p - p^\infty \exp\left(2\sigma_{vl}\Omega/rkT\right)}{(2\pi mkT)^{\frac{1}{2}}}\right\}^n \left(1 + \frac{\lambda}{r}\right) \tag{4.46}$$

In any case, the compensation for curvature in p_r is very similar to Equation 4.5, which provides a relation between the partial pressure in the system and the droplet curvature. From Equations 4.45 and 4.46, diameter dependency of growth rate is very similar to the previous model. For thicker whiskers, the Gibbs–Thomson effect can be neglected, in which case thinner wires grow faster than thicker wires following Equation 4.44. For thinner whiskers, the dependence of p_r on r becomes dominating, which causes the thinner whiskers to grow slower than thicker whiskers.

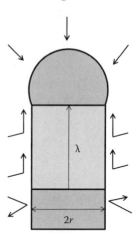

FIGURE 4.27
Schematic of the model with a whisker of diameter $2r$ where material is being added from the vapor at the top and by adatom diffusion at the sides. (Reproduced from Seifert, W. et al., *J. Cryst. Growth*, 272, 211, 2004. With permission.)

4.3.5 Role of Surface Diffusion on the Metal Droplet

In the majority of cases of VLS nanowire growth modeling, the role of diffusion on the wire surface and on the substrate is considered but the role of diffusion on the catalyst droplet surface has not been treated with proper attention. The role of liquid droplet as a catalyst or sink for preferential material incorporation and precipitation site has already been recognized. Species on the droplet can either diffuse through the bulk of the droplet or travel on the droplet surface via diffusion and add at the growth interface on the outer surface. Treating the process as 1-D diffusion problem as depicted in Figure 4.28a [29], the expressions for mass transfer rate through the liquid–solid interface in the bulk is given as

$$q_s = -\pi d D_s \frac{dC_S}{dy} \tag{4.47}$$

where
    d is the diameter of the whisker
    D_S is the surface diffusion coefficient
    C_S is the concentration of growth species on the liquid surface

Since the comparison of mass transfer rates is between the interface transfer and the surface transfer, the surface mass transfer rate is given by

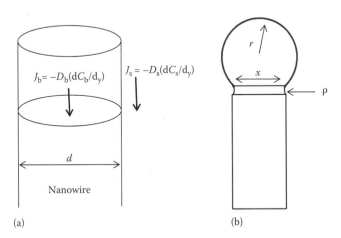

(a)   (b)

FIGURE 4.28
(a) Schematic of the mass transfer model for diffusion indicating the material flux on the wire and on the outer surface. (b) Schematic showing the curved neck region along with the radius of curvatures r, x, and ρ of the droplet and neck regions. (Reproduced from Hongyu, W. and Gary, S.F., *J. Appl. Phys.*, 76, 1557, 1994. With permission.)

$$q_b = -\pi \left(\frac{d}{2}\right)^2 D_b \frac{dC_S}{dy} \tag{4.48}$$

where

D_b is the bulk liquid diffusion coefficient
C_b is the concentration of growth species in the liquid droplet

It has been observed that the relative dominance of the two mass transport processes is dependent on the nanowire diameter, concentration of growth species in the liquid droplet, and the ratio of individual diffusivities. It can be noted from Equations 4.47 and 4.48 that mass transport by surface diffusion becomes more important at smaller nanowire diameters, lower solubility of growth species in the droplet, and larger ratio of diffusivity on the surface to diffusivity at the interface. In addition to the driving forces created by the concentration gradient, the change in curvature at the interface can also affect the diffusion on the droplet. A neck region has been observed between the droplet and the whisker, and Figure 4.28b shows a schematic of the concave neck region. From the Kelvin equation, since the neck has a lower chemical potential than the flat surface, the chemical potential decrease is expressed as

$$\Delta\mu_1 = \gamma_{lv}\Omega\left(\frac{1}{\rho} + \frac{1}{x}\right) \tag{4.49}$$

where

γ_{lv} is the surface energy of the liquid–vapor interface
v is the molar volume
ρ and x are the principal radii of curvature of the interface

Since the droplet has a higher chemical potential than the neck, the increase is given by

$$\Delta\mu_2 = \gamma_{lv}\Omega\frac{2}{r} \tag{4.50}$$

where r is the radius of the droplet. For the transport of species from droplet to the neck, chemical potential decrease results in a driving force for diffusion. Diffusion on the droplet surface can also play an important role in nanowire growth and morphology and needs to be considered for an exhaustive model for nanowire growth.

4.3.6 Role of Interwire Spacing

Another very important aspect in nanowire growth that has not been discussed so far is the interwire spacing. The importance of vertical-aligned nanowire architectures for device applications makes is very essential to control the growth process. In case of the GaP nanowire growth with different catalyst droplet sizes and interwire spacing, three different regimes were observed [30] as shown in Figure 4.29. The first is materials competition regime; here, for very small interwire spacing, the source supply is shared, which causes thinner wires to grow faster than thicker wires. With an increase in the spacing, linear growth rate is observed due to material addition without competition.

For larger spacing, diffusion is not the controlling criterion and the wires act as independent isolated island and hence the name independent regime. Here, the growth rate is independent of diameter, while at smaller spacing, the growth rate increases with diameter until material competition regime is reached. For intermediate spacing (synergetic regime), growth rate decreases with increasing interwire separation. In this region, growth rate is also observed to be influenced by the nearest neighbors as shown in Figure 4.30. In this region, the nanowires do not share any surface collection regions but interactions play a part. The species decompose at the metal droplet surface proportional to the size of the droplet and the material is shared by the neighboring wires. In the synergetic growth regime, thinner wires that are in proximity to thicker wires grow faster.

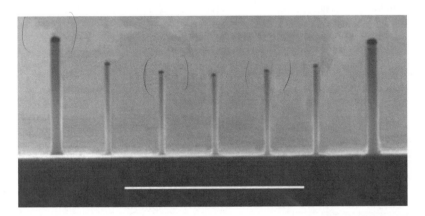

FIGURE 4.29

SEM micrographs of nanowires grown using different sizes of gold particles (25 and 100 nm catalyst droplets and 300 nm spacing). Clearly, wires closer to thicker wires are taller than ones which are further away. This shows that growth rate of wires in enhanced by the presence of neighbors and it depends on the size of the catalyst droplet. (Scale bar = 1 μm.) (Reproduced from Borgstrom, M.T., *Nat. Nanotechnol.*, 2, 541, 2007. With permission.)

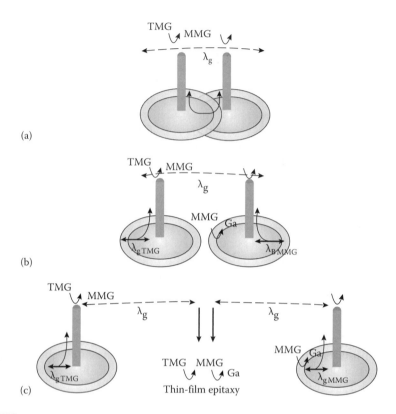

FIGURE 4.30

Schematic of the various regimes of the nanowire growth model. (a) *Competitive regime*, where surface collection areas are defined by the surface diffusion length of the species and partly decomposed species overlap. Most of the material decomposes through the catalytic particles. (b) *Synergetic regime*, partly overlapping with a where wire-to-wire gas-phase interaction occurs and surface collection areas become fully separate. The species in vapor phase crack at the droplet. Parts of the growth species desorb and are shared with neighboring wires, enhancing growth of wires in the proximity. (c) *Independent regime*, where the growth rate is determined by material surface diffusion to the wire. Species landing outside the respective surface collection areas of the wires contribute to competitive substrate epitaxy. Bent arrows with primary and secondary species indicate decomposition events, solid arrows surface diffusion and broken arrows gas-phase diffusion. (Reproduced from Borgstrom, M.T., *Nat. Nanotechnol.*, 2, 541, 2007. With permission.)

References

1. R. S. Wagner and W. C. Ellis, *Appl. Phys. Lett.* 4, 89 (1964).
2. T. Y. Tan, N. Li, and U. Gösele, *Appl. Phys. Mater. Sci. Process.* 78, 519 (2004).
3. Y. W. Wang, V. Schmidt, S. Senz, and U. Gosele, *Nat. Nanotechnol.* 1, 186 (2006).
4. Y. Yao and S. Fan, *Mater. Lett.* 61, 177 (2007).

5. A. M. Morales and C. M. Lieber, *Science* 279, 208 (1998).
6. R. S. Wagner, W. C. Ellis, K. A. Jackson, and S. M. Arnold, *J. Appl. Phys.* 35, 2993 (1964).
7   S. Sharma, H. Li, H. Chandrasekaran, R. C. Mani, and M. K. Sunkara, *Encyclopedia of Nanoscience and Nanotechnology*, H. S. Nalwa (Ed.), Vol. 10, The American Scientific Publishers, Los Angels, CA, 2003, p. 327.
8. G. Bhimarasetti and M. K. Sunkara, *J. Phys. Chem. B* 109, 16219 (2005).
9. S. Vaddiraju, M. K. Sunkara, A. H. Chin, C. Z. Ning, G. R. Dholakia, and M. Meyyappan, *J. Phys. Chem. C.* 111, 7339 (2007).
10. S. Vaddiraju, A. Mohite, A. Chin, M. Meyyappan, G. Sumanasekera, B. W. Alphenaar, and M. K. Sunkara, *Nano Lett.* 5, 1625 (2005).
11. H. Li, A. H. Chin, and M. K. Sunkara, *Adv. Mater.* 18, 216 (2006).
12. H. Chandrasekaran, G. U. Sumanasekara, and M. K. Sunkara, *J. Phys. Chem. B* 110, 18351 (2006).
13. H. Chandrasekaran, PhD Dissertation, Rationalizing Nucleation and Growth in the Vapor-Liquid-Solid (VLS) Methods, University of Louisville (2006).
14. Database of Pb-free soldering materials, surface tension and density, experiment vs. modeling, *Data Sci. J.* 4, 195 (2005)
15. S. Vaddiraju, PhD Dissertation, Chemical and Reactive Vapor Transport Methods for the Synthesis of Inorganic Nanowires and Nanowire Arrays, University of Louisville (2006).
16. C. Pendyala, J. H. Kim, J. Jacinski, Z. Chen, and M. K. Sunkara, "Self-catalysis" schemes for Group III-antimonide nanowires: Antimony vs. Group-III metal. *Semicond. Sci. Tech.* submitted. (2009).
17. V. A. Nebol'sin and A. A. Shchetinin, *Inorg. Mater.* 39, 899 (2003).
18. D. N. McIlroy, A. Alkhateeb, D. Zhang, D.E. Aston, A. C. Marcy, and M. G. Norton, *J. Phys.: Condens. Matter.* R415 (2004).
19. F. M. Ross, J. Tersoff, and M. C. Reuter, *Phys. Rev. Lett.* 95, 146104 (2005).
20. V. Nebol'sin, A. Shchetinin, A. Korneeva, A. Dunaev, A. Dolgachev, T. Sushko, and A. Tatarenkov, *Inorg. Mater.* 42, 339 (2006).
21. M. K. Sunkara, PhD Dissertation, Monte-Carlo Simulation of Diamond Nucleation and Growth, Case Western Reserve University, Cleveland, Ohio (1993).
22. V. Schmidt, S. Senz, and U. Gosele, *Nano Lett.* 5, 931 (2005).
23. E. I. Givargizov, *J. Cryst. Growth.* 31, 20 (1975).
24. V. Dubrovskii, N. Sibirev, and G. Cirlin, *Tech. Phys. Lett.* 30, 682 (2004).
25. J. Johansson, C. P. T. Svensson, T. Martensson, L. Samuelson, and W. Seifert, *J. Phys. Chem. B* 109, 13567 (2005).
26. L. E. Jensen, M. T. Bjork, S. Jeppesen, A. I. Persson, B. J. Ohlsson, and L. Samuelson, *Nano Lett.* 4, 1961 (2004).
27. L. Schubert, P. Werner, N. D. Zakharov, G. Gerth, F. M. Kolb, L. Long, U. Gosele, and T. Y. Tan, *Appl. Phys. Lett.* 84, 4968 (2004).
28. W. Seifert, M. Borgström, K. Deppert, K. A. Dick, J. Johansson, M. W. Larsson, T. Mårtensson, N. Sköld, C. Patrik T. Svensson, B. A. Wacaser, L. Reine Wallenberg, and L. Samuelson, *J. Cryst. Growth.* 272, 211 (2004).
29. W. Hongyu and S. F. Gary, *J. Appl. Phys.* 76, 1557 (1994).
30. M. T. Borgstrom, G. Immink, B. Ketelaars, R. Algra, and E. Bakkers, *Nat. Nanotechnol.* 2, 541 (2007).

5

Modeling of Nanowire Growth

5.1 Introduction

The physical properties of nanowires can depend on size, surface faceting, and growth direction [1,2]. As discussed in Chapter 4, the size of resulting nanowires can be controlled by controlling the size of the metal clusters in the gold-catalyzed schemes [3] and by nucleation in the case of spontaneous nucleation from low-melting metals [4]. However, the factors that control the nanowire growth direction and corresponding surface faceting in any vapor–liquid–solid (VLS) technique have not been clearly identified with a very few exceptions. For example, in the case of gallium nitride (GaN), control of growth direction was achieved using heteroepitaxy onto single-crystal templates with defined orientation [5]. In the case of amorphous substrates, control on the growth direction of GaN nanowires was obtained by changing the process conditions [6].

Givargizov considered the free energies of different planes and concluded that the growth direction of a nanowire is along the plane with the lowest free energy [7]. It is interesting to note that the typical growth direction of silicon (Si) nanowires in many of the gold-catalyzed schemes is along <111> direction, which has the lowest free energy [8,9]. Givargizov also proposed that the growth direction of the whiskers is related to their diameter using thermodynamic arguments [7]. Recently, Gosele and coworkers have provided similar experimental evidence using epitaxial growth conditions, suggesting that the growth direction of nanowires is controlled primarily by their diameter, i.e., the smaller diameter Si nanowires grow preferentially in <110> direction compared to <111> direction [10]. They used free energy minimization to predict that the growth direction undergoes transition from <111> plane to <110> plane at nanowire diameters of 20 nm [11]. Lieber and coworkers have also made similar arguments about size dependence on nanowire growth direction [12]. Further understanding on the factors controlling growth direction and faceting can benefit from modeling to provide insight into mechanisms.

There are two classes of theoretical approaches that have been proposed to understand the nanowire growth processes. The first class techniques involve the density functional theory for studying the energetics of nanowire structures with different growth directions and different surface faceting. The second type techniques are based on stochastic modeling approaches and implemented using either kinetic Monte Carlo (kMC) or lattice gas dynamics simulations for modeling the kinetic aspects of nanowire growth. This chapter discusses the above techniques using Si nanowire growth as the model system for understanding growth direction and the resulting surface faceting under equilibrium and kinetic situations.

5.2 Energetics of Stable Surface Faceting: Silicon Nanowire Example

It has been determined experimentally that the predominant growth directions in the case of oxide-assisted growth (OAG) method are the <112> and <110> [13], while the catalyst-assisted VLS mechanism produces nanowires growing in the <110>, <112>, and <111> directions [12]. There is no clear understanding of the correlation between the process conditions and the resulting growth directions, but the experimental results show that higher fraction of smaller diameter wires grow in directions other than <111> in the catalyst-assisted growth methods, suggesting that there is a diameter dependence on growth direction. This idea is addressed in detail later in Section 5.3 using both simulations and experiments.

Lee and coworkers have studied the hydrogen-terminated surface structures and energetics of Si nanowires using density-functional tight-binding method (DFTB) [14,15] to determine the energy gap as a function of both the nanometer size and growth direction. The cohesive energy was also determined as a function of surface faceting. Figure 5.1 shows the low-index surface cross-sections possible for different growth directions of <100>, <110>, <111>, and <112> in Si and growth of related materials. All the directions are enclosed by the low Miller index surfaces with lower symmetries excluded from the analysis. It is interesting that the <112> direction has only one cross-section with low indexes whereas the <100>, <110>, and <111> have many different possible configurations. In these calculations, the cohesive energy per atom for the nanowires was determined by removing the H contributions. It is shown that the cohesive energy decreases with an increase in the cross-sectional area of the nanowire.

From the cohesive energy calculations, the surface energies of the facets with H-terminated surfaces are in the order of $\gamma(111) < \gamma(110) < \gamma(100)$, different from the order for clean Si facets $\gamma(111) < \gamma(100) < \gamma(110)$ [16]. It can be seen that the <112> Si nanowire (rectangular cross-section) enclosed by two (100) facets and two (111) facets is the most favorable growth direction energetically. The differences in the energies can be attributed to the surface-to-volume ratio (svr) of the nanowires. For example, the <112> Si nanowire has lower svr compared to a <110> nanowire with four (111) surfaces.

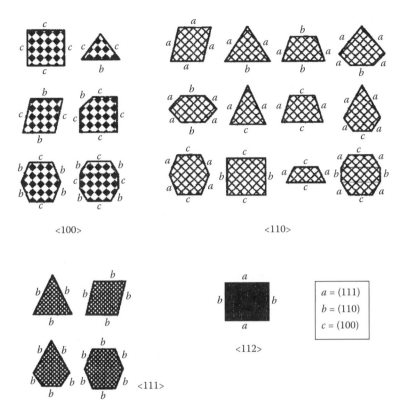

FIGURE 5.1
Schematics illustrating the possible configurations in <100>, <110>, <111>, and <211> growth directions. (Reproduced from Zhang, R.Q. et al., *J. Chem. Phys.*, 123, 144703, 2005. With permission.)

The stability of the Si nanowires arises as a result of a balance between optimum surface energy and low svr. All the above calculations indicate that <112> growth direction is the most stable among all of them and the most abundant in the OAG processes [13,17].

The energy band gaps of the different growth directions were also calculated but are less dependent on the facets compared to the cohesive energy. They follow the order <110> < <111> < <112> < <100> with a general dependence on the cross-sectional area of the wire and consistent with the earlier calculations [18] and the experimental data. Based on the understanding so far, the energetics cannot predict the surface faceting of a nanowire grown in a particular direction. So, one will have to rely on both experimental results and simulations involving growth dynamics to predict the surface faceting as a function of process variables, nanowire size, and growth direction.

5.3 Simulation of Individual Nanowire Growth

In any VLS technique, the interfacial contact area between the liquid droplet and the growing nanowire tends to be circular and is always constant, thus enforcing the one-dimensional (1-D) growth. This is illustrated in Figure 5.2 for two distinct cases of VLS growth. In the first case, the size of molten metal droplet limits the interfacial contact area depending upon the interfacial energy. In the second case, the solid nucleus maintains a limited contact area with the large molten metal droplet surface due to nonwetting conditions. The modeling of crystal growth under these circumstances needs to take care of this constraint without fixing the growth surface a priori. Typically, the crystal growth has been modeled by considering various site- and time-dependent processes of adatom adsorption, desorption, and diffusion on plane surfaces using periodic boundary conditions and on 3-D crystals without the use of any specific periodic boundary conditions [19]. The first case using a known surface with periodic boundary conditions maintains the surface with the same orientation. The second case with simulation of 3-D crystals is somewhat relevant but needs to be modified for simulating 1-D growth here.

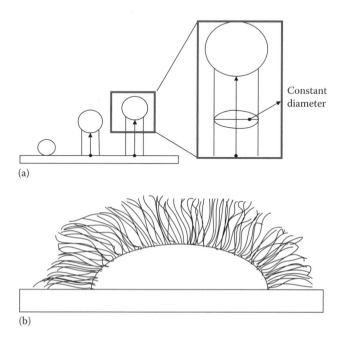

Constant diameter

(a)

(b)

FIGURE 5.2

(a) Schematic representation of traditional VLS method. (b) Schematic representation of VLS method with multiple nucleation and growth of nanowires from large droplet.

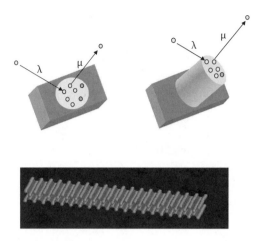

FIGURE 5.3
(a) Schematic illustrating the starting seed surface, (b) schematic illustrating the simulated nanowire growth surface, and (c) ball-and-stick model of the {100} seed surface.

In order to simulate the actual condition of VLS growth, a new condition is needed that can maintain a constant diameter for growing nanowires without fixing the growth direction a priori. The actual crystal growth dynamics are dependent upon the material and the corresponding details on site-dependent chemistry are not clear for any crystal growth system. Nevertheless, the site-dependent adsorption and desorption of adatoms needs to be included. In the case of nanowire growth, such chemistry must be considered only on the growth interface. Figure 5.3a shows the seed surface used for performing the chemisorptions and desorption of atoms. Figure 5.3b indicates the nanowire growth surface along the growth axis.

5.3.1 Simulation Methodology

Here, a simulation strategy developed by Sunkara and coworkers [19] is described. First, a circular seed surface is chosen as the growth surface. Figure 5.3c shows a typical starting seed surface. A circle on the seed surface mimics the contact area between the droplet and the growing nanowire. In the subsequent growth, the constant diameter constraint is implemented by constantly checking whether the newly added atom falls within the active growth surface. This is done by determining whether it is at a distance greater than the chosen diameter from any other sites on the active growth surface. The active growth surface is constantly updated. Actual growth is simulated by growth/etching of adatoms to the growth surface using a set of chemisorptions and desorption reactions whose rates are site and configuration dependent. The overall algorithm starts with a circular area chosen on a seed surface as shown in Figure 5.3c.

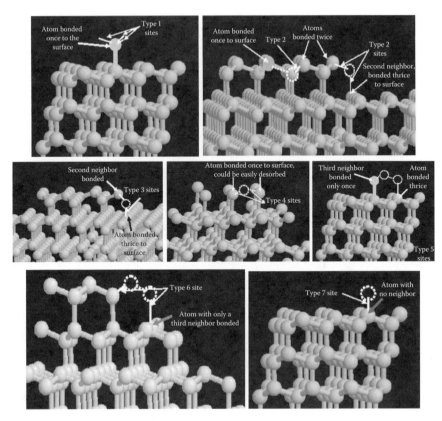

FIGURE 5.4
Ball-and-stick models of seven types of sites.

The reaction rates for all site-dependent chemisorption and desorption are determined by multiplying the rate constant with the concentration of available surface sites. The surface sites are divided into seven types using ball-and-stick models for chemisorption and desorption of atoms as shown in Figure 5.4. The rate constants for both chemisorption and desorption reactions on each surface site type can be varied. Each rate constant is estimated in relative proportion to chemisorption rate constant, λ_1 for adatoms onto bare {111} sites. This rate constant could be directly related to the supersaturation of solute species within the molten metal droplet. Similarly, the desorption rate constant (dissolution rate constant) could also be proportional to both the supersaturation multiplied by the vibrational frequency. As the diameter of the nanowire is small, the diffusion of solute species will not be limiting and does not have to be considered. This assumption is quite valid as the average diffusion distances of solute species within the molten metal would exceed the diameter of the simulated nanowire. The implementation of the above modeling strategy can be done using a kMC algorithm as detailed in Figure 5.5. The major steps within the main algorithm are

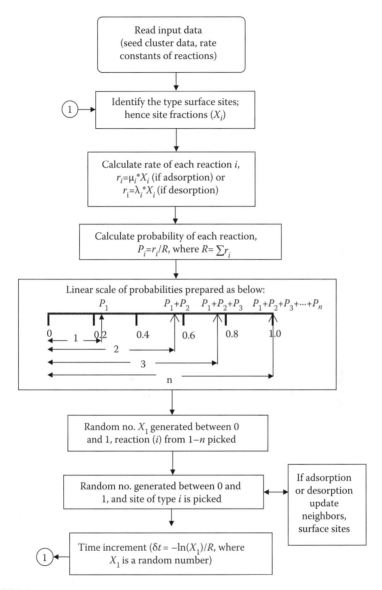

FIGURE 5.5
Flow chart of the kMC algorithm.

1. The fractions of each site type are calculated.
2. The rates of reaction, r_i for all N total reactions are calculated.
3. The probability of a reaction i occurring is $P_i = r_i/R$, where $R = \sum_{j=1,N} r_j$.
4. A linear scale of these rates is prepared. A random number is generated. If this number happens to be less than r_1, then reaction 1 occurs, if the number is less than $r_1 + r_2$, then reaction 2 occurs, and so on.

5. Once the reaction type is affirmed, the type of site is randomly picked from the cluster of all sites present at that instance on the surface.

6. Time is incremented dynamically and stochastically as $\delta t = -\ln(X)/R$, where X is a random number.

7. The simulations run until a desired number of atoms for the growing cluster are attained.

The programs were developed using C++ and were implemented on a personal computer running MS Windows. The results are viewed using free-domain software package called RASMOL. A separate module was developed to monitor the growth direction of the nanowires by determining the vectors between consecutive points on the medial axis. The angle between the medial axis and the vector perpendicular to the seed surface was also monitored.

5.3.2 Kinetic Monte Carlo Simulation Results

Simulations were performed by varying the rate constants for all the site types over a range of values as shown in Table 5.1. In each particular simulation, the growth surface site fractions are determined and grouped into three basic surface site types ({111}, {110}, {100}) and are monitored as a

TABLE 5.1

The Range of Rate Constant Values Used in the kMC Simulations

No.	Site-Specific Reaction	Range of Rate Constant Values
1.	Site 1 → Adsorption	N/A
2.	Site 1 → Desorption	μ_1:1–20
3.	Site 2 → Adsorption	λ_2:1–20
4.	Site 2 → Desorption	μ_2:0.01–0.2
5.	Site 3 → Adsorption	λ_3:1–10
6.	Site 3 → Desorption	N/A
7.	Site 4 → Adsorption	λ_4:1–5
8.	Site 4 → Desorption	μ_4:1–20
9.	Site 5 → Adsorption	λ_5:1–5
10.	Site 5 → Desorption	N/A
11.	Site 6 → Adsorption	λ_6:1–10
12.	Site 6 → Desorption	N/A
13.	Site 7 → Adsorption	λ_7:1–10
14.	Site 7 → Desorption	N/A

Note: All values are normalized with respect to chemisorptions rate constant on a site on {111} surface.

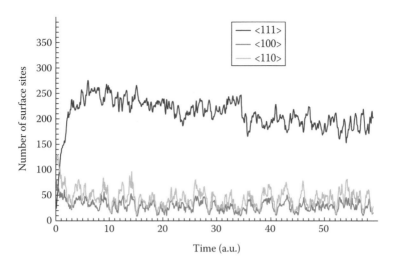

FIGURE 5.6
Plot showing the site concentrations of a crystal resulting with rate constants $\mu_1 = 5$, $\lambda_2 = 10$, $\mu_2 = 0$, $\lambda_3 = 1$, $\lambda_4 = 1$, $\mu_4 = 5$, $\lambda_5 = 1$, $\lambda_6 = 5$, $\lambda_7 = 1$.

function of the total size of the growing crystal. The plot in Figure 5.6 for a set of parameters shows that the site fractions reach a steady state quickly and remain at that point for the remainder of the simulation. In each of the simulations, the crystal size (or total number of atoms) grew linearly with time indicating 1-D growth under the kinetically controlled regime (or the precipitation kinetics limited regime in VLS scheme). Figure 5.7a shows the angles between the median axis of the nanowires and the seed crystal surface, plotted as a function of the total nanowire size for a particular simulation. The data in Figure 5.7a show that the growth direction of simulated crystal reaches a steady state after the crystal reaches 20,000 atoms in size. Figure 5.7b shows the growth kinetics of nanowires grown using a set of rate constants shown in Table 5.1. Under these conditions, the growth directions were found to be close to directions of {111}, {110}, and {100}, respectively. The growth kinetics shown in Figure 5.7b seems to follow a generic relationship as

$$v_{111} < v_{110} < v_{100} \tag{5.1}$$

As illustrated above, the slow kinetics process conditions will yield growth in the <111> direction. Similarly, the fast kinetics changes the growth direction from <111> to <100>. The simulations suggest that fastest growth direction becomes the final growth direction of nanowires. In 3-D crystal growth, the slow growing surfaces determine the final faceting of crystals [20]. In that case, the growth velocities normalized with their interplanar distance should be compared. Similarly, one can suggest that the growth velocities have to follow the relationship below for respective growth directions.

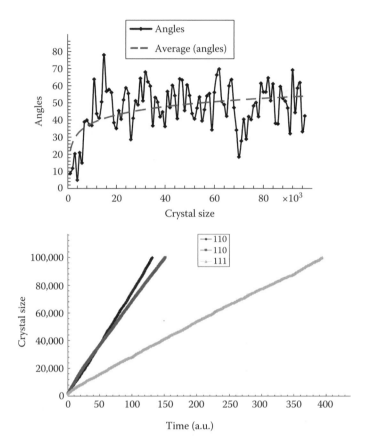

FIGURE 5.7
(a) The angle of medial axis of the simulated nanowire with respect to the seed direction. (b) The growth kinetics of the resulting nanowires with growth in <111>, <110>, and <100> directions.

$$1.732\, v_{111} < 1.414\, v_{110} < v_{100} \tag{5.2}$$

It can be seen from Equation 5.2 that if the growth rate is higher than $\sqrt{3}$ times that of <111>, it can lead to <110> or a <100> growth directions. However, it should be noted that the range of values for which a particular direction is predominant have yet to be identified. This trend seen above gives a quantitative range of growth rates at which a particular growth direction is significant. Even in the kMC simulations as shown in Figure 5.7b, it is shown that a little change in the growth rate can alter the growth direction from <110> to <100>.

The visuals of computer-grown nanowires in the three directions are presented in Figures 5.8 through 5.10. Figure 5.8 shows the ball-and-stick model of nanowires grown in <111> direction. Initially, the wire changes the

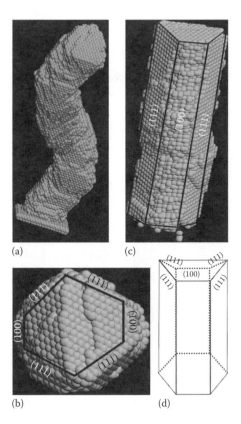

FIGURE 5.8
(a) Space-filled model of 1-D crystal simulated using the rate constants $\mu_1 = 5$, $\lambda_2 = 10$, $\mu_2 = 0$, $\lambda_3 = 1$, $\lambda_4 = 1$, $\mu_4 = 5$, $\lambda_5 = 1$, $\lambda_6 = 5$, $\lambda_7 = 1$. (b) Tip of the crystal simulated in (a). (c) Side facets of the nanowire. (d) The 3-D sketch of the nanowire faceting.

direction in space while keeping the <111> growth direction (axial vector). As shown in the image, the growth surface is a relatively smooth <111> surface with a truncated, trigonal shape with three small additional facets. Figure 5.9 is the morphology of the <100> growth direction nanowire, which becomes straight after reaching steady state. The surface faceting of the nanowire is square shaped with the tip being pyramidal, maintaining the symmetry. The observed square type surface faceting of the nanowires is consistent with that expected for <100> direction nanowires from theoretical considerations [15]. However, the pyramidal-shaped tip may be a computational artifact, which might arise from the small size of the nanowire. The morphology of nanowire grown in between {100} and {111} directions is close to {110} direction and is shown in Figure 5.10. The surface faceting seems to be composed of four {111} and two {100} facets similar to the cross-sections seen in <110> grown wires in both oxygen-assisted and gold-catalyzed growth techniques as described in Section 5.2.

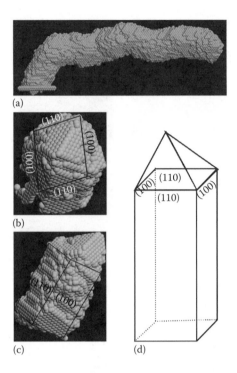

FIGURE 5.9

(a) Space-filled model of 1-D crystal simulated using the rate constants $\mu_1 = 15$, $\lambda_2 = 10$, $\mu_2 = 2$, $\lambda_3 = 5$, $\lambda_4 = 5$, $\mu_4 = 20$, $\lambda_5 = 3$, $\lambda_6 = 6$, $\lambda_7 = 6$. (b) Tip of the crystal simulated in (a). (c) Side facets of the nanowire. (d) The 3-D sketch of the resulting nanowire faceting.

In summary, the simulation methodology presented here can easily be extended to other material systems for understanding the relationship between growth kinetics, growth direction, and surface faceting. Though the direct correlation of nanowire growth kinetics to process conditions is somewhat difficult, it can be done by supersaturation within the molten metal droplet or by the dissolution/arrival flux of adatoms to the tip.

5.3.3 Experimental Results on Growth Direction and Surface Faceting

A set of careful experiments were needed specifically to validate the findings from kMC simulations. These experiments differ from the ones that exist in the literature where smaller nanowires tend to grow faster than the larger diameter wires due to surface diffusion fluxes. Such specific experiments were devised in which faster growth kinetics may occur with larger diameter nanowires. These experiments were performed using Ge nanowires growing from large Ga droplets. The dissolution of Ge into Ga droplets was controlled using the substrate temperature. In these experiments, higher substrate temperature results in higher dissolution rate and corresponding

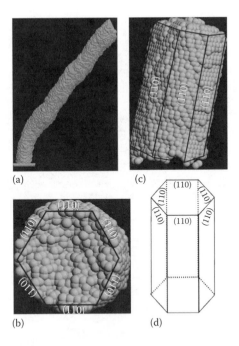

FIGURE 5.10
(a) Space-filled model of 1-D crystal simulated using the rate constants $\mu_1 = 5$, $\lambda_2 = 15$, $\mu_2 = 0$, $\lambda_3 = 15$, $\lambda_4 = 15$, $\mu_4 = 5$, $\lambda_5 = 15$, $\lambda_6 = 15$, $\lambda_7 = 15$. (b) Tip of the crystal simulated in (a). (c) Side facets of the nanowire. (d) The 3-D faceting of nanowire faceting.

higher growth kinetics. The Ge nanowires were grown by spreading a thin film of molten Ga on a clean <100> undoped, n-type Ge single-crystal wafer and exposing it to microwave plasma containing H_2 diluted in N_2 ($H_2{:}N_2$ 10:100 ratio) in a ASTeX 5010 reactor at a reactor pressure of 30 Torr. Several experiments were performed in the microwave power range of 400–680 W.

The growth rate at each experimental condition was determined by performing experiments using different growth durations and measuring the lengths of resulting nanowires. As expected, the nanowire growth rate increases with an increase in microwave power (or the substrate temperature) as seen in Table 5.2. At higher substrate temperatures, as expected,

TABLE 5.2

Growth Directions of the Nanowires Listed along with the Growth Rate and Nanowire Diameters

Nanowire Diameter (nm)	Growth Rate (nm/min)	Growth Direction
7	1.25	<112>
11	2.25	<110>
22	3.5	<100>

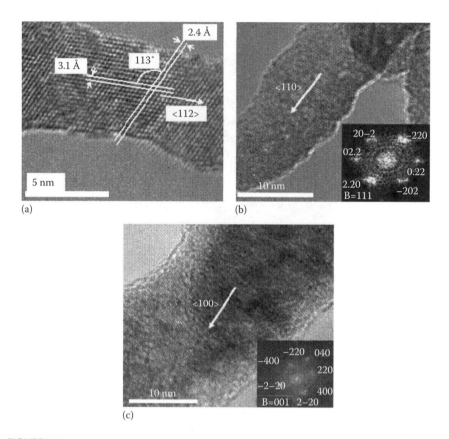

FIGURE 5.11
HRTEM micrographs of (a) 7 nm Ge nanowire, (b) 11 nm Ge nanowire, and (c) 22 nm Ge nanowire.

the resulting nanowires have larger diameters. In these set of experiments, nanowires with larger diameters grew faster than the smaller diameter nanowires. The results also show that the faster growing nanowires yielded <100> direction even though their size is larger than those grown in <211> direction as seen in the high-resolution transmission electron microscopy (HRTEM) images in Figure 5.11. Figure 5.11a shows a HRTEM image of 7 nm Ge nanowire synthesized at 500 W microwave power. Similarly, Figure 5.11b is a HRTEM image of a 11 nm Ge nanowire synthesized at 600 W microwave power with a <110> growth direction; Figure 5.11c is a 22 nm Ge nanowire with a <100> growth direction (680 W microwave power). This observation is consistent with the kMC results. As the wire diameter is less than 30 nm, it is difficult to observe the surface faceting or morphology closely. Nevertheless, the evidence presented here clearly indicates that the growth kinetics determines the growth direction of the nanowires. This evidence is contrary to the report that the growth direction of the nanowires is a function of size, i.e., thinner wires are known to grow along the fastest growth planes [12,13]. However, irrespective of the synthesis route, the faster growth rate promotes growth along the faster growing planes.

5.4 Modeling of Multiple Nucleation and Growth of One-Dimensional Structures

As described in Chapter 4, the use of low-melting metals can lead to high density of nucleation and growth of nanowires from a large droplet. The scheme described in Section 5.3 describes the growth of only one nanowire. It is quite possible to apply a kMC algorithm to model both nucleation and growth of multiple nanowires from large droplets. Such a model needs to incorporate the deposition and diffusion of the solute atoms into the solution phase along with the precipitation and diffusion of the solute atoms in the solid phase. The diffusion of the solid atoms also accommodates for the minimization of the interfacial energy of the cluster. However, the diffusion events being usually fast are separated from the growth processes to track the simulation in a large time frame. A 2-D system reduces the computational intensity but extension to a 3-D analysis is not simple. White and Welland [21] have considered a system with atoms having potential energies specific to their state and local configuration in a square lattice. It is imperative here that the atoms undergoing the transition to a solid phase are the ones in the solution and the solid atoms on cluster surface. Similar continuous time algorithm (CTA) can be implemented as that used in Section 5.3 using a time step defined as [21]:

$$\Delta t = -\frac{\ln R}{\sum_{t=1}^{N} W_i} \qquad (5.3)$$

where R is a random number between 0 and 1. The denominator represents the sum of all events possible at the current simulation step. One of the important steps in the process is the atom transition between different phases for which the rate of transition, W_{ij}, can be calculated using [21]

$$W_{ij} = v \exp\left(-\frac{\mu^* - \mu_i}{k_B T}\right) \qquad (5.4)$$

where
   μ^* is the intermediate-state potential
   μ_i is the initial potential of the reacting species, which is state- and configuration-dependent

The attempt frequency v is taken as 10^{13} s^{-1} and k_B is the Boltzmann constant. It is well known that the supersaturated species in a solution drives the precipitation thermodynamically. So, the driving force for the precipitation or the chemical potential of solute species within supersaturated solution can be expressed as [21]

$$\mu_1 = kT \ln\left(C/C_\infty\right) \tag{5.5}$$

where
C is the local solute concentration
C_∞ is the equilibrium solute concentration obtained from the phase diagram

For a single atom attachment at the cluster boundary, the transition rate is calculated by setting $\mu_i = \mu_1$ in Equation 5.4. For an atom that sublimes or melts, the transition rate is calculated by using $\mu_i = \mu_s$ in Equation 5.4, where as the chemical potential (μ_s) of the solid atom is written as follows [21]:

$$\mu_s = \mu_i^I + \mu_i^C + \mu_i^S - \mu_i^B \tag{5.6}$$

where the contributing potentials are related to the atom's excess interfacial energy (I), the excess energy due to curvature of the cluster (C), the excess energy due to strain from the impurity atoms (S), and the bonding potential (B). The bonding potential can be determined using the bond counting method, i.e., the bonding potential of an atom is based on the number of nearest bonds and next nearest neighboring bonds. Other excess energies can be easily defined through solid–vapor and solid–liquid interfacial energies. The concentration profile (diffusion) of the solute atoms within the solution can be generated using the average diffusion length for the atoms over a specific diffusion time and the variable increment time Δt, used for the growth step.

The simulations using the above methodology were carried out on a square lattice measuring 300 × 50 lattice sites. The formation of the clusters is a multistep process: the process starts with a high initial solute concentration at the liquid–vapor interface. After an initial period of deposition, the solvent is supersaturated and a stable nucleus precipitates and grows isotropically. As the critical size of the nucleus is approached, it interfaces with the vapor and at this point, solute diffusion at the interface plays a significant role in the further growth process. Hence, the surface diffusivity was varied between 10^{-8} and 10^{-5} m^2/s in increments of 10 and the resulting morphologies are shown in Figure 5.12. At slow diffusion rates, the clusters undergo a lateral 2-D growth evident in Figure 5.12a and b. For higher diffusivities, the axial growth dominates the lateral growth giving rise to 1-D and tapered morphologies seen in Figure 5.12c and d. The equivalent reactant vapor pressure and the growth temperature for the simulations were set to 100 Torr and 1100°C, respectively. The results shown in Figure 5.12 are all obtained after

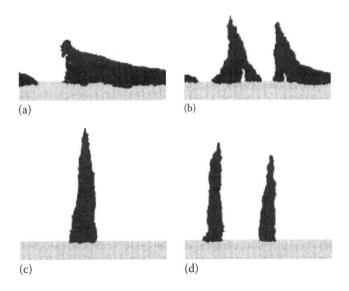

(a) (b) (c) (d)

FIGURE 5.12
The simulated structural morphology of precipitated clusters at the liquid–vapor interface of a supersaturated solution droplet with diffusion coefficients of (a) 10^{-8} m^2 s^{-1}, (b) 10^{-7} m^2 s^{-1}, (c) 10^{-6} m^2 s^{-1}, and (d) 10^{-5} m^2 s^{-1}. (Reproduced from White, R. and Welland, M.E., *J. Appl. Phys.*, 102, 104301, 2007. With permission.)

1.2 µs of simulated growth time. The different growth morphologies can be explained by the phenomena of rate of attachment. For slow diffusivities, the residence time of an atom is high because of which the attempts for attachment to a site increase, which in turn increases the attachment probability and the lateral growth rate. In the case of higher diffusion rates, the residence time is low and therefore, low lateral growth and predominant axial growth for 1-D nanostructures. This study also shows that an increase in pressure increases the solute surface concentration available for attachment as a result of which the lateral growth rate increases. Nevertheless, the above simulation strategy is useful in understanding the origin of experimentally observed tapered morphologies of nanostructures grown using a variety of techniques that involve solute precipitation from large, molten metal pools or solid metal foils.

5.5 Modeling Nanowire Array Growth

The growth of an ensemble of nanowires is a self-assembling process. The growth modeling of such a phenomenon has both scientific and technological importance. Scientifically, the modeling will yield insight into surface adsorption and desorption processes, surface diffusion, reaction kinetics, and chemical bond braking and formation events. It also serves as a gateway to study wire diameter, spacing, and the nanowire height as a function of the process parameters such as reactant pressure, temperature, gas flow rates, etc.

As described in Sections 5.3 and 5.4, the kMC simulations are event-driven and the most successful for surface processes. The time step in kMC is small in larger systems; as a result, the possible process count increases as well as the likelihood that more than one process happens in the given time step, making the computations intensive. Also, different processes cannot be easily parallelized using kMC. Gerisch et al. [22] described an approach based on lattice gas cellular automata (LGCA) method in which any number of adatoms can undergo surface diffusion at any given time and adatoms can hop from one site to another with finite probability. In these systems, some atomistic processes are on the order of nanoseconds and several other diffusion and growth processes require timescales on the order of seconds or even minutes. So, a time step is explicitly determined by the atomic jump of nearest neighbor, which is of the order of seconds and minutes, and hence, the faster atomistic processes of bond formation and breaking are ignored. LGCA adds natural parallelism as well as simple and efficient computational approach compared to kMC. In spite of some differences, both these methods share a common idea that local properties determine the adatom diffusion process. Miller and Succi [23] proposed a 1-D LGCA for the growth of nanowhiskers in which each lattice site is occupied by an integer number of adatoms. Adatoms interact with the uppermost atoms with no overhanging.

If b_k represents the number of dangling bonds, then $1 \leq b_k \leq 3$, as an adatom can grow only on the top of another adatom. Let b_0 be the number of free dangling bonds of an atom in its present location and "$b\pm$" the number of dangling bonds an atom encounters by moving right (+) or left (−). So, the rate of a particular process is then given by [23]:

$$R_k = \frac{1}{\tau} e^{-E_b (b_k - b_0)/kT} \tag{5.7}$$

where E_b is the corresponding binding energy and

$$\tau = \frac{h}{kT} e^{\frac{E_c}{kT}} \tag{5.8}$$

where
  h is the Planck constant
  E_c is the surface corrugation energy specific to a material

The time constant represents the adatom hopping duration between sites on a flat surface ($b_k = b_0$).

The probability of a particular event P_k is then calculated as [23]:

$$P_k = \frac{r_k}{1 + r_- + r_+} \tag{5.9}$$

with the moves designated as 0 (no move), left (+), and right (−), and the reduced rates designated as $r_k = R_k/R_0$. In this model, all sites are parallelized using a global time step dt. The growth equation for the height profile at lattice site, $x = i dx$ and time, t is as follows [23]:

$$h(i, t + dt) = h(i, t) + \left[\varphi^\uparrow + (i - 1, t) + \varphi^\uparrow (i + 1, t) + \varphi^\uparrow - (i, t) - \varphi^\uparrow + (i, t) \right] dt$$

$$+ D(i, t) dt \tag{5.10}$$

where $D(i,t)$ is the local deposition rate. To preserve the integer units for the change in the site height, the following rule is employed:

$$0 < \xi < p_0 : \Phi_0 = 1, \qquad \Phi_\pm = 0$$

$$p_0 < \xi < p_0 + p_- : \Phi_- = 1, \qquad \Phi_0 = \Phi_+ = 0$$

$$p_0 + p_- < \xi < 1 : \Phi_+ = 1, \qquad \Phi_0 = \Phi_- = 0$$

where ξ is a random number that lies between 0 and 1. The algorithm above was used to predict the qualitative features of the whickers as a function of deposition flux, corrugation, and bond energies. The deposition process ("rain") is implemented by allowing one drop to fall every n_{rain} steps on a randomly chosen site from N_x sites. The n_{rain} is defined as $(1/N_x D dt)$, where D is the monolayers per second (ML/s). Using a time step t and the algorithm described here, Miller and Succi studied the height of whiskers (whisker growth rates) and the whiskering efficiency (ratio of atoms in whiskers versus total number of deposited atoms) as a function of bond energy, E_b. It is clearly shown here that smaller the dangling energy (E_b), the lower the spatial resolution and hence, faster growth with less background thin film deposition.

In another model, adatoms were allowed to diffuse along the whisker rims but were restricted to a direction in which there were no adatoms as neighbors. Hopping to another adatom was not allowed. These restrictions prohibited the growth of the whisker diameter and only allowed for the growth in whisker height. In this model, the corrugation energy E_c was the same for surface distribution on substrate as well as the whisker's rim diffusion eliminating the time scale problems. An adatom at the top of the rim can enter the whisker with a probability P_{diss}, which was an input for the study. Using these simulations, the whisker height profiles (growth kinetics) were deduced as a function of dangling bond energy (E_b) as well as P_{diss}, which is a sensitive parameter. The LGCA model clearly simulated the experimentally observed whisker growth features with a trench-like background in Figure 5.13

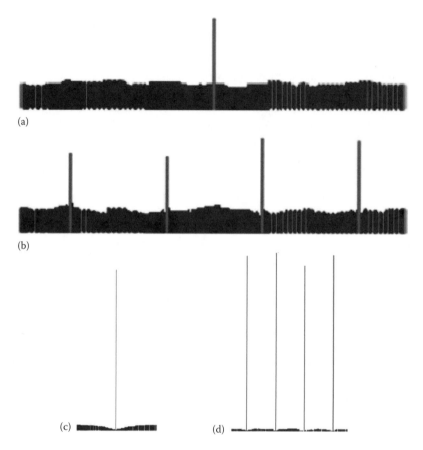

FIGURE 5.13
Spatial distribution of $h(i)$ for (a) $N_w = 1$ and (b) $N_w = 4$ using $E_b = 1.0\,\text{eV}$. (c) $N_w = 1$ and (d) $N_w = 4$ using $E_b = 0.5\,\text{eV}$. (Reproduced from Miller, W. and Succi, S., *Phys. Rev. E*, 76, 031601, 2007. With permission.)

using E_b values of 1.0 and 0.5 eV with $E_c = 0.88\,\text{eV}$ and $D = 2000\,\text{ML/s}$. It can also be seen in Figure 5.13 that the background thin film deposition was much lower for $E_b = 0.5\,\text{eV}$, because the high dissolution of particles leads to high flux from the sides to the whisker leaving trenches around them. The simulation results also predicted that the growth kinetics under the diffusion-induced processes can be fitted with m values ranging from −0.3 to −0.86 using various parameters within the following relationship [23]:

$$h_{\text{whisker}} \propto d_{\text{whisker}}{}^{m} \tag{5.11}$$

It is essentially demonstrated here that a simple 1-D LGCA model can be used to capture essential aspects of nanowire array growth.

References

1. Y. Cui, Z. Zhong, D. Wang, W. U. Wang, and C. M. Lieber, *Nano Lett.* 3, 149 (2003).
2. S. Gradečak, F. Qian, H. G. Park, and C. M. Lieber, *Appl. Phys. Lett.* 87, 173111 (2005).
3. Y. Cui, L. J. Lauhon, M. S. Gudiksen, J. Wang, and C. M. Lieber, *Appl. Phys. Lett.* 78, 2214 (2001).
4. H. Chandrasekaran, G. U. Sumanasekera, and M. K. Sunkara, *J. Phys. Chem. B* 110, 18351 (2006).
5. T. Kuykendall, P. J. Pauzauskie, Y. Zhang, J. Goldberger, D. Sirbuly, J. Denlinger, and P. Yang, *Nat. Mater.* 3, 524 (2004).
6. H. Li, A. H. Chin, and M. K. Sunkara, *Adv. Mater.* 18, 216 (2006).
7. E. I. Givargizov, *J. Crystal Growth* 31, 20 (1975).
8. A. M. Morales and C. M. Lieber, *Science* 279, 208 (1998).
9. Y. Yao, F. Li, and S. T. Lee, *Chem. Phys. Lett.* 406, 381 (2005).
10. V. Schmidt, S. Senz, and U. Gösele, *Nano Lett.* 5, 931 (2005).
11. T. Y. Tan, N. Li, and U. Gösele, *Appl. Phys. A* 78, 519 (2004).
12. Y. Wu, Y. Cui, L. Huynh, C. J. Barrelet, D. C. Bell, and C. M. Lieber, *Nano Lett.* 4, 433 (2004).
13. C. P. Li, C. S. Lee, X. M. Ma, N. Wang, R. Q. Wang, and S. T. Lee, *Adv. Mater.* 15, 607 (2003).
14. Th. Frauenheim, G. Seifert, M. Elstner, Z. Hajnal, G. Jungnickel, D. Porezag, S. Suhai, and R. Scholz, *Phys. Status Solidi. B* 217, 41 (2000).
15. R. Q. Zhang, Y. Lifshitz, D. D. D. Ma, Y. L. Zhao, F. Th, S. T. Lee, and S. Y. Tong, *J. Chem. Phys.* 123, 144703 (2005).
16. D. J. Eaglesham, A. E. White, L. C. Feldman, N. Moriya, and D. C. Jacobson, *Phys. Rev. Lett.* 70, 1643 (1993).
17. D. D. D. Ma, C. S. Lee, F. C. K. Au, S. Y. Tong, and S. T. Lee, *Science* 299, 1874 (2003).
18. B. Delley and E. F. Steigmeier, *Phys. Rev. B* 47, 1397 (1993).
19. R.C. Mani and M. K. Sunkara, *Dia. Relat. Mater.* 12, 324 (2003).
20. J. C. Angus, Y. Wang, and M. Sunkara, *Ann. Rev. Mater. Sci.* 21, 221 (1991).
21. R. White and M. E. Welland, *J. Appl. Phys.* 102, 104301 (2007).
22. A. Gerisch, A. T. Lawniczak, R. A. Budiman, H. E. Ruda, and H. Fukś, in Proceedings of CCECE 2003-CCGEI 2003, 2, 1413 (2003).
23. W. Miller and S. Succi, *Phys. Rev. E* 76, 031601 (2007).

6

Semiconducting Nanowires

6.1 Introduction

Semiconducting nanowires are of great interest due to their potential in electronics, optoelectronics, and sensor applications. Silicon nanowires (SiNWs) have received the most attention due to the historical role of silicon in the integrated circuits (IC) industry. Continued high performance from silicon may require alternatives to thin film based planar CMOS transistors in the form of novel devices, three-dimensional structures, innovative architectures, and integration with other functional components [1]. SiNWs may provide new avenues in these directions. Silicon becomes a direct band gap semiconductor at nanoscale dimensions [2] and thus may have interesting optoelectronics applications. The interest in optical interconnects beyond the copper era in IC manufacturing and the desire to have all silicon-based process (instead of using III–V materials for the optical interconnect part) may benefit from mature nanowire technology if and when it becomes available. More recently, SiNWs have also been used in the fabrication of DNA sensors [3]. In many of the applications, Ge provides a competitive avenue and thus received serious consideration in the nanowire community. Nanowires of III–V semiconductors such as GaAS, $Al_xGa_{1-x}As$, InP, InAs, and wide band gap III–V nitrides have also been reported. As in the case of all nanowires discussed in this book, a wide variety of techniques including the VLS approach, laser ablation, template-guided synthesis, oxide-assisted growth, and others have been used to grow semiconducting nanowires.

6.2 Silicon Nanowires

Investigation of silicon whiskers and nanowires dates back to the pioneering work on vapor–liquid–solid (VLS) approach by Wagner and Ellis [4,5] as discussed in Chapter 2 and has continued since then using a variety of techniques [6–43]. Most of the studies in the literature report SiNWs grown randomly on the substrate regardless of the synthesis approach while some attempted vertically oriented nanowires [7,30,31,35,43]. Since the primary motivation for SiNWs is electronics, VLS-based growth is discussed first in this chapter; also it is closer to the chemical vapor deposition (CVD) widely used in the IC industry. Two common chemistries in silicon epitaxy involve $SiCl_4/H_2$ and SiH_4/H_2, which are also used in SiNW growth. Variations of the above such as dichlorosilane, trichlorosilane, disilane, etc. found in IC industry have not been used to date in nanowire preparation.

6.2.1 SiCl$_4$/H$_2$ System

$SiCl_4$ has been commonly used in thin film epitaxy of silicon and therefore has been popular also in the SiNW literature. A detailed description of SiNW growth process using this system is given below. The growth reactor consists of a quartz tube heated by a programmable furnace, which can heat up to 1100°C and maintain within ±1°C. A thermocouple located in the middle of the reactor monitors the quartz wall temperature during growth. The substrate is located at the thermocouple to ensure accurate measurement. Process gases include $SiCl_4$ (99.998% purity, Aldrich), Ar (99.999%, Scott Specialty Gas), and H_2 (99.99%, Air Gas). A dilution line purges the system to minimize impurities such as water and oxygen before introducing $SiCl_4$. Gaseous and solid impurities are minimized through the use of suitable materials and isolation valves. Perfluoroalkoxy (PFA), an acid-resistant plastic, is the material of choice for the valves and tubing, and the valves isolate the liquid source from ambient water and air when the system is idle. The $SiCl_4$ vapor is introduced into the system by flowing argon through the liquid $SiCl_4$ reservoir. Transport of the $SiCl_4$ vapor source occurs by saturation of the Ar carrier gas, as it bubbles through a sparger. The amount of $SiCl_4$ depends on the amount of carrier gas and bath temperature, and Raoult's law can be used to determine the amount of $SiCl_4$ delivered. The premixed process gas is fed to the quartz tube and flows toward the substrate, which is located in a quartz boat in the middle of the heater. As it travels downstream, the gas is heated by flowing through 7 in. of heated quartz reactor tube before arriving at the boat.

The substrates are n-type [111] Si that is 4° off normal with a specified resistivity of 0.001–0.005 Ω-cm. The substrate surface treatment includes removing impurities that could be incorporated into the nanowires and eliminating the outer oxide layer that could be a barrier against Au alloying with Si. The cleaning processes included the following steps [43,44]:

1. Sonicating wafers in double distilled water for 15 min
2. Rinsing with double distilled water
3. Sonicating with acetone for 15 min
4. Rinsing with double distilled water
5. Soaking the wafer in 10% HF for 2 min to remove SiO_2 surface layer
6. Rinsing with double distilled water and drying with Ar

In principle, the by-product HCl from the $SiCl_4 + H_2$ reactions can be expected to etch the oxide layer, thus obviating the need for an extra step incorporated above [31]. However, the amount of HCl generated depends on process conditions and it is a good practice to use an oxide-removing step commonly used in silicon processing. After the cleaning steps, a thin film of gold (5 nm) is sputtered onto the wafer using ion-beam sputtering and then the wafer is cleaved into small pieces. The heating during the growth process above the eutectic point converts the thin film to small droplets or beads. For example, after heating to 925°C with 180 sccm of Ar and 20 sccm of H_2 for 15 min, the catalyst bead distribution from SEM images (not shown here) shows an average of 48.4 nm with a standard deviation of 24.5 nm. Of course, the actual droplet size distribution during growth may be different depending on local temperature, droplet mobility, agglomeration, etc.

The wire growth processes consist of the following procedure. The temperature is first ramped up to the growth temperature over 15 min and the system is purged with Ar and H_2 to remove oxygen and moisture. Hydrogen is used to reduce residual oxygen not removed by purging. With the Ar and H_2 still flowing, the temperature is stabilized in the next 5 min. In this stage, Au alloys with the Si substrate at the operating temperature. Next, $SiCl_4$ is introduced for 15 min to supply material for wire growth. When growth is over, the $SiCl_4$ flow is stopped while maintaining the temperature and flowing the dilution gases—Ar and H_2—for 5 more minutes. Finally, with Ar still flowing, the system is slowly cooled over 15 min to minimize instabilities that can arise from steep temperature gradients and an unstable temperature.

The exploration matrix shown in Table 6.1 helps to understand the effects of temperature (T), residence time (τ, given by reactor volume divided by flow rate), H_2 concentration, and $SiCl_4$ concentration on wire morphologies and orientation [44]. The range of parameters here is restricted to only those values that yield SiNWs, thus eliminating particulates and thin films. The SiNWs produced by these conditions can be classified into five different types of nanowires as summarized in Table 6.2. Most of the results are not ideal for device fabrication; nevertheless they are presented here because this is precisely what happens when one sets out to conduct growth experiments. Almost all journal publications reveal only successful runs, resulting in nice SEM images, thus leaving it to the readers to learn the nuances and difficulties by themselves.

Type I wires, as shown in Figure 6.1, are the only wires which are vertically aligned. These wires are crystalline as confirmed by the TEM images in Figure 6.2 and have average diameters less than 100 nm. According to Figure 6.2a and c, regular crystal lattices, which are indication of crystallinity, are observed on the wire body with the exception of a thin outer layer. The outer sheath is likely to be amorphous silicon dioxide (SiO_x), which forms when the samples are exposed to oxygen. Crystallinity is also confirmed by the diffraction pattern on Figure 6.2d. The tip, however, is noncrystalline

TABLE 6.1

Experimental Matrix Showing the Parametric Variables in SiNW Growth Using $SiCl_4/H_2$

Test	T (°C)	Q_{H_2} (sccm)	Q_{tot} (sccm)	Q_{SiCl_4} (sccm)	[$SiCl_4$] (mol%)	[H_2] (mol%)	τ (s)
E1	925	20	500	2	0.04	4	2.9
E2				5	0.10	4	2.9
E3			200	2	0.10	10	7.3
E4				5	0.25	10	7.3
F1		50	500	2	0.04	10	2.9
F2				5	0.10	10	2.9
F3			200	2	0.10	25	7.3
F4				5	0.25	25	7.3
G1	1000	20	500	2	0.04	4	2.8
G2				5	0.10	4	2.8
G3			200	2	0.10	10	6.9
G4				5	0.25	10	6.9
H1		50	500	2	0.04	10	2.8
H2				5	0.10	10	2.8
H3			200	2	0.10	25	6.9
H4				5	0.25	25	6.9

Source: Courtesy of A. Mao.

TABLE 6.2

Summary of Wire Types Observed Corresponding to Conditions in Table 6.1

Wire Type	Cond.	T (°C)	[$SiCl_4$] (%)	τ (s)	[H_2] (%)	Wires Characteristics and Surface Morphology
I	E1	925	0.04	2.9	4	Most ideal wires
	F1		0.10	7.3	10	Straight and aligned
	E3				10	Average diameter < 100 nm
	F3				25	Greater than 3 μm long
II	E2	925	0.10	2.9	4	Thin and curvy
	F2		0.25	7.3	10	Densest of all growth
	E4				10	
	F4				25	
III	G3	1000	0.10	6.9	10	Straight, long but
	G4		0.25		25	unaligned wires
	H3				10	Larger than Type I wires
	H4				25	
IV	G1	1000	0.04	2.8	4	Thin, long, curvy wires
	H1				10	Pitted surfaces
V	G2	1000	0.10	2.8	4	Combination of Types
	H2				10	III and IV wires

Source: Courtesy of A. Mao.

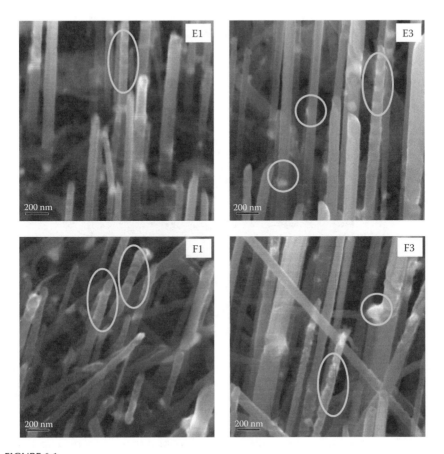

FIGURE 6.1
Type I wires produced by conditions E1, E3, F1, and F3 (Table 6.1). Circles indicate possible catalyst beads and ovals indicate screwlike patterns on wires. Wires are viewed at 45° and have a common scale of 200 nm. (Courtesy of A. Mao.)

as indicated by Figure 6.2b. The noncrystallinity could be attributed to the changes in temperature and concentration, which occur after growth stoppage. The catalyst particle, normally seen at the tip in VLS growth, is missing in Figure 6.2b.

These aligned wires have average wire diameters of 68.3–96.4 nm. The average diameters of SiNWs from this group are larger than the average catalyst particle size of 48 nm. As mentioned earlier, particle size characterization is done on the cooled sample taken out from the reactor after exposing to the growth temperature. This distribution may not be representative of the dynamic situation during growth wherein the droplets are mobile and coalesce to form larger droplets. Also, the droplets can undergo swelling as silicon diffuses into them [31] until reaching a critical supersaturation level to begin nanowire growth. Wires smaller than 50 nm are not observed in these results, which suggests that the effective chemical potential is insufficient

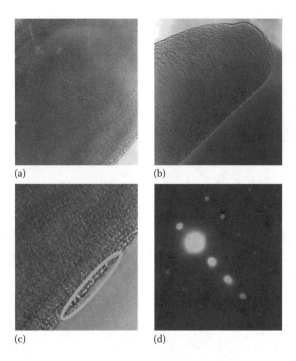

(a)   (b)

(c)   (d)

FIGURE 6.2
TEM images for straight wire produced by condition E3. Crystal lattice of (a) wire body, (b) wire tip, and (c) wire body (circle indicates noncrystalline outer SiO_x sheath). (d) Diffraction pattern of wire body. (Courtesy of A. Mao.)

for formation of nanowires smaller than 50 nm and that wires are selectively formed on catalyst beads with a diameter larger than the minimum radius (R_{min}) determined by the prevailing process conditions. R_{min} is given by $R_{min} = 2\sigma_{LV}V_L/RT\ln\sigma$ where σ_{LV} is the liquid–vapor interfacial energy, V_L is the liquid molar volume, T is the temperature, and σ is the supersaturation determined by the excess of partial pressure over the equilibrium vapor pressure, $(P - P_{eq})/P_{eq}$. For the conditions employed here, R_{min} is estimated to be in the range of 50–100 nm.

Though the growth densities are high (3.9–4.7 per μm^2), it is less than 8% of the catalyst beads created by the 5 nm Au films. The location of the catalyst beads, however, cannot be substantiated on the SEM images. Analysis of Figure 6.1 reveals two interesting features: shiny beads and screwlike features on the wire side. The shiny beads, indicated by circles, are observed on the side of the wire. These beads could be the missing catalyst particles, but their composition was not determined. A screwlike pattern, as marked by ovals in Figure 6.1, is also observed on approximately 10% of the wires. This pattern is independent of the wire diameter, length, and alignment. According to the right-hand rule, the screw directions are either into or out of the substrate. It cannot be determined whether the screwlike pattern originates from the substrate or catalyst bead instability. Several researchers had provided support for catalyst bead instability; both Wagner [5] and Givargizov [45] describe the catalyst bead as a high interfacial energy system that is very unstable and these instabilities can lead to undesirable morphologies.

Type I wires are consistent with the VLS mechanism even though no catalyst bead on the tip can be confirmed in the electron microscopy images. The relatively long wires suggest that the material incorporation rate into the catalyst for wire formation is much greater than that on the bare Si substrate. This is possible only if the catalyst is present to increase the deposition rate, otherwise nodules or films would be produced [5,45]. Finally, the wire cross-sectional areas remain constant throughout the length of the wires. Indeed, there have been other reports on VLS-grown SiNWs with missing catalyst particles on the tip [17]. In all these cases, it is possible to have the base growth mode wherein the catalyst particle is at the base of the nanowire instead of at the tip; such occurrence is common, for example, in carbon nanotube (CNT) growth—which also follows VLS mechanism—when the catalyst particle adhesion to the substrate is very strong.

Type II wires are curvy and unaligned as illustrated in Figure 6.3 and summarized in Table 6.2 along with the growth conditions. These wires are extremely dense ($\sim$17/μm^2), thin ($\sim$15 nm), short ($\sim$1.5 μm), and very uniform in diameter and length. Compared to the conditions for Type I wires, the supersaturation here is higher with increased $SiCl_4$ concentrations, which results in smaller diameter wires. Bright spots are observed at the tip and within the wires, which could be the catalyst beads, but the composition is undetermined without further analysis. The curviness of these wires could be due to the fact that thinner wires are in general less stable. Instability of thinner wires is consistent with Givargizov's observation [45] that stability decreases with decreasing wire size. The Type II wires also have features that are consistent with the VLS mechanism. The high length-to-diameter ratio suggests that material is incorporated through the growing tip; without the catalyst, polycrystalline nodules or film would have formed instead.

Type III wires are long, straight, and unaligned as shown in Figure 6.4. These wires, grown at 1000°C (see Table 6.2 for conditions), are very similar to Type I wires in straightness but not vertical; they are less densely grown (0.2–1.5 per μm^2), longer (12.9–43.8 μm), and thicker (163.7–278.9 nm). The extremely long lengths indicate rapid growth rates estimated to be between 0.86 and 2.92 μm/min. Even at these rapid growth rates, the wires remain straight with a constant cross-sectional area. The Type III wires are found to have wide diameter distributions ranging from 50 to 960 nm, and have densities as low as 0.5% on the substrates. The catalyst beads must be larger to generate larger wires. The lower densities, larger wires, and unaccounted catalyst beads suggest that smaller catalyst beads could agglomerate to form larger beads when the temperature is increased. As expected, the catalyst particle distribution is dynamic and affected by the process conditions.

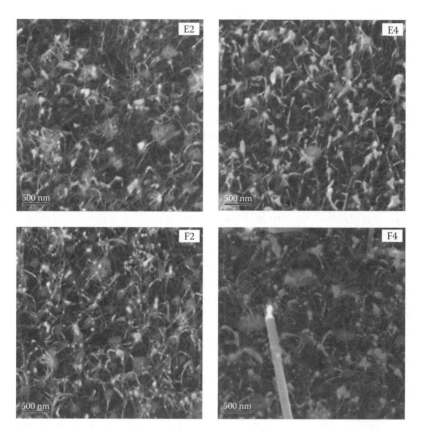

FIGURE 6.3
Type II wire samples generated by conditions E2, E4, F2, and F4 (Table 6.1). Specimens are viewed at 45° with a common scale of 500 nm. (Courtesy of A. Mao.)

Type IV wires (Figure 6.5) illustrate the detrimental effect of the combination of high temperature and low $SiCl_4$ concentration. These wires are long, thin, and curvy, but the surfaces are marked by triangular pits that are 3 µm long and 500 nm deep. Within some of these pits, there are straight and large wires approximately 200 nm in diameter and 2 µm in length. Unlike the long, thin, curvy unaligned wires found outside the pit, these wires inside the pits are larger, shorter, and aligned. The larger wires within the pits could be caused by the pits acting like wells to accumulate and enlarge the catalyst beads. The improved alignment of the larger wires may possibly be attributed to the stability of larger wires, as described by Givargizov [45] and Westwater et al. [7]. Finally, Type V wires appear to be a combination of Types III and IV when the $SiCl_4$ concentration was increased to 0.1% at 1000°C.

In addition to the qualitative observations, the computed wire diameter, wire length, growth rate, and wire density can be correlated to the input process parameters through a design of experiment (DOE) analysis.

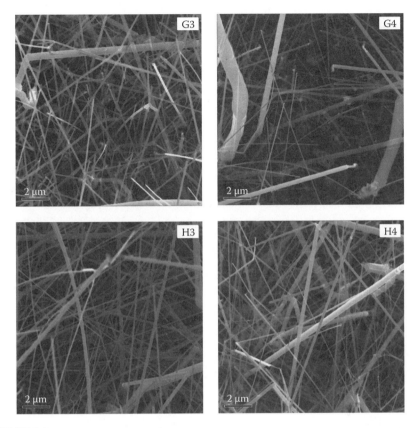

FIGURE 6.4
Type III wires produced by conditions G3, G4, H3, and H4. Specimens are viewed at 45° with a common scale of 2 μm. (Courtesy of A. Mao.)

The uncertainty in estimating the above four figures of merit from numerous images is 8.7%, 8.7%, 8.7%, and 5.0%, respectively. The results of this analysis are discussed in the following [44].

The most important effect of H_2 mole fraction is on the wire alignment and is best illustrated by the Type I wires in Figure 6.1. For a fixed $SiCl_4$ concentration, an increase in H_2 concentration by 2.5 times (E1 versus F1, E3 versus F3) leads to misalignment. Nevertheless, even the nonnormally oriented wires are not necessarily randomly directed; rather, most of them appear to have three major directions that are approximately 60° from each other. This observation is rationalized by the preferred SiNW growth direction, which is along the <111> direction. The directionality of the unaligned wires and the preferred <111> growth direction suggest that the wires are likely to originate from other {111} surfaces, which are not uniformly and normally directed; these new surfaces could have been created by surface etching. A possible etchant could be HCl, which is a by-product of $SiCl_4$ reduction by H_2. The etching is slight since the wire tip and base have similar diameter;

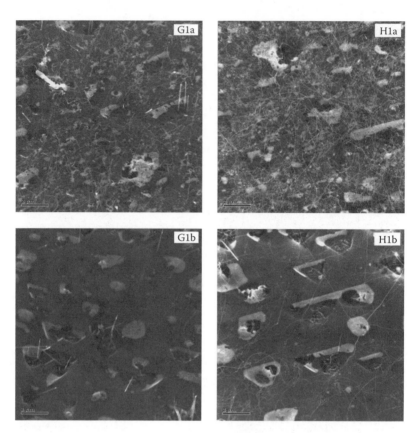

FIGURE 6.5
Growth results from conditions G1 and H1. Samples are viewed at 45° with a common scale of 2 μm. (Courtesy of A. Mao.)

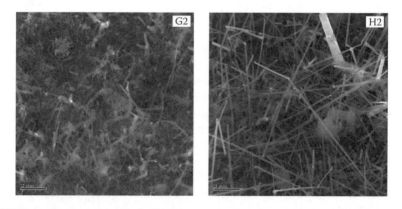

FIGURE 6.6
Growth results from conditions G2 and H2. Specimens are viewed at 45° with a common scale of 2 μm. (Courtesy of A. Mao.)

if the etching was aggressive, the wire base would be thinner than the top. Other effects of H_2 mole fraction tend to be temperature-dependent and more pronounced at 1000°C. Both the wire length and wire density increase with H_2 concentration. This is consistent with the observation by Zhang et al. [46] reporting an increasing yield of silicon whiskers with low molar ratios of $SiCl_4/H_2$.

An increase in temperature in general changes the wire morphologies and surfaces radically. As noted in Type IV wires, pitting occurs on the surface. In general, the wires are longer, thicker, and unaligned when the temperature is increased. One may expect an increase in vapor–solid epitaxy with increased temperature, which could possibly explain the formation of thicker wires. If this were the case, the base would be larger than the tip. All the results here indicate a constant cross-section over the entire length of the wires, thus eliminating vapor–solid epitaxy. As mentioned earlier, it is entirely likely that catalyst particles agglomerate when the temperature is increased to form larger beads that lead to larger wires. Increased agglomeration can reduce the number of the catalyst beads, thus reducing the wire density, which is the case observed here with increased temperature. The wire density and diameter appear to have an inverse relationship from these results. Another effect of temperature is to increase the length of SiNWs, which may be explained by an increase in the conversion rate of the endothermic reactions (that produce silicon).

$SiCl_4$ concentration, as expected, affects wire morphologies and surface topologies and this effect itself is strongly dependent on the temperature level. An increase in $SiCl_4$ concentration, in the range investigated for nanowire production here, causes a reduction in wire diameter and an increase in wire misalignment. Higher concentrations lead to an increase in supersaturation and the wire radius is inversely proportional to supersaturation [5]. Note that much higher concentrations beyond the levels in Table 6.1 result in particulates and coating-like materials as reported in Ref. [42].

Hochbaum et al. [31] reported SiNW growth using $SiCl_4/H_2$ system and gold colloids instead of gold thin film for catalyst. First, 0.1 wt% poly-L-lysine is deposited on a Si(111) wafer and then nanoclusters of gold of specified size are deposited by immersing the substrate containing the gold particle solution (10^{10}–10^{11} particles/cc). The negatively charged gold particles stick to the positively charged poly-L-lysine. Without this procedure, the gold colloids do not stick to the silicon wafer. Growth of vertical nanowires was reported at a growth temperature of 800°C–850°C. The use of monodispersed colloids appears to give a narrow distribution of wire diameters. Colloids of nominally 20, 30, and 50 nm in diameter (with a standard deviation of about 10%) yielded nanowire diameters of 39 ± 3.7, 43 ± 4.4, and 93 ± 7.4 nm, respectively. Swelling of the seed droplets until reaching a critical supersaturation concentration was suggested as the reason for the deviation between the seed size and nanowire diameter. However, the above disparity between the seed and nanowire sizes is much higher than that reported for the silane/H_2 system using monodispersed colloids [16] where the nanowires were on the average only 1–2 nm larger than the colloids. This suggests that in the $SiCl_4/H_2$ system, the high growth temperatures (>800°C) may lead to surface migration and agglomeration of the gold particles, resulting in larger droplets.

6.2.2 Silane Feedstock in VLS Growth

The use of silane as feedstock allows lower growth temperatures than for $SiCl_4$ which is well known in silicon microelectronics industry. Therefore, this has been a popular approach in VLS growth of SiNWs [7,15–17,20,21,26,32–34,36–42]. Though, pure silane feedstock has been used occasionally [20,21,41], it is common to dilute it with H_2 [26,36–40], He [7,15–17], or argon [32–34,42]. Westwater et al. [7] provided a comprehensive analysis of crystalline SiNW growth characteristics using silane, similar to the discussion for $SiCl_4$ system in Section 6.2.1. Their findings are summarized in Figure 6.7, which correlates

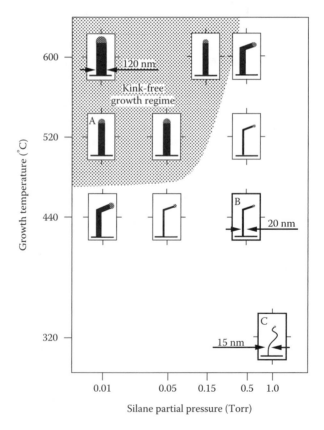

FIGURE 6.7

Summary of SiNW morphology for various growth temperatures and silane partial pressures. (Reproduced from Westwater, J. et al., *J. Vac. Sci. Technol. B* 15, 554, 1997. With permission.)

growth morphology to silane dilution and temperature. Here, gold thin film (0.6 nm) is used to catalyze SiNW growth. Silane is diluted with He, varying the partial pressure from 0.01 to 1 Torr, and the growth temperature ranges between 320°C and 600°C. As with the case of SiCl$_4$, low partial pressure of the silicon source vapor is necessary to obtain vertical nanowires. When the silane partial pressure is increased, the growth rate increases accompanied by a tendency toward kinking, and the nanowires are thin and curvy under these conditions. The lower end of the temperature range also leads to defects as seen in Figure 6.7. Vertical wires are obtained at temperatures of 500°C–600°C and the diameter increases with temperature.

In contrast to the SiCl$_4$ system, much thinner wires are possible with silane, and the results summarized in Figure 6.7 include nanowires as small as 10 nm. Even smaller nanowires (~2 nm) have been obtained using mono-dispersed gold colloids. Cui et al. [16] reported mean nanowire diameters of 6.4 ± 1.2, 12.3 ± 2.5, 20 ± 2.3, and 31.1 ± 2.7 nm for gold clusters of nominal size 5 nm (4.9 ± 1.0), 10 nm (9.7 ± 1.5), 20 nm (19.8 ± 2.0), and 30 nm (30.0 ± 3.0), respectively. It is remarkable that the nanowire diameter on the average is only 1–2 nm larger than the colloid size. This indicates that catalyst particle agglomeration is not an issue at the growth temperature of 440°C and the small disparity in sizes can be attributed to the gold droplet swelling under supersaturation as well as possible native oxide on the nanowire surface upon exposure to air. The general growth characteristics, for the conditions of 10% silane in He and 440°C used here, are consistent with the summary in Figure 6.7.

All other published works using silane in VLS growth also confirm the general trends with respect to the effects of partial pressure and temperature in Figure 6.7. Dense growth of SiNWs happens around 500°C, which seems to be the optimum temperature. Typically no growth is found at temperatures below 300°C regardless of the type of catalyst. Very high temperatures ($T > 700°C$) induce radial epitaxy [36], resulting in tapered structures with a wide base. Undiluted silane always gives very long, high-aspect-ratio nanowires [21]. The growth rate, as expected, increases with temperature and silane partial pressure. Sample A in Figure 6.7 has a growth rate of 0.15 nm/s at 520°C with 0.01 Torr partial pressure of silane [7]. In the case of tapered structures mentioned above, the vertical growth rate is about 3.1 nm/s with a sidewall deposition rate of 0.1 nm/s at 700°C [36]. When the temperature is lowered to 500°C, the vertical growth decreases to 0.1 nm/s with complete elimination of radial epitaxy. Much higher growth rates of up to 17 nm/s (~1 μm/min) have also been reported for 2.5% silane in H$_2$ at 500°C–650°C using gold thin film catalyst [40].

6.2.3 Other Sources

Besides gas/vapor sources for silicon as discussed in Sections 6.2.1 and 6.2.2, silicon powder can be used as well. Yu et al. [8,9] mixed high-purity silicon and Fe powders (95:5) and formed a plate by hot pressing at 150°C. This plate is placed in a typical VLS reactor described in Chapter 3 and Section 6.2.1 and heated to 1200°C under a flowing carrier gas at reduced pressures (100 Torr). This approach produces SiNWs with diameters of 10 nm and lengths of tens to hundreds of microns. The nanowires exhibit a crystalline silicon core (10 nm diameter) with an overcoating of 2 nm thick amorphous silicon oxide. Here the iron acts as the catalyst seed and iron particles are found in the body of the wires instead of at the tips. In the absence of Fe powder, SiNWs are not formed. When the pressure is increased, SiNW diameter increases, likely due to the formation of larger catalyst droplets through coalescence. Since Fe is capable of introducing impurity states undesirable in photoluminescence, alternatives such as Ge, B, Ga, or P can be mixed with silicon powder to facilitate nanowire growth. The major disadvantage of this approach is the very high growth temperature.

6.2.4 Oxide-Assisted Growth

In the approaches discussed thus far, metal particles act as seeds in a VLS growth mechanism. An alternative without metal catalysts is the oxide-assisted growth to produce SiNWs [10–14,28]. This approach involves laser ablation of a target placed in a quartz tube heated by a furnace. The target is a pressed disk consisting of high-purity Si and SiO_2 powders (90:10). An inert gas flow is maintained through the system, which is at a temperature of 1200°C near the target. This produces long wires (~10 μm) with diameters in the range of 6–20 nm. All nanowires consist of a crystalline silicon core with an amorphous SiO_2 sheath. The silicon core shows many lattice defects including twins on the {111} planes and stacking faults.

Lee and coworkers [10–14] proposed a mechanism to explain the oxide-assisted growth, which centers on the formation of SiO nanoparticles on the surface of SiO matrix particles. Silicon is precipitated due to the disproportion or oxidation–reduction reaction of amorphous SiO at temperatures of 950°C–1250°C. Si–O clusters are continuously fed to the amorphous nanoparticles at the tip of the elongating nanowire, which, following phase separation and precipitation, results in a nanowire with a silicon core and oxide sheath.

The pressed disk of Si/SiO powders can be heated to above 1100°C in flowing argon instead of laser ablation to produce SiNWs. The typical yield with or without laser is about 10% of the source material. If a closed system is used instead, wherein the inert gas is sealed inside the chamber, the yield nearly doubles to about 20% [12]. Several variations of the oxide-assisted technique have also been explored [14,22]. One of them involves the use of gold catalyst film on a silicon substrate with a source SiO powder. In this approach [14], the nanowire diameter is largely determined by the gold particle diameter as in VLS growth. While the source vapor generation from SiO requires 1300°C, the growth zone can be at a lower temperature (800°C) with the aid of the catalyst. This combined approach also allowed the growth of oriented SiNWs [14].

Another variation involves carbothermal reduction of the silica layer covering silicon [22]. Here silicon oxide is not used explicitly in the powder form as before. Instead silicon powder is mixed with activated carbon at a ratio of 1:1 and heated to 1200°C in a flowing stream of argon and hydrogen. SiNWs of diameter 75–350 nm and length of few microns are obtained. While a surface oxide layer appears on all the nanowires, no carbides are formed. When amorphous carbon is sputtered onto a silicon substrate instead of using a mixture of the two powders, SiNWs are obtained again. In all cases, the absence of carbon does not produce any nanowires. Gundiah et al. [22] explain the mechanism as carbothermal reduction of the oxide layer converting the silicon.

$$Si_xO_2 + C \rightarrow Si_xO + CO \; (x > 1)$$

$$Si_xO \rightarrow Si_{x-1} + SiO$$

$$2\,SiO \rightarrow Si + SiO_2$$

In the last step above, crystalline silicon nucleates and grows normal to the (111) direction. The major drawback of the oxide-assisted growth is that it is a high-temperature (1200°C) process.

6.2.5 Template-Assisted Synthesis

Templates have been advocated for the growth of CNTs and various nano-wires as a means to produce vertical structures with controlled diameter as discussed in Chapter 3. The most popular template is anodized alumina membrane, which is typically prepared by the anodic oxidation of an electropolished aluminum plate or foil. The cell voltage, acid concentration, temperature, and oxidization time can be varied to control the opening size of the pores. Lu et al. [47] used an anodized alumina template without any catalyst to grow SiNWs using 50:50 silane in H_2 at 900°C. Their results indicate nanowires of 50 nm in diameter, which is consistent with the pore diameter of the template. A silicon surface film is present at the bottom of the nanowires. TEM characterization reveals that the nanowires exhibit a smooth surface and are virtually defect-free with no kinks or dislocations. Though there is no explicit use of a catalyst, the anodized surface has a large number of surface sites of Lewis acid nature, which possess an intrinsic catalyst activity [47] leading to a VLS-like growth.

Others have explicitly added a gold thin film to the anodized surface to promote VLS growth [27,35]. Depending on the growth time, the grown nanowires can be entirely within the pores or they can protrude outside the template. A careful template removal process is used to release the nanowires for further use. There is no information currently on the surface quality of the wires after the template removal, how they compare against as-grown nanowires without the template and their suitability for device fabrication. If device quality nanowires are desired, then an extremely gentle process to remove the template is critical. The template preparation by Bogart et al. [27] showed a pore size variation of 136 ± 27 nm with a nanowire diameter variation of 138 ± 13 nm if the nanowires are entirely within the pores. The diameter variation for nanowires protruding above was 158 ± 9 nm. The growth conditions were 5% silane in H_2 at a growth temperature of 500°C. These results do not show any better diameter control than the gold colloid approach used by Cui et al. [16]; as discussed before, the disparity between the nominal colloid size and the nanowire diameter in Ref. [16] was very small at 1–2 nm. Hence, it is not clear if the extra efforts expended in preparation and removal of the template pay off, and much more work is needed to draw conclusions. In any case, at least one unique advantage of using a porous template appears to be the ability to grow along [100] direction on Si(100) substrates [35] as seen in Figure 6.8 showing vertical SiNWs. Without the template, SiNWs are grown most commonly in <111>, <112>, and <110> directions. A detailed discussion related to nanowire orientation, specifically in the context of SiNW growth, was provided in Chapter 4.

6.2.6 Plasma Enhancement

In most growth systems reported for nanowire preparation, the source generation and actual growth proceed at two different temperatures by separating these activities into two different zones. The growth temperature is then determined by the choice of the catalyst (as in VLS process) as will be discussed later. The source generation temperature is dependent on the choice of precursor, for example, $SiCl_4$ versus silane. Whereas the homemade growth systems feature two zone furnaces in a hot wall system, if one were to use a typical cold wall commercial CVD system common in IC industry, then the growth temperature will be dictated by the precursor dissociation even with a low-melting catalyst (such as In or Ga). This creates a need for a

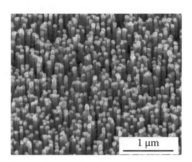

FIGURE 6.8
Vertical SiNWs grown along [100] direction in a Si (100) substrate using an anodized alumina template. (Reproduced from Shimizu, T. et al., *Adv. Mater.*, 19, 917, 2007. With permission.)

process capable of growing SiNWs at temperatures below 400°C in order to protect other layers present in the processed wafer. This need prompts the exploration of low-temperature plasmas in SiNW growth [19,20,38]. Hoffman et al. [20] used a parallel plate 13.56 MHz radiofrequency (rf) plasma reactor with a 100% silane feedstock and gold catalyst. This produced very long, thin, curvy SiNWs consistent with the observations in Table 6.2. Since the early findings by Westwater et al. [7], it is well known that higher partial pressures of silane yield thin, curvy wires. Here the pure silane well dissociated by the plasma also produces enough active radicals depositing silicon radially on the axially elongating nanowire out of the seed. As a result, the resulting structures tend to be tapered.

Aella et al. [38] also used a rf plasma reactor to grow SiNWs on Si (100) surfaces but with a 10% silane in hydrogen and gold thin film as catalyst. Here in the absence of plasma stimulation, a majority of the nanowires grew along [111] orientation whereas plasma enhanced growth primarily yielded wires along [110] direction. They also found nanowires to be tapered with a wider diameter at the base. The growth rate with plasma stimulation was found to be larger by an order of magnitude. The plasma is efficient in breaking the feedstock silane into SiH_3, SiH_2, and SiH radicals due to electron impact dissociation. Therefore precursors available at the growth front are substantially different with and without the plasma. Though detailed modeling and diagnostics of these systems for nanowire growth are not available, thermal and plasma dissociation of common precursors in the growth of Si, SiO_2, Si_3N_4, etc. are well known. The differences seen here under plasma enhancement are somewhat analogous to the observations in CNT growth using thermal versus plasma CVD. For example, in thermal CVD of CNTs, the growth temperature is lower than the pyrolysis temperature of the hydrocarbon feedstock. Therefore gas phase diagnostics using mass spectrometry and modeling show nothing beyond the input hydrocarbon gas above the wafer [48,49]. All feedstock dissociation seems to be restricted to the catalyst surface. In contrast, modeling and diagnostics showed under PECVD conditions that a pure methane feedstock is converted to CH_3, CH_2, CH, C_2H_4, C_2H_6, and a variety of ions [49,50]. Such a disparity in the available species at the growth front affects the growth rate, morphology, diameter, orientation, defects, growth temperature, radial epitaxy, leading to tapered structures, etc. A microwave plasma of 2% silane in hydrogen with molten Ga droplets as catalyst produced SiNWs but these do not appear to be tapered [19]. Well-diluted feedstock, low-plasma power, and perhaps even a remote plasma (wherein the plasma generation and growth are done in two separate zones) may be the prerequisites for controlled growth of nanowires using plasma enhancement.

6.2.7 Doping of SiNWs

Doping of the nanowires is desirable to tune the electronic properties suitable for fabricating diodes and transistors. Typically boron can be used as p-type dopant and phosphorous or Bi as n-type dopant. Doping can be achieved in situ during the nanowire growth or as a postgrowth step. Boron doping is normally done through the addition of B_2H_6 to the source gas flow (such as silane) [15,41]. Interestingly, the addition of small amounts of B_2H_6 to silane (<1%), while achieving the intended doping levels, also affects the morphology of the SiNWs depending on the growth catalyst [41]. When gold is used as a catalyst, the eutectic temperature of Au–B (1064°C) is higher than both the Au–Si eutectic temperature and typical growth temperatures (400°C–500°C). This means the unlikelihood of the formation of the Au–B liquid alloy. Since the dissociation energy of B_2H_6 is lower than that of silane, all the boron unused in the doping process appears as a layer around the nanowire and the catalyst tip. This typically interferes with the silicon diffusion into the catalyst, thus reducing growth rate. In addition, as known from silicon thin film growth, B_2H_6 increases the epitaxial rate. These phenomena yield conical SiNWs during in situ doping process as evidenced by Whang et al. [41]. In contrast, the eutectic temperature of Al–Si alloy is 577°C and when Al is used as a catalyst, Al–B liquid alloy forms at 659°C promoting the incorporation of boron inside the growing nanowire. The morphology of the Al-catalyzed SiNWs appears to be unaffected by the in situ doping process.

The undesirable effects on morphology during in situ doping such as the case described above can be eliminated in postgrowth doping techniques. Whang et al. [41] also studied postgrowth doping of SiNWs by exposing them to a B_2H_6 plasma at 440°C. X-ray photoelectron spectroscopy (XPS) and secondary ion mass spectroscopy (SIMS) are normally used to establish the doping concentration and depth of penetration. Byon et al. used a vapor phase postgrowth doping technique to dope SiNWs with bismuth vapor, which has a relatively high vapor pressure [29]. For in situ n-type doping, PH_3 can be added to the reactant flow during the nanowire growth.

6.2.8 Properties of SiNWs

Quantum size effect in nanowires will have an impact on electronics and photonics applications. Quantum confinement occurs when the nanomaterial dimensions approach the size of an exciton in bulk crystal, called the exciton Bohr radius. This leads to an increase in band gap with a decrease in size of the nanomaterial. Ma et al. [51] performed careful experiments to map the band gap variation for SiNWs smaller than 10 nm in diameter. The nanowires were prepared by an oxide-assisted technique and the surface oxide on the wires was removed by a HF acid treatment. Scanning tunneling spectroscopy measurements done on these nanowires allowed determination of the band gap for various wires as plotted in Figure 6.9. The band gap of a 7 nm wire is unchanged from that of bulk silicon at 1.1 eV. The increase in band gap for smaller diameter is at first gradual, with a steep increase to 3.5 eV finally for a 1.3 nm diameter wire.

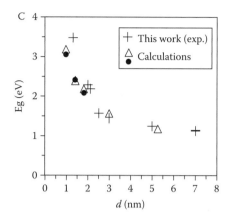

FIGURE 6.9
Band gap of SiNW as a function of nanowire diameter as determined by scanning tunneling spectroscopic measurements. (Reproduced from Ma, D.D.D. et al., *Science*, 299, 1874, 2003. With permission.)

Thermal conductivity of SiNWs has been extensively studied [52–56] and both theoretical predictions and experimental measurement show this property to be size-dependent at the nanoscale. Figure 6.10 shows measured thermal conductivity values for SiNWs of 22, 37, 56, and 115 nm diameter [52]. The nanowires were produced by a VLS process and found to have a well-defined crystalline order. The nanowire thermal conductivities are about two orders of magnitude smaller than that of bulk silicon. A reduction in thermal conductivity accompanies the decrease in nanowire diameter. These observations were ascribed to an increase in phonon boundary scattering at

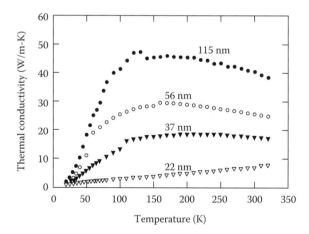

FIGURE 6.10
Thermal conductivity measurements as a function of temperature for SiNWs of various diameters. (Reproduced from Li, D. et al., *Appl. Phys. Lett.*, 83, 2934, 2003. With permission.)

the nanoscale [52]. Direct molecular dynamics simulations [56] confirm the decrease in thermal conductivity with size as a result of increased surface scattering effects. However, when the diameter reaches very small values (below 1.5 nm), phonon confinement effects increase the thermal conductivity.

Photoluminescence (PL) studies of SiNWs grown by a physical evaporation of silicon and Fe powders showed a stable blue luminescence visible to the naked eye with two PL peaks at 2.4 and 2.95 eV [8]. The diameter of the nanowires in this study is 15 ± 3 nm and thus is too big to exhibit quantum confinement effect. The origin of the luminescence may be an overlayer of silicon oxide on the nanowire. This conclusion was also verified by Qi et al. [23] who observed PL peaks at 455 and 525 nm, which quenched rapidly with an increase in temperature. Guichard et al. [26] presented evidence for light emission due to quantum confined excitons in SiNWs. They grew their nanowires using $TiSi_2$ catalyst to avoid the difficulties with gold, which, as a deep-level trap, causes a fast nonradiative decay of excited carriers. The intentionally thermal-oxidized nanowires had a core diameter comparable to the silicon Bohr radius. The PL emission data showed a blue shift and a reduction in PL lifetime with decreasing diameter.

Room temperature electroluminescence is also possible with SiNWs [24]. The emission spectrum (not shown here) shows a peak around 600 nm (~2 eV), which appears to originate from nanowires of about 4 nm in diameter. The EL demonstration was done on SiNWs intercalated with a HfO_2 dielectric, which may have traps or interface states at the interface. The electrons from the traps tunnel into the direct band gap of the nanowires under the applied ac field. The electron–hole pairs produced, upon recombination, are responsible for the light emission [24].

6.3 Germanium Nanowires

Germanium is an important semiconductor with a direct band gap of 0.8 eV and an indirect band gap of 0.66 eV. It has been receiving much attention recently either as a pure element or as an alloy with silicon (Si_xGe_{1-x}) for logic and other devices. Germanium offers higher intrinsic carrier mobilities than silicon:

$$\left(\mu_n = 3900, \mu_p = 1900 \, cm^2/Vs \right) \text{ versus } \left(\mu_n = 1500, \mu_p = 450 \, cm^2/Vs \right)$$

at room temperature, enabling faster switching and higher frequency devices. Other advantages include higher intrinsic carrier concentration, $2.4 \times 10^{13} \, cm^{-3}$ versus $1.45 \times 10^{10} \, cm^{-3}$; larger bulk excitonic Bohr radius, 24.3 nm versus 4.7 nm for more prominent quantum confinement effects even at larger material dimensions; and compatibility with high-k dielectrics enabling integration to current semiconductor processing technology. Germanium is also compatible with III–V materials, for example, good lattice matching with GaAs. Germanium oxide exhibits interesting optical properties suitable for interesting integrated optoelectronic circuits. All these features have prompted extensive investigations of germanium nanowire (GeNW) growth, characterization, and application development [57–78].

6.3.1 Synthesis Using Germanium Powder

First, a simple VLS process using Ge powder as source material is described below [60]. The feedstock consists of a 1:1 weight ratio of germanium powder (Alfa Aesar, 99.999% metal basis) and graphite powder (Alfa Aesar, 99.9995% metal basis). A quartz boat containing this mixture is placed upstream in a quartz tube VLS reactor described in Chapter 3 and throughout this book. The addition of graphite enhances the surface area for the evaporation of germanium and allows control of the germanium partial pressure at a given carrier gas flow and source temperature. Graphite also reduces the native germanium oxide present on the germanium powder by carbothermal reduction. This mixture approach does not seem to produce germanium carbide nanowires but only crystalline GeNWs for the following reasons [79]. The heats of formation of gaseous carbon and germanium species are substantially different; the Ge–C bond is relatively less stable compared with the C–C bond, and carbon is relatively insoluble in crystalline germanium at a wide range of temperatures and pressures. An intrinsic germanium substrate with a thin film of gold deposited by ion-beam sputtering is placed downstream in the reactor. The temperature upstream in the two-zone furnace system is 1020°C–1030°C with the substrate downstream maintained at 470°C–480°C. A carrier gas flow of 100–140 sccm argon (Scott Specialty Gases, 99.999% pure) and 50–80 sccm hydrogen (Scott Specialty Gases, 99.999% pure) is maintained. The hydrogen flow prevents oxidation of gas-phase and solid-phase germanium by the residual oxygen in the system.

Figure 6.11 shows an SEM image of GeNWs of uniform diameter (42 ± 10 nm) and length ($1 \pm 0.2\,\mu$m). TEM analysis indicates that the nanowires grow in the [111] direction and exhibit a hemispherical gold catalyst bead on top of a germanium receptacle, followed by the GeNW stem. The nanowire body is characterized with a uniform diameter and smooth surface as seen in Figure 6.11b. High-resolution TEM images reveal well-defined lattice fringes of the (111) and (11$\bar{1}$) planes, along with a smooth surface, as shown in Figure 6.11c. The lattice fringe is measured to be 3.26 Å, matching well with x-ray powder diffraction data of bulk germanium (04–0545). An oxide layer of ~1–2 nm thickness is usually observed on the nanowires. Selected area electron diffraction pattern is resolved to show spots in the (200), (220), and (222) plane families, with interplanar distances of 2.83, 2.00, and 1.63 Å, respectively. These nanowires are crystalline germanium of the cubic diamond structure, with the preferred growth direction of [111].

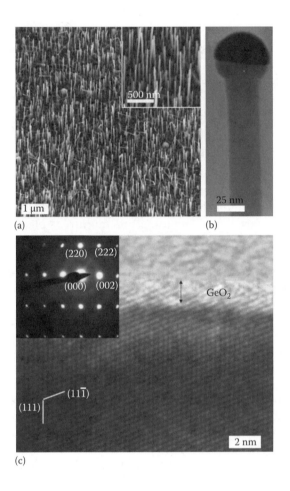

FIGURE 6.11
GeNWs discussed in Section 6.3.1. (a) SEM image of an array of nanowires grown on a Ge(111) substrate. The inset shows a higher magnification. (b) TEM image of a single nanowire showing the gold catalyst bead at the top. (c) HRTEM image showing the (111) lattice fringes and a native oxide layer. The inset shows the electron diffraction pattern indicating high crystallinity. (Courtesy of P. Nguyen.)

Compositional analysis using TEM energy dispersive spectrometry (EDS) and XPS in Figure 6.12 shows germanium L- and K-lines along with a small oxygen $K\alpha_1$ peak and a small copper $K\alpha_1$ peak. The copper signal originates from the TEM grids. Area calculation indicates a ~40 at% of oxygen, arising from the thin germanium oxide sheath. The XPS analysis also confirms this with a molar composition of 83% Ge and 17% germanium oxide. The germanium peak is fitted to a single component at 29.38 eV (FWHM = 1.77 eV), which corresponds to germanium in the germanium matrix. The second peak in Figure 6.12b is also fitted to a single component at 32.58 eV (FWHM = 1.44), which corresponds to Ge^{4+} in germanium oxide. The inset in this figure is

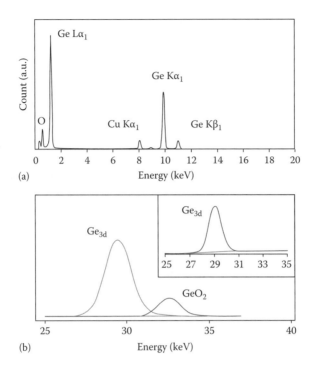

FIGURE 6.12
Spectral analysis of the GeNWs in Figure 6.11. (a) TEM-EDS spectrum and (b) XPS spectrum showing peaks of Ge and germanium oxide. (From Nguyen, P. et al., *Adv. Mater.*, 17, 549, 2005.)

the spectrum after the sample is cleaned up by sputtering using an argon beam at 5 mA and 5 keV. The germanium oxide peak is completely eliminated, leaving only the germanium peak at 29.4 eV (FWHM 1.18) and no other contaminants, indicating high-density growth of GeNWs with high purity and high crystallinity.

Sun et al. [59] reported a similar process as above using sublimation of Ge powder at 950°C without the addition of graphite powder. They used 2 nm gold particles produced in situ as catalyst and carried out growth at 600°C on silicon, alumina, and ceramic substrates. The GeNWs were on the average 30 nm in diameter and tens of microns long. When thicker gold films (30 nm) are used, the resulting particle size is in the range of 80–100 nm [62] and with such large particles, higher temperatures ($T \sim 700$°C) are required to attain the supersaturation necessary for nanowire growth. At a temperature of 400°C, which is just above the eutectic temperature, only Ge nanodots are produced with large gold catalyst particles and at 550°C, the structures are mostly nanorods [62]. Instead of sublimation or carbothermal reduction of Ge powder, the latter mixed with gold particles can be pressed into a disk and ablated using an appropriate laser to produce the vapor source [30,57,58].

6.3.2 Germane and Related Sources

Similar to the use of silane in SiNW growth, germane and related precursors have been used in GeNW preparation [64–70] as these precursors are well known in microelectronics industry. Only a limited temperature range (320°C–380°C) yields GeNWs using germane [64] on silicon substrates. At higher temperatures, the germane decomposition at the gold catalyst particle surface is faster than the diffusion of Ge atoms to the growth interface and as a result, three-dimensional gold-containing Ge structures are formed. Even at the optimum temperature of 320°C, the nanowires grown along the <111> direction appear to be slightly tapered [64]. Adikhari et al. [65] also found tapering as an issue at 350°C using GeH_4/H_2 (5 and 550 sccm respectively) on germanium substrates. The use of Ge (111) substrate or epitaxially grown Ge layer on Si (111) substrate does not make any difference to the growth characteristics in terms of tapering or nanowire length. The nanowires grow vertically along the <111> direction. When the temperature is lowered, the nanowires are not tapered but no longer maintain the epitaxial relation with the substrate. Therefore, Adikhari et al. [65] introduced a two-temperature process wherein a quick 2 min heating to facilitate nucleation followed by growth at 280°C yields vertical nanowires of uniform diameter along the axis while epitaxially aligned to the <111> direction of the substrate.

Sidewall or radial epitaxy leading to tapering is also dependent on pressure besides temperature (>350°C). Jin et al. [70] showed that only Ge nanocones are possible at 10 Torr even at a temperature of 290°C. Raising the pressure to 50 Torr at the same temperature appears to provide perfectly cylindrical nanowires of ~20 nm diameter. Another cause of tapering involves in situ doping attempts using B_2H_6 as a dopant source, as was discussed earlier in Section 6.2.7 under doping of SiNWs. Under common growth conditions for GeNWs such as 10% germane in helium, 285°C and 5 Torr, B_2H_6 dramatically enhances lateral growth rate [67] in GeNWs aligned with the [111] direction of the silicon substrate. This work does not have a parametric study including higher pressures to eliminate pressure as a variable causing the tapering as in the case of Jin et al. [70]. However, at 5 Torr pressure, the tapering appears to decrease with the boron concentration. Even undoped nanowires have a slight tapering at 5 Torr, which may be the effect of low pressure. Whereas diborane serves as a p-type dopant, addition of PH_3 to germane allows to produce n-type GeNWs [68]. However, identical growth conditions with PH_3 addition does not cause the tapering problems found with diborane.

It is possible to grow SiGe nanowires by mixing the appropriate source gases as demonstrated by Qi et al. [69]. They mixed disilane (10% in Ar) and germane (10% in He) and used 1–2 nm gold thin film or 20 nm gold colloids as catalyst. Obtaining smooth nanowire surface is found to be restricted to Si:Ge = 1:1 while departure from this ratio only yields rough surfaces.

6.4 Catalyst Choice

Interestingly, the VLS growth is not actually a catalytic process and the seed metal is not really a catalyst. The metal droplet just receives the source material and when supersaturation is reached, the excess material precipitates out of the droplet in the form of a nanowire. Thus, the seed metal is only a "soft template" [80] to collect the material and facilitate nanowire precipitation, guidance, and elongation in the axial direction. Johannsson et al. [81] provide a more rigorous evidence to prove that the metal is not a catalyst. A catalyst, as well known in chemistry literature and chemical industry, is a material that increases the rate of a chemical reaction while remaining intact in the process. The activation energy for SiNW growth using gold seeds and silicon thin film growth in microelectronics industry (with no seeds, of course) is about the same at ~130 kJ/mol [81], indicating that gold does not aid in increasing the reaction rate. Similarly, MOCVD of GaAs nanowires with the aid of gold is in the range of 67–75 kJ/mol, which compares with the 67 kJ/mol value for GaAs thin film epitaxy [81]. In spite of all this evidence, we have still adopted the commonly used term "catalyst" throughout this book, with the reservation noted above.

It is clear by now that gold is the most favored seed metal in the literature for nanowire growth. However, gold is viewed as a contaminant in semiconductor processing and offers no useful properties to the nanowire after growth. It can modulate carrier recombination in both n-type and p-type materials because high-mobility interstitial gold atoms can transform into electrically active low-mobility substitutional sites. Gold is not desirable also in optoelectronics applications since, as a deep level trap in silicon, it causes fast nonradiative decay of excited carriers [26]. For this reason, SiNWs with gold catalyst are not capable of light emission whereas the nanowires produced with other seed metals or oxide-assisted growth are known to produce light emission in the visible and near IR ranges. Another disadvantage is the high eutectic composition of Au–Si system at 371°C (31 at% of silicon), which would make it impossible to obtain abrupt heterojunctions in a nanowire by rapidly switching the gas phase precursors. For these reasons, several alternatives to gold have been investigated in SiNW and GeNW growth [18,19,26,36,39,41, 63,76,77,80]. Gallium is a metal with low melting point which forms a eutectic at a low temperature (29.8°C) with a low silicon content (only 5×10^{-8} at%). Germanium forms a eutectic with Ga at 30°C at a Ge atomic composition of 5×10^{-5}%. This allowed Sunkara et al. [18,19,76] to grow dense SiNWs and GeNWs using gallium droplets with a variety of elemental sources such as bulk silicon, silane plasma, Ge powder, etc.

$TiSi_2$ has been successfully used as a catalyst for SiNW growth [26]. Titanium sputtered onto a silicon wafer forms small islands of $TiSi_2$ when annealed at 900°C in hydrogen for 5 min. Similarly, a thin layer of Pt on silicon wafer obtained by physical vapor deposition forms PtSi when annealed at high temperatures. PtSi, commonly used for device contacts, has also been used in SiNW growth [36]. Copper, which is commonly used as an interconnect, can also serve as a seed metal for SiNW growth [39] though primarily through the vapor-solid-solid (VSS) process at temperatures below 800°C. The Cu-Si phase diagram indicates a higher temperature range of 802°C–1084°C for VLS growth of silicon using copper. Compared to this, Al (which is also an interconnect material) has a eutectic temperature of 577°C for Al-Si alloys and therefore can be used as a viable seed metal [41]. However, since Al oxidizes easily even with trace amounts of oxygen, growth under vacuum conditions will be necessary.

Iron, nickel, and titanium also have been used to seed nanowire growth. These are however the so-called back-end materials in semiconductor processing, which may potentially lead to device contamination. Therefore, Sun et al. [63] used indium and antimony in GeNW growth both of which are front-end materials commonly used for channel or shallow junction doping in CMOS device fabrication. Indium, just like Ga, has a low melting point (156°C), which is also the eutectic temperature with Ge. Sb has a melting point of 631°C and the eutectic temperature of Sb-Ge alloy system is 587°C. The solid solubility of Ge in both In and Sb is low, which reduces the time needed for supersaturation and facilitating axial growth. The low melting point of In has one disadvantage in that very thin films (1–2 nm) are not useful for nanowire growth as the metal is lost quickly through evaporation at the growth temperatures. At least a 10 nm thick In layer is needed to compensate for the metal evaporation loss. Nguyen et al. [80] studied the effectiveness of 16 different metals (Ti, Cr, Fe, Co, Ni, Cu, Ni, Mo, Pd, Ag, Ta, W, Ir, Pt, Au, and Al) in the growth of tin oxide and SiNWs. While all these metals except Ir and Pt have been successful in seeding nanowire growth, the only apparent correlation they could find was between the bulk melting point and growth density. These two have an inverse relation with the nanowire density increasing with a decrease in melting point. As a soft template guiding the growth, better seed formation from the thin film precursor and the fluidity of the molten seed appear to be two important parameters in facilitating nanowire growth.

6.5 III–V Nanowires

III–V compound semiconductor materials are of great interest in electronics and optoelectronics for applications in high-power or high-frequency transistors, light emitting diodes, lasers, resonant tunneling diodes, solar cells, photodetectors, and others. Their usefulness in the above applications has been well established in the last two decades through thin films and quantum wells of GaAs, $Al_xGa_{1-x}As$, $In_xGa_{1-x}As$, InP, GaP etc grown using techniques such as metal organic chemical vapor deposition (MOCVD) and molecular beam epitaxy (MBE). Recently, these materials have also been grown as one-dimensional nanowires since quantum confined low-dimensional structures have the potential to provide superior performance in all of the above applications. Most of the reported growth studies use the VLS approach for growing III–V nanowires along with template-guided synthesis drawing some attention. Under VLS growth, further classification is done based on the source generation. The MOCVD approach uses well known vapor sources such as AsH_3, PH_3, TMGa, TMIn, etc. in conventional commercial reactors for growth on wafers coated with a thin film of catalyst (invariably gold). In the MBE approach, both solid sources or gas sources have been used, again with gold catalyst film. Both MOVCD and MBE growth systems and procedures are along the lines of III–V thin film growth perfected over a couple of decades, with the only exception of introducing catalyst particles for guiding the nanowire growth along the axis. A third approach involves a pulsed laser deposition system that uses a target of the III–V material mixed with gold catalyst [82]. All three approaches appear to be capable of synthesizing III–V nanowires of 3–100 nm in diameter and various lengths, compositional control, and heterostructure formation through fast switching of sources.

6.5.1 GaAs Nanowires

Gallium arsenide is a direct band gap semiconductor with a band gap of 1.424 eV, electron saturation velocity of 10^7 cm/s, and carrier mobility of 8500 and 450 cm^2/V s for electrons and holes, respectively. It has been grown in the nanowire form by laser-assisted catalytic growth [82,83], MBE [84–87], MOCVD [81,83,88–94] and template-guided approach [95]. In a typical MOCVD run in Ref. [89], trimethylgallium (TMGa) and arsine (ASH_3, 10% in hydrogen) are used as source gases with the ASH_3 to TMGa molar flow ratio maintained between 1.1 to 125. Growth temperatures between 380°C and 520°C are explored to understand the impact on morphology. The pressure inside the MOCVD reactor is 2×10^4 Pa. Undoped or Si-doped GaAs(111) substrates are used with gold deposited as a thin layer (0.1, 1.0, and 10 nm) by vacuum evaporation. Figure 6.13 shows growth results [89] from the above wherein the images on the left (a, b and c) present particle size distribution after the gold film is heated to 500°C for 10 min to allow the film to break into droplets. The SEM images on the right present the corresponding GaAs nanowire morphology for growth conditions of 420°C and AsH_3/TMGa flow ratio of 16. For a gold film thickness of 0.1 nm, the particle size ranges from 8 to 30 nm and the nanowires appear to be like needles with the diameter at the midpoint approximately 20–30 nm. The nanowires are not vertical, instead appear to be brush-like. When the Au film thickness is increased to 1 nm, the particle size distribution is 10–40 nm and the nanowires are cylindrical with a diameter of 70–80 nm. Finally, for a 10 nm thick gold film, the resulting particles are fairly large at 80–500 nm, yielding correspondingly thick

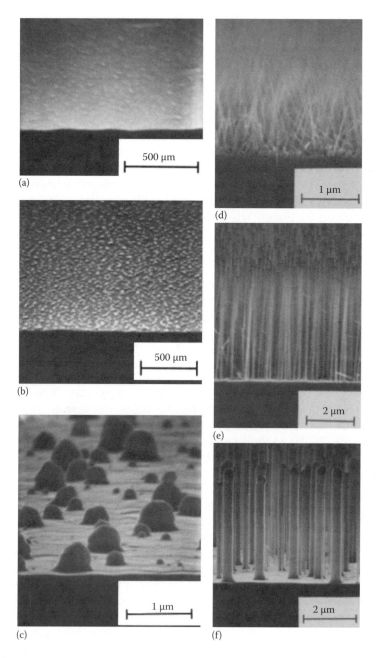

FIGURE 6.13

GaAs nanowires by MOCVD. Growth conditions: 420°C, AsH_3/TMGa flow ratio = 16. The effect of gold film thickness on growth morphology is shown here. The images on the left show particle size distribution for gold films of thickness: (a) 0.1 nm, (b) 1 nm, and (c) 10 nm. The images on the right show the corresponding nanowire growth results. (Reprinted from Hiruma, K. et al., *Nanotechnology*, 17, S369, 2006. With permission.)

nanowires of 70–400 nm in diameter. In this case as well as for 1 nm film, the nanowires are vertical on the substrate. In Figure 6.13f, the presence of gold particles at the tip is evident, which confirms the VLS mechanism. EDX analysis confirms the tip to be composed of gold and TEM/EDX together confirm the semiconductor to be stoichiometric GaAs along the entire length.

When the growth temperature is increased from 420°C to 500°C for the case of 1 nm gold film (Figure 6.13b and e), the nanowire shape changes from cylindrical to conical (not shown here). The reason for this is the radial or sidewall epitaxy adding material along the axially growing wire. Similar observations have been made by others as well. Paiano et al. [90] used Au colloids of ~60 nm and AsH_3/TMGa molar ratio of 7 and found that the nanowires grown above 425°C have substantial tapering while the wires grown at 400°C are cylindrical. Interestingly, in laser-assisted growth utilizing solid targets mixed with gold, tapering does not seem to be an issue since radial epitaxy from feedstock does not occur at elevated temperatures as in MOCVD [82]. This approach has also produced the smallest GaAs nanowires with 3–5 nm diameter. With such small wires, room temperature PL measurements show a strong blueshift relative to bulk GaAs, indicating radial confinement of excitons in the nanowires [82].

MBE also produces tapered nanowires under unfavorable conditions. Ihn et al. [87] examined growth temperatures between 480°C and 640°C and V/III ratios between 3 and 20. At low temperatures (480°C–500°C) and low flux ratios (less than 9), the GaAs nanowires grown on Si(111) and Si(001) substrates are tapered and randomly oriented on the substrate. The tapering at low temperature here is attributed to the reduced surface diffusion of the growth species on the sidewall. At higher temperatures and flux ratios, the GaAs nanowires are cylindrical. Hammand et al. [84], on the other hand, reported tapering as a problem occurring when the nanowire is too long. In excessively long nanowires, the last 4–5 μm appear to be tapered with the diameter decreasing at a rate of about 25 nm per micron length. The analysis of their results indicates that Ga adatoms migrate along the sidewalls of the nanowire with a mean length of 3 μm at 590°C. When the length exceeds 3 μm, some of the Ga atoms migrating on the sidewall are lost due to desorption before reaching the gold particle at the top. This reduces the growth rate with a concomitant gradual reduction of the particle size, leading to a reduced diameter.

Several of the applications using compound semiconductors involve heterojunctions and both MOCVD and MBE are amenable to achieve abrupt or intentionally graded heterostructures by controlling the switching of the group V or III sources as warranted. Borgstrom et al. [88] studied the interface sharpness of GaP-GaAs nanowires as a function of temperature and MOCVD source flux ratios. Fixed conditions in their study include 50 mbar total pressure, TMGa mole fraction of 7.3×10^{-5} and total gas flow rate of 6.0 λ/min including the hydrogen carrier gas flow. Both PH_3 and AsH_3 mole fractions are varied between 7.5×10^{-4} to 500×10^{-4} in a growth temperature range of 400°C–520°C. In general, the interface is sharper at low growth temperatures and low group V mole fractions.

6.5.2 InAs Nanowires

InAs exhibits a small energy band gap (0.35 eV), high electron mobility (33000 cm²/V s), and a large quantum confinement energy. There have been several reports on the preparation of InAs nanowires [96–103]. Mandl et al. [97] presented InAs growth by MOCVD with a thin layer of SiO serving as catalyst on InP(111), InAs(001), and Si(001) substrates. Figure 6.14 shows SEM images of InAs nanowires on InP substrates at temperatures 540°C–660°C. The total pressure is 10 kPa with TMIn and AsH₃ mole fractions of 2×10^{-6} and 2×10^{-4} respectively under a 6 L/min H₂ flow. In all cases, the wires are normal to the substrate and exhibit constant diameter along the entire height without any tapering. An increase in temperature results in an increase in wire diameter and decrease in density. This indicates formation of larger clusters perhaps due to coalescence at high temperatures, resulting in larger diameter wires grown on fewer clusters. When using gold colloids as catalyst, Chuang et al. [100] found the existence of a critical diameter for particle size to enable nanowire growth which is inversely proportional to the lattice mismatch. Analyzing nanowire/substrate combination cases of In As/Si, InP/Si, InP/GaAs, GaP/Si, and InP/InP grown by MOCVD, they deduced that the critical diameter for InAs/Si (11.6% mismatch) is 26 nm.

6.5.3 InP Nanowires

Indium phosphide is a direct band gap semiconductor with a gap of 1.34 eV that has been widely investigated and used for electronics and optoelectronics applications. It has been grown in the form of nanowires using laser-assisted

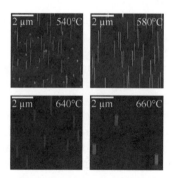

FIGURE 6.14

InAs nanowires grown on InP(111)B by MOCVD at various temperatures. Growth conditions: pressure = 10 kPa, TMIn mole fraction = 2×10^{-6}, AsH₃ mole fraction = 2×10^{-4}, H₂ carrier gas glow = 6 L/min. (Reprinted from Mandl, B. et al., *Nano Lett.*, 6, 1817, 2006. With permission.)

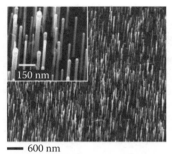

FIGURE 6.15

InP nanowires grown by MOCVD using 20nm gold colloids. TBP/TMIn ratio = 120 and 430°C growth temperature. (Reprinted from Bhunia, S. et al., *Appl. Phys. Lett.*, 83, 3371, 2003. With permission.)

growth [104,105], sublimation of InP powder [106], MBE [107–110], and MOCVD [111–115]. A typical MOCVD process for growing InP nanowires is described below. The process uses TMIn and phosphine (or tertiary butyl phosphine, TBP) as sources with hydrogen as carrier gas [111]. Gold colloidal particles of average size 20nm are applied from solution on InP wafers. First, the substrates are annealed at 500°C under a constant TBP flow. Then a V/III flow ratio of 120 is maintained at 430°C during a growth period of 1 min. Figure 6.15 shows results from this run, which provides vertical, uniform diameter nanowires. Room temperature PL measurements (not shown here) indicate a blueshift of 32 meV due to the quantum confinement of carriers [111], which corresponds to a nanowire diameter of 18 nm. Cathode luminescence spectra of these nanowires [112] show a broader than expected peak of 1.6 eV for a 18–20 nm wire with quantum confinement, which may be the result of carrier scattering by lattice defects.

Variations in growth conditions appear to give a widely varying morphologies of InP nanowires. Whereas tapering of III–V nanowires have been discussed in the cases of GaAs, InAs, and GaP with typically a thicker base, Novotny and Yu [113] showed evidence that InP nanowires can exhibit "reverse tapering," i.e., thinner base. The major difference in their work involves the use of indium as "self-catalyst." InP wafers are annealed first when surface reconstruction results in an In-rich surface. The TMIn source vapor striking this surface creates an excess of In adatoms, which causes a liquid phase separation providing the droplets for seeding. When the V/III ratio is as low as 2, the flux of P atoms to the seed is not enough and the In atoms from both the source vapor and the adatoms migrating on the surface upward are in excess to cause the swelling of the seed. This causes the head of the nanowire to be larger than the base and up to a certain height, this inverse structure is stable enough not to fall off. When the V/III ratio is raised to 25 and above, the tapering decreases as shown in Figure 6.16. When the growth temperature is raised to 450°C for the molar ratios in Figure 6.16b

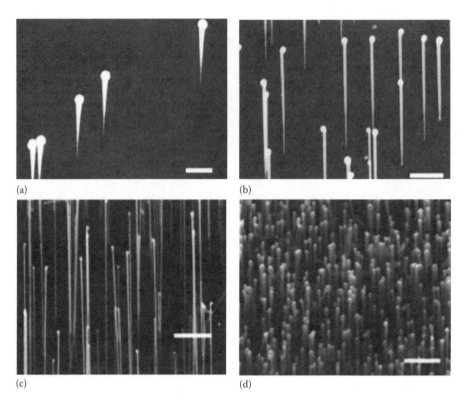

FIGURE 6.16
InP nanowires grown by MOCVD with In serving as self-catalyst at various V/III ratios: (a) 2, (b) 25, and (c) 125. Growth temperature = 350°C (d) VI/III ratio of 25 at a higher temperature of 450°C. (Reprinted from Novotny, C. and Yu, P.K.L., *Appl. Phys. Lett.*, 87, 203111, 2005. With permission.)

and c, a growth density of 10^9 wires/cm^2 is obtained with a diameter distribution of 25–50 nm.

6.5.4 GaP Nanowires

Gallium phosphide is a wide band gap semiconductor with an indirect gap of 2.26 eV. The electron mobility of bulk GaP is 160 cm^2/V.s. GaP is preferred for high-temperature electronics applications and light emission in the visible range. In laser catalytic growth of GaP [116], a GaP solid target is laser ablated to provide the source vapor and the use of gold colloids appears to provide a tight diameter distribution of grown nanowires. In general, the mean nanowire diameter is only 1–2 nm larger than that of the colloids. Sublimation of GaP powder has also been used to generate the vapor source [117–120]. Typically the temperature is above 1000°C and gold, nickel, or other particles are used as catalysts. Flowing ammonia during growth enables N-doping of the GaP nanowires [117]. The nitrogen doping makes GaP appear with a bright orange color instead of the typical greenish yellow color for the undoped wires and allows production of yellow LEDs. The simple sublimation process yields stoichiometric GaP nanowires and typically, they are randomly oriented on the substrate. MOCVD using TMGa and PH_3 sources yields more vertically oriented nanowires on Si(111) substrates [121].

References

1. J. Hutchby, G.I. Bourianoff, V.V. Zhirnov, and J.E. Brewer, *IEEE Cir. Dev.* 28 (2002).
2. D.P. Yu, Z.G. Bai, J.J. Wang, Y.H. Zou, W. Qian, J.S. Fu, H.Z. Zhang, Y. Ding, G.C. Xiong, L.P. You, J. Xu, and S.Q. Feng, *Phys. Rev. B* 59, R2498 (1999).
3. Z. Li, Y. Chen, X. Li, T.I. Kamins, K. Nauka, and R.S. Williams, *Nano Lett.* 4, 245 (2004).
4. R.S. Wagner and W.C. Ellis, *Appl. Phys. Lett.* 4, 89 (1964).
5. R.S. Wagner, VLS mechanism of crystal growth, in *Whisker Technology*, ed A.P. Levitt, John Wiley & Sons, New York (1970).
6. E.E. Givargizov and N.N. Sheftal, J. *Cryst. Growth* 9, 326 (1971).
7. J. Westwater, D.P. Gosain, S. Tomiya, and S. Usui, *J. Vac. Sci. Technol. B* 15, 554 (1997).
8. D.P. Yu, Z.G. Bai, Y. Ding, Q.L. Hang, H.Z. Zhang, J.J. Wang, Y.H. Zou, W. Qian, G.C. Xiong, H.T. Zhou, and S.Q. Feng, *Appl. Phys. Lett.* 72, 3458 (1998).
9. H.Z. Zhang, D.P. Yu, Y. Ding, Z.G. Bai, Q.L. Hang, and S.Q. Feng, *Appl. Phys. Lett.* 73, 3396 (1998).
10. Y.F. Zhang, Y.H. Tang, H.Y. Peng, N. Wang, C.S. Lee, I. Bello, and S.T. Lee, *Appl. Phys. Lett.* 75, 1842 (1999).
11. S.T. Lee, N. Wang, Y.F. Zhang, and Y.H. Tang, *MRS Bull.*, 36 (1999).
12. X.H. Fan, L. Xu, C.P. Li, Y.F. Zheng, C.S. Lee, and S.T. Lee, *Chem. Phys. Lett.* 334, 229 (2001).
13. Z. Zhang, X.H. Fan, L. Xu, C.S. Lee, and S.T. Lee, *Chem. Phys. Lett.* 337, 18 (2001).
14. Y. Yao, F. Li, and S.T. Lee, *Chem. Phys. Lett.* 406, 381 (2005).
15. Y. Cui, X. Duan, J. Hu, and C.M. Lieber, *J. Phys. Chem. B* 104, 5213 (2000).
16. Y. Cui, L.J. Lauhon, M.S. Gudiksen, J. Wang, and C.M. Lieber, *Appl. Phys. Lett.* 78, 2214 (2001).
17. J.Y. Yu, S.W. Chung, and J.R. Heath, *J. Phys. Chem. B* 104, 11864 (2000).
18. M.K. Sunkara, S. Sharma, R. Miranda, G. Lian, and E.C. Dickey, *Appl. Phys. Lett.* 79, 1546 (2001).
19. S. Sharma and M.K. Sunkara, *Nanotechnology* 15, 130 (2004).
20. S. Hoffman, C. Duacti, R.J. Neill, S. Piscanec, A.C. Ferrari, J. Geng, R.E. Dunin-Borokowski, and J. Robertson, *J. Appl. Phys.* 94, 6005 (2003).
21. N. Sakulchaicharoen and D.E. Resasco, *Chem. Phys. Lett.* 377, 377 (2003).
22. G. Gundiah, F.L. Deepak, A. Govindraj, and C.N.R. Rao, *Chem. Phys. Lett.* 381, 579 (2003).
23. J. Qi, J.M. White, A.M. Belcher, and Y. Masumoto, *Chem. Phys. Lett.* 372, 763 (2003).

24. J. Huo, R. Solanki, J.L. Freeouf, and J.R. Carruthers, *Nanotechnology* 15, 1848 (2004).
25. M.S. Islam, S. Sharma, T.I. Kamins, and R.S. Williams, *Nanotechnology* 15, L5 (2004).
26. A.R. Guichard, D.N. Barsic, S. Sharma, T.I. Kamins, and M.L. Brongersma, *Nano Lett.* 6, 2140 (2006).
27. T.E. Bogart, S. Dey, K.K. Lew, S.E. Mohney, and J.M. Redwing, *Adv. Mater.* 17, 114 (2005).
28. H. Pan, S. Lim, C. Poh, H. Sun, X. Wu, Y. Feng, and J. Lin, *Nanotechnology* 16, 417 (2005).
29. K. Byon, D. Tham, J.E. Fischer, and A.T. Johnson, *Appl. Phys. Lett.* 87, 193104 (2005).
30. Y. Wu, R. Fan, and P. Yang, *Nano Lett.* 2, 83 (2002).
31. A.I. Hochmaum, R. Fan, R. He, and P. Yang, *Nano Lett.* 5, 457 (2005).
32. V. Schmidt, S. Senz, and U. Gosele, *Nano Lett.* 5, 931 (2005).
33. Th. Stelzner, G. Andra, E. Wendler, W. Wesch, R. Scholz, U. Gosele, and S. Christiansen, *Nanotechnology* 17, 2895 (2006).
34. S. Christiansen, R. Schneider, R. Scholz, U. Gosele, Th. Stelzner, G. Andra, E. Wendler, and W. Wesch, *J. Appl. Phys.* 100, 084323 (2006).
35. T. Shimizu, T. Xie, J. Nishikawa, S. Shingubara, S. Senz, and U. Gosele, *Adv. Mater.* 19, 917 (2007).
36. T. Baron, M. Gordon, F. Dhalluin, C. Ternon, P. Ferret, and P. Gentile, *Appl. Phys. Lett.* 89, 233111 (2006).
37. A. Lugstein, M. Steinmair, Y.J. Hyun, E. Bertagnolli, and P. Pongratz, *Appl. Phys. Lett.* 90, 023109 (2007).
38. P. Aella, S. Ingole, W.T. Petuskey, and S.T. Picraux, *Adv. Mater.* 19, 2603 (2007).
39. J. Arbiol, B. Kalache, P.R. Cabarrocas, J.R. Morante, and A.F. Morral, *Nanotechnology* 18, 305606 (2007).
40. A.F. Morral, J. Arbiol, J.D. Prades, A. Cirera, and J.M. Morante, *Adv. Mater.* 19, 1347 (2007).
41. S.J. Whang, S. Lee, D.Z. Chi, W.F. Yang, B.J. Cho, Y.F. Liew, and D.L. Kwong, *Nanotechnology* 18, 275302 (2007).
42. J. Kikkawa, Y. Ohno, and S. Takeda, *Appl. Phys. Lett.* 86, 123109 (2005).
43. A. Mao, H.T. Ng, P. Nguyen, M. McNeil, and M. Meyyappan, *J. Nanosci. Nanotech.* 5, 831 (2005).
44. A. Mao, M.S. Thesis, San Jose State University (2004).
45. E.I. Givargizov, *J. Crystal Growth* 20, 217 (1973).
46. Y. Zhang, Q. Zhang, N. Wang, Y. Yan, H. Zhou, and J. Zhu, *J. Crystal Growth* 226, 185 (2001).
47. M. Lu, M.K. Li, L.B. Kong, X.Y. Guo, and H.L. Li, *Chem. Phys. Lett.* 374, 542 (2003).
48. N.R. Franklin and H. Dai, *Adv. Mater.* 12, 890 (2000).
49. D.B. Hash and M. Meyyappan, *J. Appl. Phys.* 93, 750 (2003).
50. B.A. Cruden, A.M. Cassell, D.B. Hash, and M. Meyyappan, *J. Appl. Phys.* 96, 5284 (2004).
51. D.D.D. Ma, C.S. Lee, F.C.K. Au, S.Y. Tong, and S.T. Lee, *Science*, 299, 1874 (2003).
52. D. Li, Y. Wu, P. Kim, L. Shi, P. Yang, and A. Majumdar, *Appl. Phys. Lett.* 83, 2934 (2003).
53. D. Li, Y. Wu, R. Fan, P. Yang, and A. Majumdar, *Appl. Phys. Lett.* 83, 3186 (2003).
54. N. Mingo, L. Yang, D. Li, and A. Majumdar, *Nano Lett.* 3, 1713 (2003).

55. H. Scheel, S. Reich, A.C. Ferrari, M. Contoro, A. Colli, and C. Thomsen, *Appl. Phys. Lett.* 88, 233114 (2006).

56. I. Ponomareva, D. Sreevastava, and M. Menon, *Nano Lett.* 7, 1155 (2007).

57. A.M. Morales and C.M. Lieber, *Science*, 279, 208 (1998).

58. Y. Wu and P. Yang, *Chem. Mater.* 12, 605 (2000).

59. X.H. Sun, C. Didychuk, T.K. Sham, and N.B. Wong, *Nanotechnology* 17, 2925 (2006).

60. P. Nguyen, H.T. Ng, and M. Meyyappan, *Adv. Mater.* 17, 549 (2005).

61. H.D. Park and S.M. Prokes, *Appl. Phys. Lett.* 90, 203104 (2007).

62. K. Das, A.K. Chakraborty, M.L. Nandagoswami, R.K. Shingha, A. Dhar, K.S. Coleman, and S.K. Ray, *J. Appl. Phys.* 101, 074307 (2007).

63. X.H. Sun, G. Galebotta, B. Yu, G. Selvaduray, and M. Meyyappan, *J. Vac. Sci. Technol. B*, 25, 415 (2007).

64. T.I. Kamins, X. Li, R.S. Williams, and X. Liu, *Nano Lett.* 4, 503 (2004).

65. H. Adhikari, A.F. Marshall, C. Chidsey, and P.C. McIntyre, *Nano Lett.* 6, 318 (2006).

66. H. Jagannathan, M. Deal, Y. Noshi, J. Woodruff, C. Chidsey, and P.C. McIntyre, *J. Appl. Phys.* 100, 024318 (2006).

67. E. Tutuc, S. Guha, and J.O. Chu, *Appl. Phys. Lett.* 88, 043113 (2006).

68. E. Tutuc, J.O. Chu, J.A. Ott, and S. Guha, *Appl. Phys. Lett.* 89, 263101 (2006).

69. C. Qi, G. Goncher, R. Solanki, and J. Jordan, *Nanotechnology* 18, 075302 (2007).

70. C.B. Jin, J.E. Yang, and M.H. Jo, *Appl. Phys. Lett.* 88, 193105 (2006).

71. T. Hanrath and B.A. Korgel, *J. Am. Chem. Soc.* 124, 1424 (2002).

72. T. Hanrath and B.A. Korgel, *Adv. Mater.* 15, 437 (2003).

73. N. Zaitseva, J. Harper, D. Gerion, and C. Saw, *Appl. Phys. Lett.* 86, 053105 (2005).

74. L.T. Ngo, D. Almecija, J.E. Sader, B. Daly, N. Petkov, J.D. Holmes, D. Erts, and J.J. Boland, *Nano Lett.* 6, 2964 (2006).

75. S. Mathur, H. Shen, V. Sivakov, and U. Werner, *Chem. Mater.* 16, 2449 (2004).

76. H. Chandrasekaran, G.U. Sumanasekara, and M.K. Sunkara, *J. Phys. Chem. B* 110, 18351 (2006).

77. B. Yu, X.H. Sun, G.A. Calebotta, G.R. Dholakia, and M. Meyyappan, *J. Cluster Sci.* 17, 579 (2006).

78. X. Wang, A. Shakouri, B. Yu, X.H. Sun, and M. Meyyappan, *J. Appl. Phys.* 102, 014304 (2007).

79. A.G. Cullis, L.T. Canham, and P.D.J. Calcott, *J. Appl. Phys.* 82, 909 (1997).

80. P. Nguyen, H.T. Ng, and M. Meyyappan, *Adv. Mater.* 17, 1773 (2005).

81. J. Johansson, B.A. Wacaser, K.A. Dick, and W. Seifert, *Nanotechnology* 17, S355 (2006).

82. X. Duan, J. Wang, and C.M. Lieber, *Appl. Phys. Lett.* 76, 1116 (2000).

83. A.L. Roset, M.A. Verhiejen, O. Wunnicke, S. Serafin, H. Wondergem, and E. Bakkers, *Nanotechnology* 17, S271 (2006).

84. J.C. Harmand, G. Patriarche, N. Pere-Laperne, M.N. Merat-Combes, L. Travers, and F. Glas, *Appl. Phys. Lett.* 87, 203101 (2005).

85. M. Tchernycheva, J.C. Harmand, G. Patriarche, L. Travers, and G.E. Cirlin, *Nanotechnology* 17, 4025 (2006).

86. F. Martelli, S. Rubini, M. Piccin, G. Bais, F. Jabeen, S. de Franceschi, V. Grillo, E. Carlino, F. D'Acapito, F. Buscherini, S. Cabrini, M. Lazzarino, L. Businaro, F. Romanato, and A. Franciosi, *Nano Lett.* 6, 2130 (2006).

87. S.G. Ihn, J.I. Song, T.W. Kim, D.S. Leem, T. Lee, S.G. Lee, E.K. Koh, and K. Song, *Nano Lett.* 7, 39 (2007).

88. M.T. Borgstrom, M.A. Verheijen, G. Immink, T. de Smet, and E. Bakkers, *Nanotechnology* 17, 4010 (2006).

89. K. Hiruma, K. Haraguchi, M. Yazawa, Y. Madokoro, and T. Katsuyama, *Nanotechnology* 17, S369 (2006).

90. P. Paiano, P. Prete, N. Lovergine, and A.M. Mancini, *J. Appl. Phys.* 100, 094305 (2006).

91. Y. Kim, H.J. Joyce, Q. Gao, H.H. Tan, C. Jagadish, M. Paladi, J. Zou, and A.A. Suvorova, *Nano Lett.* 6, 599 (2006).

92. H.J. Joyce, Q. Gao, H.H. Tan, C. Jagadish, Y. Kim, X. Zhang, Y. Guo, and J. Zou, *Nano Lett.* 7, 921 (2007).

93. P. Parkinson, J. Lloyd-Hughes, Q. Gao, H.H. Tan, C. Jagadish, M.B. Johnston, and L.M. Herz, *Nano Lett.* 7, 2162 (2007).

94. M. Paladugu, J. Zou, Y.N. Guo, G.J. Auchterlonie, H.J. Joyce, Q. Gao, H.H. Tan, C. Jagadish, and Y. Kim, *Small*, 3, 1873 (2007).

95. J. Noborisaka, J. Motohisa, and T. Fukui, *Appl. Phys. Lett.* 86, 213102 (2005).

96. H.J. Parry, M.J. Ashwin, and T.S. Jones, *J. Appl. Phys.* 100, 114305 (2006).

97. B. Mandl, J. Stangl, T. Martensson, A. Mikkelsen, J. Eriksson, L.S. Karlsson, G. Bauer, L. Samuelson, and W. Seifert, *Nano Lett.* 6, 1817 (2006).

98. K. Dick, K. Deppert, L.S. Karlsson, W. Seifert, L.R. Wallenberg, and L. Samuelson, *Nano Lett.* 6, 2842 (2006).

99. X. Zhou, S.A. Dayeh, D. Wang, and E.T. Yu, *Appl. Phys. Lett.* 90, 233118 (2007).

100. L.C. Chuang, M. Moewe, C. Chase, N.P. Kobayashi, C. Chang-Hasnain, and S. Crankshaw, *Appl. Phys. Lett.* 90, 043115 (2007).

101. E.C. Heeres, E.P.A.M. Bakkers, A.L. Roest, M. Kaiser, T.H. Oosterkemp, and N. deJonge, *Nano Lett.* 7, 536 (2007).

102. H.D. Park, S.M. Prokes, M.W. Twigg, R.C. Cammarata, and A.C. Gaillot, *Appl. Phys. Lett.* 89, 223125 (2006).

103. T. Martensson, J.B. Wagner, E. Hilner, A. Mikkelsen, C. Thelander, J. Stangl, B.J. Ohlsson, A. Gustafsson, E. Lungren, L. Samuelson, and W. Seifert, *Adv. Mater.* 19, 1801 (2007).

104. X. Duan, Y. Huang, Y. Cui, J. Wang, and C.M. Lieber, *Nature*, 409, 66 (2001).

105. M.S. Gudiksen, J. Wang, and C.M. Lieber, *J. Phys. Chem. B* 105, 4062 (2001).

106. C. Tang, Y. Bando, Z. Liu, and D. Goldberg, *Chem. Phys. Lett.* 376, 676 (2003).

107. P.J. Poole, J. Lefebure, and J. Fraser, *Appl. Phys. Lett.* 83, 2055 (2003).

108. A.I. Persson, M.T. Bjork, S. Jeppesen, J.B. Wagner, L.R. Wallenberg, and L. Samuelson, *Nano Lett.* 6, 403 (2006).

109. A. Fuhrer, L.E. Froberg, J.N. Pedersen, M.W. Larsson, A. Wacker, M. Pistol, and L. Samuelson, *Nano Lett.* 7, 243 (2007).

110. D.M. Cornet, V.G.M. Mazzetti, and R.R. LaPierre, *Appl. Phys. Lett.* 90, 013116 (2007).

111. S. Bhunia, T. Kawamura, Y. Watanabe, S. Fujikawa, and K. Tokushima, *Appl. Phys. Lett.* 83, 3371 (2003).

112. N. Yamamoto, S. Bhunia, and Y. Watanabe, *Appl. Phys. Lett.* 88, 153106 (2006).

113. C. Novotny and P.K.L. Yu, *Appl. Phys. Lett.* 87, 203111 (2005).

114. P. Mohan, J. Motohisa, and T. Fukui, *Nanotechnology* 16, 2903 (2005).

115. M. Mattila, T. Hakkarainen, H. Lipsanen, H. Jiang, and E.I. Kauppinen, *Appl. Phys. Lett.* 90, 033101 (2007).

116. M.S. Gudiksen and C.M. Lieber, *J. Am. Chem. Soc.* 122, 8801 (2000).

117. H.W. Seo, S.Y. Bae, J. Park, M. Kang, and S. Kim, *Chem. Phys. Lett.* 378, 420 (2003).

118. S.C. Lyu, Y. Zhang, H. Ruh, H.J. Lee, and C.J. Lee, *Chem. Phys. Lett.* 367, 717 (2003).

119. J.R. Kim, B.K. Kim, J.O. Lee, J. Kim, H.J. Seo, C.J. Lee, and J.J. Kim, *Nanotechnology* 15, 1397 (2004).

120. B. Liu, L. Wei, Q. Ding, and J. Yao, *Nanotechnology* 15, 1745 (2004).

121. K. Tateno, H. Hibino, H. Gotoh, and H. Nakano, *Appl. Phys. Lett.* 89, 033114 (2006).

7

Phase Change Materials

7.1 Introduction

Chalocogenide materials such as $Ge_2Sb_2Te_5$ undergo reversible phase change between crystalline and amorphous states under thermal stimulation. The crystalline phase exhibits a long-range atomic order, high free-electron density, low activation energy, and low resistivity. In contrast, the amorphous phase is characterized by a short-range atomic order, low free-electron density, high activation energy, and high resistivity. The phase change between the orderly single crystalline or polycrystalline phase (c-phase) and the less orderly amorphous phase (α-phase) is accompanied by a change in reflectivity and resistivity [1]. Indeed, the reflectivity change in $Ge_2Sb_2Te_5$ (GST) under thermal stimulation has long been used in mainstream optical data storage such as CDs and DVDs [2–5]. The resistance of the chalgogenides also changes substantially between the phases—enough to represent two logic states—which can be exploited to develop an electrically operated phase-change random access memory (PRAM) [6–12]. Common phase change materials (PCMs) in the above applications include binary materials such as GeTe (GT), InSb, InSe, SbTe, and GeSb; ternary materials such as GST, InSbTe, GaSeTe, SnSbTe, and InSbGe; and quaternary materials such as AgInSbTe, (GeSn) SbTe, GeSb (SeTe), and TeGeSbS. Relevant properties of some of these materials are presented in Table 7.1. Critical parameters to select a candidate material for PRAM operation include melting point, electrical resistivity, thermal conductivity, and specific heat as they determine phase change energy threshold, heating efficiency, and programming current/power.

PRAM technology offers several attractive features: faster write/read speeds, improved endurance, higher scalability, and simpler fabrication than conventional transistor-based nonvolatile memories. These advantages, combined with the scaling issues currently facing FLASH memory, have renewed interest in PRAM technology. While the optical data storage technology using phase change is highly commercial today, the PRAM commercialization needs to overcome a few difficulties to become competitive. A typical programmable cell consists of a thin film of a PCM sandwiched between two electrodes. The PCM is heated by Joule heating from the applied power leading to the phase transition. The primary limitation has been the large programming current needed for the thermal stimulation. In addition, intercell thermal interferance may be an issue, that is, heating of one bit inadvertently

TABLE 7.1

Properties of Various PCMs

Materials	Melting Point (°C)	Specific Heat (J/g K)	Thermal Conductivity (W/m K)	Electrical Resistivity (Ωcm)
GeTe	725	0.26	4.4	4.78×10^{-4}
In_2Se_3	890	0.27		6.7
InSb	527	0.2	180.0	0.03
Sb_2Te_3	620	0.2	1.9	2.5×10^{-4}
$Ge_2Sb_2Te_5$	632	0.202	0.46	4.16×10^{-4}

Source: Courtesy of B. Yu.

influencing the neighboring bit; this will prevent scale up efforts in the quest for ultrahigh density storage. These are precisely the areas where PCM nanowires can offer solutions. First, as mentioned in Chapter 1, the melting point of semiconductors and metals at nanoscale is substantially reduced below the corresponding bulk values. Thermal conductivity and other thermal properties at nanoscale are also favorable. In addition, the smaller memory cell volume with a nanowire leads to a direct reduction of energy needs. The reduction in thermal budget gained from these features would reduce the programming current to acceptable levels. Finally, since there is no charge transport involved in the memory operation, the device is immune to space radiation effects.

The possibility of constructing PRAM devices has invigorated research on the growth and characterization of phase change nanowires (PCNWs) [13–19]. This chapter discusses preparation and characterization of PCM nanowires. Demonstrations of reduction in melting point are discussed, whereas reduction in programming current and device results will be discussed in Chapter 12 under electronics applications.

7.2 Phase Change Nanowire Growth

The PCNW growth reported in the literature is primarily through vapor–liquid–solid (VLS) technique with sublimation of corresponding powders to generate the vapor source. Therefore, the generic VLS reactor setup discussed in Chapter 3 is relevant here. Table 7.2 summarizes growth conditions for various PCM nanowires using the VLS approach. A two-zone furnace system is ideal, allowing high temperature generation of the source by sublimation in one zone and the nanowire growth in a low-temperature zone. To grow the GT nanowires shown in Figure 7.1, high-purity GeTe powder is placed on a quartz boat in the center of the high-temperature zone [15]. The growth substrate with the catalyst is placed downstream in the low-temperature zone.

TABLE 7.2

Growth Conditions for Various PCNWs

PCM	Source	Source Temperature	Growth Temperature	Pressure (Torr)
GT	GeTe powder	720°C	450°C	200
GST	Powders of Sb_2Te_3: GeTe = 1:2	690°C	450°C	200
In_2Se_3	In_2Se_3 powder	900°C–950°C	650°C–700°C	30–600
GeSb	Ge and Sb powders	900°C for Ge/ 500°C for Sb	300°C	250

Source: Courtesy of X.H. Sun.

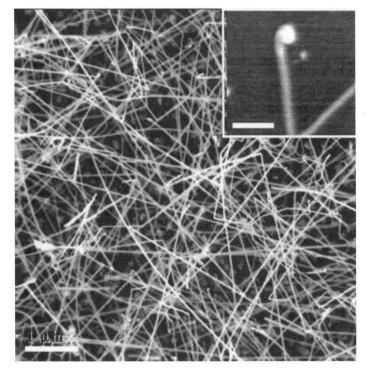

FIGURE 7.1

SEM image of GeTe nanowires on SiO_2-coated silicon (100) substrate. Scale bar = 1 μm. Inset is a close-up view of a catalyst bead at the tip of a nanowire. Scale bar (inset) = 200 nm. (Courtesy of X.H. Sun and B. Yu.)

The catalyst in this experiment consists of 20 nm gold particles on Si (100) substrate coated with SiO_2. The reactor is pumped down to a base pressure of 10^{-2} Torr using a roughing pump. A 20% of mixture of hydrogen in argon is introduced as carrier gas (25 sccm flow rate) at 200 Torr. The temperature

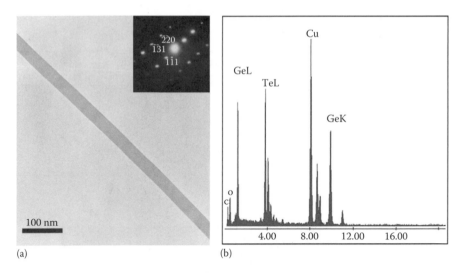

FIGURE 7.2
(a) Low-magnification TEM image of a GeTe nanowire with a diameter of 40 nm and length over 5 μm. The inset shows the SAED pattern of an fcc lattice structure. (b) EDS spectrum of the same nanowire as in (a). (Courtesy of X.H. Sun and B. Yu.)

of the upstream zone is raised to 720°C and maintained for 1 h. The growth zone temperature is maintained at 450°C. At the end of the growth run, the reactor is slowly cooled to room temperature prior to the removal of the sample.

Figure 7.1 shows an SEM image of GT nanowires from the above experiment. The growth density appears to be high and the nanowire diameter ranges from 40 to 80 nm. The 1 h growth time yields nanowire lengths of tens of microns. The gold catalyst appears as beads at the tips of nanowires providing evidence for VLS mechanism. Controlled experiments without the gold catalyst on the SiO_2 surface do not yield any GT wires [15]. Figure 7.2 shows a low-magnification TEM image revealing a 40 nm diameter GT nanowire. It is a single crystal with cubic lattice structure as indicated by selected area electron diffraction (SAED). The energy-dispersive spectroscopic (EDS) analysis of an individual nanowire with locally focused beam spot confirms the presence of only Ge and Te at a ratio of 1:1. The oxide outer layer is the source of the trace O peak in the EDS analysis. XPS analysis also confirms the chemical composition, compound nature, and atomic ratio of GT nanowires via Ge 3d and Te $3d_{5/2}$ peaks with binding energies of 30.0 and 572.4 eV, respectively.

GST nanowires can be grown by mixing Sb_2Te_3 powder with GeTe powder (mole ratio of 1:2) [19]. The source zone is heated to 690°C and all other details are as before. Figure 7.3 shows an SEM image of high-yield GST nanowires with diameters in the 20–40 nm range. Gold catalyst particles end

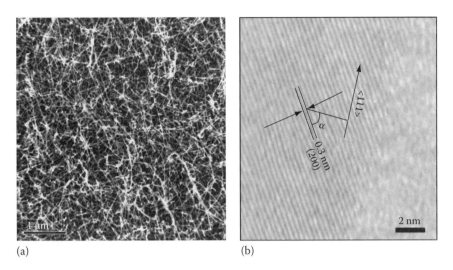

FIGURE 7.3
(a) SEM image of GST nanowires. Scale bar = 1 μm. (b) HR-TEM image of the (200) planes of GST fcc nanowire with an interplanar distance of ~0.3 nm. (Courtesy of X.H. Sun and B. Yu.)

up at the nanowire tips as seen in TEM images (not shown here) confirming VLS growth mechanism. Local energy-dispersive x-ray spectroscopic (EDX) analysis of individual wires indicates an approximate atomic ratio of 2:2:5 for Ge, Sb, and Te, respectively. The high-resolution transmission electron microscopy (HRTEM) image in Figure 7.3b shows planes with $a \sim 3.0$ Å lattice spacing corresponding to d-spacing of the (200) plane of the GST fcc structure. The 54.7° angle matches the angle between the (200) and (111) planes, which indicates a preferential <111> growth orientation.

The VLS reactor setup above can be used to grow In_2Se_3 nanowires by placing the corresponding powder in the upstream zone [16]. A temperature of 900°C–950°C is necessary to generate the vapor source. The substrate temperature is 650°C–700°C in the growth zone and other experimental details are the same as in GT and GST cases. In_2Se_3 nanowires decorated with gold tips are obtained (see Figure 7.4). These nanowires are single crystal of the β-phase hexagonal structure with lattice constants $a = 4.0$ Å and $c = 19.2$ Å. A high fraction (~90%) of the nanowires exhibits [1120] crystallographic direction with the remaining in the [0001] orientation. The indium selenide nanowires can also be grown with an indium film serving as self-catalyst, thus eliminating a contaminating metal such as gold. However, the use of relatively thick films of indium (~40 nm, which may be sputtered onto the substrate) is warranted due to the low melting point of In (156°C) and the consequent loss of substantial indium through evaporation during the pregrowth temperature ramping.

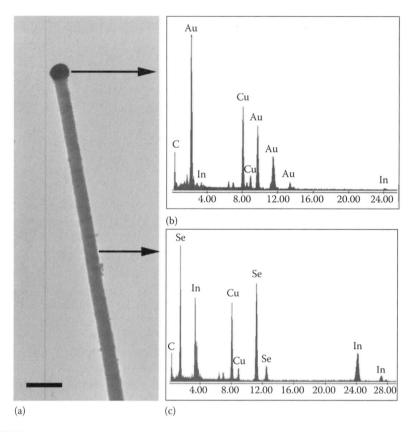

FIGURE 7.4
(a) TEM image of an In₂Se₃ nanowire, (b) EDS spectrum from the catalyst bead, and (c) EDS spectrum from the nanowire. (Courtesy of X.H. Sun.)

Synthesis of GeSb nanowires can be carried out by employing high-purity (99.999%) powders of Ge and Sb placed in the high- and low-temperature zones, respectively. The Si(100) substrate with gold nanoparticle dispersion (20 nm) is also placed downstream about one-fourth the way from the end of the reactor tube. The growth is carried out at 250 Torr with a carrier gas flow of 50 sccm Ar mixed with hydrogen (10%). The two zones are maintained at 900°C and 500°C, respectively. The GeSb nanowires are 40–100 nm in diameter and several microns long. Figure 7.5 shows a high-resolution TEM image of GeSb nanowire covered with a thin 1–2 nm amorphous layer [20]. The EDS results indicate a nanowire stoichiometry of 91%–93% Sb and 7%–9% Ge. Though the nanowire has a rich antimony content, it is still referred to as GeSb as there is no stoichiometric compound. The elemental mapping in Figure 7.6 shows that both Ge and Sb are homogeneously distributed across the nanowire.

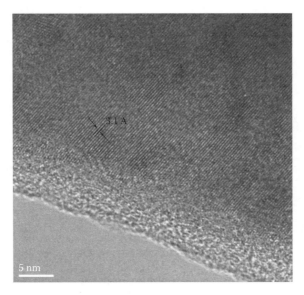

FIGURE 7.5
High-resolution TEM image of a GeSb nanowire showing a thin amorphous outer layer. (Courtesy of X.H. Sun.)

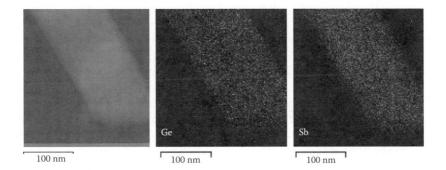

FIGURE 7.6
Elemental mapping of a GeSb nanowire. (Courtesy of X.H. Sun.)

7.3 Properties Relevant to PRAM

The melting point of the PCM is an important consideration since it determines the phase transition temperature. For example, the lower the melting point, the lower would be the thermal budget corresponding to the memory reset activity. As mentioned earlier, it is well known that materials in

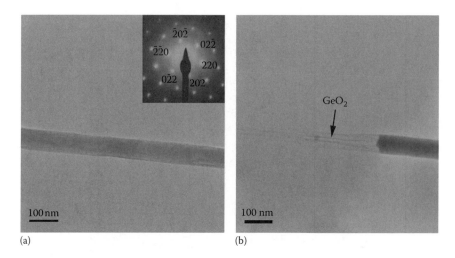

FIGURE 7.7

Measurement of GeTe nanowire melting point under real-time TEM morphology monitoring. The nanowire melts and its mass is gradually lost through evaporation, leaving an oxide nanotube as outer shell. (a) TEM image under room temperature with the electron diffraction pattern in the inset and (b) nanowire during melting. (Courtesy of X.H. Sun.)

the nanoscale exhibit much lower melting points compared to their bulk counterparts. Numerous measurements and model predictions attest to this phenomena in metals and semiconductors [21–25]. Such a reduction in melting point has also been reported for PCNWs [15,16]. The demonstration involves measurement of the nanowire melting temperature via an in situ heating experiment in a TEM chamber while monitoring its morphology in real time [15]. The sample stage is resistively heated at a rate of 10°C/min. When the nanowire starts to melt, the electron diffraction (ED) pattern would disappear and this is identified as the melting point. The process of melting and evaporation is monitored real time by TEM and recorded by a video camera in the bright field mode.

Figure 7.7 shows the TEM image of a single GT nanowire before melting commences along with its ED pattern in the inset. When melting begins, the contrast of the wire abruptly turns dark and the single-crystalline ED pattern disappears, following change from solid crystalline to liquid phase (Figure 7.7b). This process commences at 390°C, indicating a 46% reduction from the bulk melting point of 725°C. Since the nanowires have an outside oxide layer, which has a higher melting point of 1115°C, a shell of the oxide nanotube is left behind when GT is vaporized out. The presence of the oxide layer is confirmed by EDS analysis. Similarly, the In_2Se_3 nanowires melt at 680°C compared to bulk melting point of 890°C [16].

Though Figure 7.7 shows a single nanowire, it is important to point out that the melting experiment sample consists of numerous wires and hence, the observed reduction is for nanowires in the 40–50 nm diameter range. Controlled experiments of the type reported in Ref. [22] for CdS nanocrystals would show further reduction for smaller nanowires. Indeed, thermodynamic considerations and an energy balance associated with the melting process [26] yield:

$$\Delta\theta = \frac{\sigma T_b}{L\rho} \times \frac{1}{r}\left(1 + \frac{1}{AR}\right)$$

where

$\Delta\theta$ is the difference between bulk melting point T_b and the nanowire melting point

r and AR are the radius and aspect ratio (l/r) of the nanowire

L is the latent heat of fusion

ρ is nanowire density

σ is the surface tension of the nanowire

According to this relation, as the nanowire size gets smaller, the reduction in melting point is larger.

In any case, the most relevant metric is the programming current, which is established by proper electrical characterization of the nanowires. Lee et al. [17] constructed a simple memory device with GT nanowires to demonstrate the reduction in current and similar reductions for In_2Se_3 nanowire-PRAM have been reported as well [27]. These results and other relevant details will be discussed in Chapter 12.

References

1. S.R. Ovshinsky, *Phys. Rev. Lett.*, 21, 1450 (1968).
2. D. Adler, M.S. Shur, M. Silver, and S.R. Ovshinsky, *J. Appl. Phys.*, 51, 3289 (1980).
3. M. Chen, K. Rubin, and R. Barton, *Appl. Phys. Lett.*, 49, 502 (1986).
4. C.A. Volkert and J.M. Wuttig, *J. Appl. Phys.*, 86, 1808 (1999).
5. N. Yamada and T.J. Matsunaga, *J. Appl. Phys.*, 88, 7020 (2000).
6. J. Kotz and M.P. Shaw, *J. Appl. Phys.*, 55, 427 (1984).
7. S. Lai, *IEDM Tech. Dig.*, 255 (2003).
8. A. Pirovano, A.L. Lacaita, A. Benvenuti, S. Pellizzer, and R. Bez, *IEEE Trans Elec. Dev.*, 51, 452 (2004).
9. S.M. Kim, M. J. Shin, D.J. Choi, K.N. Lee, S.H. Hong, and Y.J. Park, *Thin Solid Films*, 469, 322 (2004).
10. H. Lee, Y.K. Kim, D. Kim, and D.H. Kang, *IEEE Trans. Mag.*, 41, 1034 (2005).
11. J.D. Maimon, K.K. Hunt, L. Burcin, and J. Rodgers, *IEEE Trans. Nuc. Sci.*, 50, 1878 (2003).
12. D.S. Suh, E. Lee, K.H.P. Kim, J.S. Noh, W.C. Shin, Y.S. Kang, C. Kim, and Y. Khang, *Appl. Phys. Lett.* 90, 023101 (2007).

13. D. Yu, J.Q. Wu, Q.A. Gu and H.K. Park, *J. Am. Chem. Soc.*, 128, 8148 (2006).

14. S. Meister, H.L. Peng, K. McIlwrath, K. Jarqusch, X.F. Zhang, and Y. Cui, *Nano. Lett.*, 6, 1514 (2006).

15. X.H. Sun, B. Yu, G. Ng, and M. Meyyappan, *J. Phys. Chem. C*, 111, 2421 (2006).

16. X.H. Sun, B. Yu, G. Ng, T.D. Nguyen, and M. Meyyappan, *Appl. Phys. Lett.*, 89, 233121 (2006).

17. S.H. Lee, D.K. Ko, Y. Jung, and R. Agarwal, *Appl. Phys. Lett.*, 89, 223116 (2006).

18. Y. Jung, S.H. Lee, D.K. Ko, and R. Agarwal, *J. Am. Chem. Soc.*, 128, 14026 (2006).

19. X.H. Sun, B. Yu, and M. Meyyappan, *Appl. Phys. Lett.*, 90, 183116 (2007).

20. X. Sun, B. Yu, G. Ng, M. Meyyappan, S. Ju, and D.B. Janes, *IEEE Trans. Elec. Dev.*, 55, 3131 (2008).

21. Ph. Buffat and J.P. Borel, *Phys. Rev. A*, 13, 2287 (1976).

22. A.N. Goldstein, C.M. Echer, and A.P. Alivisatos, *Science*, 256, 1425 (1992).

23. A.N. Goldstein, *Appl. Phys. A*, 62, 33 (1996).

24. F.G. Shi, *J. Mater. Res.*, 9, 1307 (1994).

25. T. Karabacak, J.S. DeLuca, P.I. Wang, G.A. Ten Eyck, D. Ye, G.C. Wang, and T.M. Lu, *J. Appl. Phys.*, 99, 064304 (2006).

26. O. Koper and S. Winecki, in *Nanoscale Materials in Chemistry*, K.J. Klabunde, ed., Wiley-Interscience, New York 2001.

27. B. Yu, X. Sun, D.B. Janes, and M. Meyyappan, *IEEE Trans. Nanotechnol.* 7, 496 (2008).

8

Metallic Nanowires

Metallic nanowires provide an interesting model system to study nanostructural effects such as quantized conductance, localization effects, and other physical phenomena. They also have applications in electronics, optoelectrtonics, magnetics, etc., due to desirable values for many physical properties such as electrical conductivity, and thermal conductivity, and excellent optical and magnetic properties. Nanowires of bismuth [1–5], silver [6–15], gold [16–22], platinum [23], copper [24–37], indium [38], iron [39], nickel [40–48], cobalt [49], palladium [50], molybdenum [51], tungsten [52], and zinc [53–56] have been grown successfully. The most common approach for growing metallic nanowires is to use some type of template in a chemical or electrochemical synthesis. The anodized alumina (AA) template is the most popular one, and other template-assisted methods to obtain ordered arrays of metallic nanowires include polycarbonate-etched ion-track membranes [27,28], polymer templates, carbon nanotubes [23], DNA [29], molecular sieves, zeolites [9,11], and block copolymers. The preparation of many of these templates has been discussed in detail in Chapter 3.

8.1 Bismuth Nanowires

Bismuth is a semimetal with a small electron effective mass, small electron density, long carrier mean free path, and highly anisotropic Fermi surface [1]. The energy overlap between the L-point conduction band and the T-point valence band is small, about 38 meV at 77 K, which can lead to a semimetal–semiconductor transition with decreasing diameter, about 60 nm at 77 K [3]. The potential in thermoelectric devices (see Chapter 16) has made the study of bismuth and its alloy with antimony attractive.

An anodized alumina membrane (AAM) can be effectively used to grow bismuth and Bi_xSb_{1-x} nanowires. The template may be prepared in a manner similar to the description in Chapter 3. For electrochemical deposition using this template [1], a 60 nm nickel is sputtered on the bottom side of the AAM to serve as a contact and a graphite plate is used as counterelectrode. The plating solution consists of 40 g/L of $BiCl_3$, 50 g/L of tartaric acid, 100 g/L of glycerol, 70 g/L of NaCl, and 1 mol/L of HCl. The pH of the electrolyte is adjusted to 0.9 by adding aqueous ammonia to prevent

(a)   (b)

FIGURE 8.1
SEM images of bismuth nanowires of different diameters: (a) 36 nm and (b) 60 nm. (Reproduced from Zhu, Y. et al., *J. Phys. Chem. B*, 110, 26189, 2006. With permission.)

corrosion of the AAM. The pulsed electrodeposition is carried out at voltages of negative 1–2 V for pulse periods of 40–80 ms. Figure 8.1 shows SEM images of bismuth nanowires (BiNWs) of 36 and 60 nm in diameter that are ordered and uniform with a high degree of pore filling of the AAM [1]. In preparing Bi_xSb_{1-x} alloys, 0.02 M/L of $SbCl_3$ is added to the mixture to enable the growth of nanowires with an antimony fraction of 17.5%–33.0% and diameters of 22–60 nm [2].

The bismuth nanowires exhibit a size-dependent melting point [1] that can be measured by differential scanning calorimetry (DSC). Figure 8.2

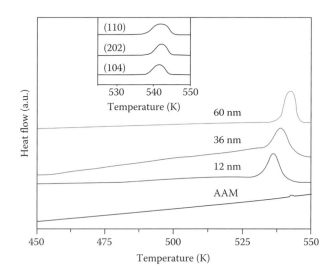

FIGURE 8.2
Results from DSC to estimate the melting points of nanowires of various diameters. Inset: DSC results for nanowires of different orientations. (Reproduced from Zhu, Y. et al., *J. Phys. Chem. B*, 110, 26189, 2006. With permission.)

shows the DSC results for nanowires of 12, 36, and 60 nm diameters. The peak that occurs in the heat flow versus temperature curves is related to the melting fusion of the material and therefore is located at the melting point of BiNW. The melting points are 536, 539, and 542 K for 12, 36, and 60 nm diameter wires whereas the bulk melting point is 544 K. This provides clear evidence that the melting point decreases with diameter. As shown in the inset of Figure 8.2, the melting point is independent of orientation for a fixed diameter wire. This melting point depression is lot smaller than for bismuth particles [57], germanium nanowires [58], and phase change nanowires such as GeTe reported in Chapter 7. The reason may be because the DSC measurements in Ref. [1] were for bismuth wires embedded in the template, not for individual, loose nanowires. Therefore, the results in Figure 8.2 can be used only to understand the relative melting points of nanowires with different diameters [1]. Further discussion of this is provided later in the context of Zn nanowires.

8.2 Silver Nanowires

Silver nanowires have received attention due to their high electrical and thermal conductivities. Silver also finds applications in electronics, photonics, photography, conductive inks, antibacterial agents, and catalysis. A simple approach to producing these nanowires involves the use of AAM and growth in the pores by electroplating [8]. A copper electrode of thickness 0.5 μm is sputtered on one side of the AAM to serve as a contact. The electroplating solution consists of 10 g/L of $AgNO_3$, 10 g/L of Na_2SO_3, and 20 g/L of CH_3COONH_4. A dc voltage of 1–4 V is used for 10 min and the length of the nanowires depends on the electroplating time. A template-free approach involves a solution-phase method [10] by reducing silver nitrate with ethylene glycol in the presence of poly(vinyl pyrrolidone) (PVP). Ethylene glycol acts both as a solvent and a reducing agent here, and the addition of nanoparticles of Pt seeds the growth of silver nanowires. When silver nitrate is reduced in the presence of these seeds, silver particles are formed, and during the refluxing, particles grow by Ostwald ripening. The presence of PVP controls the growth rates of various faces of silver and produces cylindrical wires [10]. A self-seeding approach is also possible, thus avoiding Pt seeds, which may become impurities later. The solutions mentioned above can be injected together in a controlled manner and the silver particles formed initially can act as seeds themselves. Figure 8.3 shows Ag nanowires of 60 nm in diameter, and this cross section is uniform along the axis of the nanowires. The XRD patterns indicate the face-centered cubic (fcc) structure of the nanowires.

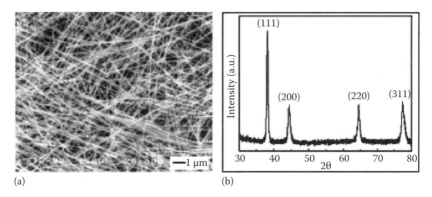

(a)　　　　　　　　　　　　　　　　(b)

FIGURE 8.3
Solution phase growth of silver nanowires. (a) SEM image. (b) XRD pattern showing the fcc structure of the nanowires. (Reproduced from Sun, Y. and Xia, Y., *Adv. Mater.*, 14, 833, 2002. With permission.)

8.3 Copper Nanowires

Copper is currently used as an interconnect material in silicon CMOS technology due to its high electrical conductivity. For the same reason, copper nanowires are of interest to examine their potential use in 32 nm generation CMOS and beyond. Though copper nanowires can be prepared with AAM as a template similar to other metallic nanowires described in this chapter, a simple solid-phase method without a template is also possible [25] and may even be preferable. A thin film of $(KI)_{1.5}(CuI)_{8.5}$ is used as a medium to transmit cuprous ions in a two-electrode structure (see Figure 8.4), and when cuprous ions are reduced at the cathode, nanowires are formed locally. The thin film above is prepared by vacuum thermal evaporation of a source mixture of (1.5:8.5) KI and CuI at 525–535 K on a glass substrate. On either side of the film, a copper film is deposited as an electrode with a gap of 1 mm. Copper

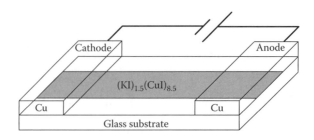

FIGURE 8.4
Schematic of a two-electrode (copper–copper) apparatus with a $(KI)_{1.5}(CuI)_{8.5}$ thin film to produce copper nanowires under a dc field. The nanowires are formed at the edge of the cathode. (Reproduced from Zhang, J. et al., *Nanotechnology*, 16, 2030, 2005. With permission.)

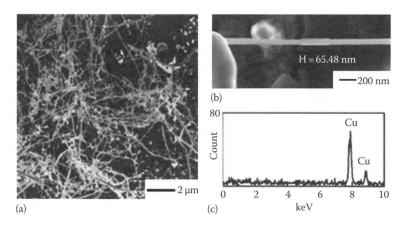

FIGURE 8.5
Copper nanowires grown from a mixture of $(KI)_{1.5}(CuI)_{8.5}$ under a dc field. (a) SEM image of nanowires at the cathode. (b) SEM image of an individual copper nanowire. (c) EDS results of an as-prepared copper nanowire. (Reproduced from Zhang, J. et al., *Nanotechnology*, 16, 2030, 2005. With permission.)

nanowires form at the edge of the cathode at a dc electric field of 1.5×10^4 V/m. Copper atoms lose electrons in this process and become cuprous ions at the anode. These ions move through the $(KI)_{1.5}(CuI)_{8.5}$ film under the electric field and get reduced at the edge of the cathode. When this goes on for a certain length of time, the locally congregating atoms form nanowires of up to $6\,\mu m$ length over 30 min. Figure 8.5 shows an SEM image of these nanowires, and individual wires are uniform in diameter along the axis [25]. The EDS analysis confirms that the structure is pure copper. Interestingly, when a pure CuI thin film is used, only curled nanowires are formed at the edge of the cathode instead of the straight wires seen in Figure 8.5a.

Polycarbonate membranes are also useful to produce size-controlled copper nanowires [27]. When polycarbonate foils are irradiated with ions of energy ~10 MeV, the ions create cylindrical tracks; these tracks are chemically etched in a solution of 6 N NaOH with 10% methanol at 50°C to create nanopores. On one side of the membrane, a metal layer such as gold is needed to serve as the contact for the potentiostatic growth of copper nanowires using an aqueous solution of $CuSO_4$ and sulfuric acid [27].

Characterizing the resistivity of copper nanowires is of great interest due to their potential to serve as interconnects. Practical measurements are always complicated due to poor contacts. It is commonly found that when nanowires are transferred onto prepatterned contacts for electrical measurements, the resistance of Cu nanowires is of the order of gigaohms [27]. This is due to an adsorbate formation between the contact and the nanowire. In contrast, the contacts deposited above the nanowire lying on a substrate appear to be the more favorable scheme to make resistance measurements. Even here, the specific resistivity of copper nanowire is $17.1\,\mu\Omega$ cm, which is an order of magnitude higher than that of copper thin film (1.75 $\mu\Omega$ cm). Since careful preparation of the contacts eliminates the parasitic contribution of the contacts to the observed resistance, other likely contributions can be surface scattering of electrons within the nanowires and copper oxidation. Continuous monitoring of the current–voltage characteristics of a 50 nm copper wire over a 12 h period shows a gradual increase of resistance from 640 Ω by six orders of magnitude, clearly revealing the effect of oxidation [27]. Since nanowires have a high surface-to-volume ratio, they are more susceptible to oxidation. Room temperature oxidation behavior of copper nanorods [32] shows that the resistance to oxidation is a lot lower than that of bulk copper. The oxide layer on the wires and rods is porous and unstable. In contrast, when copper nanowires are within 230 nm deep SiO_2 trenches, oxidation does not appear to be an issue and the resulting resistivity is only a factor of 2.6 higher (for a 40 nm wire) than the bulk value [36]. This increase is contributed by two factors. The first, nonspecular surface scattering at the nanowire surface, is important when the nanowire dimensions are comparable to or smaller than the mean free path of conduction electrons. The latter is 40 nm for copper at room temperature. The second contribution is from the scattering of conduction electrons at the grain boundaries, which depends on the average grain size along the direction of the electrical current.

Another important property of interest is the thermal expansion coefficient. For copper nanowires, this coefficient is an order of magnitude smaller than for bulk copper ($1.2\times10^{-6}K^{-1}$ versus $148\times10^{-6}K^{-1}$) [26]. Typically when metal particles are implanted in a ceramic to improve the toughness, the larger thermal expansion coefficient of the metal relative to the ceramic may lead to the breaking of cermets at high temperatures. Therefore, the lower thermal expansion coefficient of copper nanowires is desirable for cermet preparation.

8.4 Nickel Nanowires

The interest in nickel nanowires (NiNWs) is due to their magnetic behavior and thus their potential in magnetic data storage and microsensor applications. In the AAM approach, the electrolyte for NiNW preparation consists of 100 g/L of $NiSO_4 \cdot 6H_2O$, 30 g/L of $NiCl_2 \cdot 6H_2O$, and 40 g/L of H_3BO_3 solution. The 10H of the mixture is adjusted to 2.5 by the addition of 1 M H_2SO_4. A potential of 1.35 V at 10°C enables direct current electrodeposition of the nanowires inside the pores [43]. Figure 8.6 shows SEM images of nickel nanowires of various diameters, and their sizes correspond well to the pore openings. Examination of the corresponding XRD patterns (see Figure 8.7) reveals that growth orientation is dependant on the nanowire diameter [43]. The smallest wires, 25 nm in diameter, show only one peak of (220) plane,

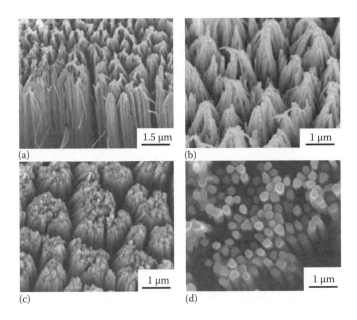

FIGURE 8.6
SEM images of nickel nanowires of different diameters: (a) 25 nm, (b) 45 nm, (c) 90 nm, and (d) 225 nm. (Reproduced from Wang, X.W. et al., *J. Phys. Chem. B*, 109, 24326, 2005. With permission.)

indicating a preferred growth orientation along the [110] direction. This trend continues with 45 nm diameter wires although it is not shown in Figure 8.7. When the diameter increases to 70 nm, a peak of a (111) plane appears in addition to the peak of the (220) plane; this suggests that some nanowires

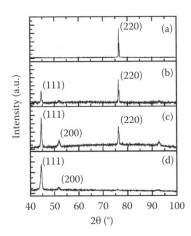

FIGURE 8.7
XRD patterns of nickel nanowires of different diameters: (a) 25 nm, (b) 70 nm, (c) 90 nm, and (d) 225 nm. (Reproduced from Wang, X.W. et al., *J. Phys. Chem. B*, 109, 24326, 2005. With permission.)

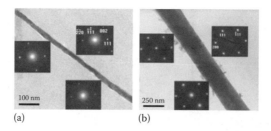

FIGURE 8.8
TEM images of nanowires. (a) 25 nm diameter and (b) 225 nm diameter. SAED patterns are also shown. (Reproduced from Wang, X.W. et al., *J. Phys. Chem. B*, 109, 24326, 2005. With permission.)

grow along the [110] direction while others follow [111] direction. The higher intensity of the (220) plane in Figure 8.7b relative to the (111) plane implies that a higher fraction of nanowires grows along the [110] direction. When the diameter reaches 90 nm, the trend reverses as seen in Figure 8.7c, with a higher fraction of nanowire growing along the [111] direction than [110]. Finally, for large diameters such as 225 nm, the most preferred growth orientation is [111]. Since the penetration of x-rays in XRD analysis is small, Wang et al. [43] repeated their analyses from the bottom side and observed the same behavior. This diameter-dependent growth orientation is also confirmed by TEM and SAED analysis shown in Figure 8.8. The SAED patterns show clearly that the orientation is [110] and [111] for 25 and 225 nm wires, respectively. The TEM images show that the diameters are uniform along the nanowire axis and the nanowires are single crystalline [43].

Nickel has been found in two different phases: a stable face-centered cubic (fcc) phase and a metastable hexagonal close-packed (hcp) phase. With heat treatment, a phase transition from an hcp to an fcc structure occurs, bringing an improved (220) orientation to the nickel nanowire arrays [41]. A study of the magnetic properties of these wires shows that the hcp structures are nonmagnetic. The magnetic behavior of nickel nanowires embedded inside an anodized alumina membrane [47] shows the coercive field to be diameter dependent. Keeping the interpore distance the same and reducing the diameter of NiNW from 55 to 30 nm leads to an increase in the coercive field from 600 to 1200 Oe.

8.5 Zinc Nanowires

Zinc nanowires exhibit superconducting properties, and their luminescence properties make them attractive in optoelectronics applications such as light emitters and lasers. Thermal evaporation of ZnS powder at 1225°C under an argon flow in a quartz tube reactor appears to be a simple approach to produce Zn nanowires [54]. This approach needs neither a catalyst nor a

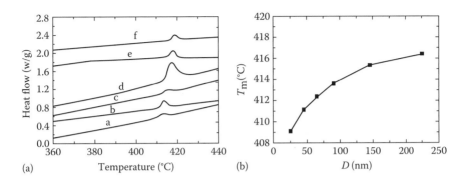

FIGURE 8.9
(a) DSC results for Zn nanowires of different diameters. The endothermic peak corresponds to the melting point. Curves a through f represent 25, 45, 65, 90, 145, and 225 nm diameter wires respectively. (b) The melting point versus diameter plot. (Reproduced from Wang, X.W. et al., *Appl. Phys. Lett.*, 88, 173114, 2006. With permission.)

template. The drawback is ZnS wires and ribbons are also produced, although at the high-temperature region of the tube. For optical characterizations and related applications, it is desirable to produce wires without oxide coverage. Thermal evaporation of pieces of pure zinc sheets under argon flow at 900°C and 500 Pa pressure produces Zn nanowires of high crystallinity and very low oxygen content [55]. Comparison of the PL spectra of these Zn nanowires along with that for a Zn sheet shows blue shifts from 393 to 346 nm, 406 to 361 nm, and 422 to 378 nm, confirming the quantum size effects arising from the size-dependent widening of the energy gap [55].

To prepare Zn nanowires with the aid of AAMs, electrodeposition into the pores is done using an electrolyte consisting of 100 g/L of $ZnSO_4 \cdot H_2O$, 40 g/L of $ZnCl_2$, and 40 g/L of H_3BO_3. These samples have been analyzed using DSC [53] to study the size dependence of the melting point, as in the case of BiNWs discussed in Section 8.1. The DSC results in Figure 8.9 show the shift in endothermic peak—located around the melting point—toward lower temperatures with a decrease in the diameter. The corresponding melting point versus diameter plot is also shown in Figure 8.9. The extent of the reduction in melting point is low as in the case of the bismuth nanowires discussed in Section 8.1, because the Zn nanowires are embedded in a matrix. The deviation of the melting point from the bulk value is known to be inversely proportional to the diameter. The nanowire melting point for a material in a confined geometry is given by

$$T_m = T_o\left(1 - \frac{4\Delta\sigma\upsilon}{\Delta H_f D}\right) \tag{8.1}$$

where
 T_o is the bulk melting point
 υ is the molar volume of a liquid at the melting point T_m

ΔH_f is the heat of fusion

$\Delta \sigma$ is the difference in interfacial energies of the solid/wall and the liquid/wall interfaces in the confined geometry

The corresponding relation for free nanoparticles and nanowires uses σ, the surface tension of the nanomaterial instead (see Section 7.3). Though this relation indicates a linear relation between T_m and $1/D$, the data generally deviate from this relationship. It is important to realize that ΔH_f also depends on the diameter as given below, which accounts for a curvilinear behavior of T_m with inverse diameter:

$$\Delta H_f = \Delta H_o \left(1 - \frac{2t_o}{D} \right)^2 \tag{8.2}$$

where

ΔH_o is the heat of fusion for the bulk material

t_o is the thickness of the liquid overlayer covering the cylindrical core at the melting temperature T_m [53]

References

1. Y. Zhu, X. Dou, X. Huang, L. Li, and G. Li, *J. Phys. Chem. B* 110, 26189 (2006).
2. X. Dou, Y. Zhu, X. Huang, L. Li, and G. Li, *J. Phys. Chem. B* 110, 21572 (2006).
3. L. Li, Y. Zhang, G. Li, and L. Zhang, *Chem. Phys. Lett.* 378, 244 (2003).
4. L. Li, Y.W. Yang, G.H. Li, and L.D. Zhang, *Small* 2, 548 (2006).
5. M. Tian, J. Wang, N. Kumar, T. Han, Y. Kobayashi, Y. Liu, T.E. Mallouk, and M.H.W. Chan, *Nano Lett.* 6, 2773 (2006).
6. X. Liu, J. Luo, and J. Zhu, *Nano Lett.* 6, 408 (2006).
7. Y.H. Cheng, C.K. Chou, C. Chen, and S.Y. Cheng, *Chem. Phys. Lett.* 397, 17 (2004).
8. Y.H. Cheng and S.Y. Cheng, *Nanotechnology* 15, 171 (2004).
9. X. Ding, G. Briggs, W. Zhou, Q. Chen, and L.M. Peng, *Nanotechnology* 17, S376 (2006).
10. Y. Sun and Y. Xia, *Adv. Mater.* 14, 833 (2002).
11. M.J. Edmondson, W. Zhou, S.A. Sieber, I.P. Jones, I. Gameson, P.A. Anderson, and P.P. Edwards, *Adv. Mater.* 13, 1608 (2001).
12. S. Liu, J. Yue, and A. Gedanken, *Adv. Mater.* 13, 656 (2001).
13. B.H. Hong, S.C. Bae, C.W. Lee, S. Jeong, and K.S. Kim, *Science*, 294, 348 (2001).
14. S. Bhattacharrya, S.K. Saha, and D. Chakravorty, *Appl. Phys. Lett.* 76, 3896 (2000).
15. G. Sauer, G. Brehm, S. Schneider, K. Nielsch, R.B. Wehrspohn, J. Choi, H. Hofmeister, and U. Gosele, *J. Appl. Phys.* 91, 3243 (2002).
16. L.K. Tan, A.S.M. Chong, X.S.E. Tang, and H. Gao, *J. Phys. Chem. C* 111, 4964 (2007).
17. J. Wang, M. Tian, T.E. Mallouk, and M.H.W. Chan, *J. Phys. Chem. B* 108, 841 (2004).

18. P.L. Gai and M.A. Harmer, *Nano Lett.* 2, 771 (2002).
19. P.A. Smith, C.D. Nordquist, T.N. Jackson, T.S. Mayer, B.R. Martin, J. Mbindyo, and T.E. Mallouk, *Appl. Phys. Lett.* 77, 1399 (2000).
20. Y. Kondo and K. Takayanagi, *Science* 289, 606 (2000).
21. B.M.I. van der Zande, L. Pages, R.A.M. Kikmet, and A. van Blaaderen, *J. Phys. Chem. B*, 103, 5761 (1999).
22. V.M. Cepak and C.R. Martin, *J. Phys. Chem. B* 102, 9985 (1998).
23. S. Lu, K. Sivakumar, and B. Panchapakesan, *J. Nanosci. Nanotechnol.* 7, 2473 (2007).
24. F. Neubrech, T. Kolb, R. Lovrincic, G. Fahsold, A. Pucci, J. Aizpurua, T.W. Cornelius, M.E. Toimil-Molares, R. Neumann, and S. Karim, *Appl. Phys. Lett.* 89, 253104 (2006).
25. J. Zhang, J. Sun, W. Liu, S. Shi, H. Sun, and J. Guo, *Nanotechnology* 16, 2030 (2005).
26. Y. Wang, J. Yang, C. Ye, X. Fang, and L. Zhang, *Nanotechnology* 15, 1437 (2004).
27. M.E.T. Molares, E.M. Hohberger, C. Schaeflein, R.H. Blick, R. Neumann, and C. Trautmann, *Appl. Phys. Lett.* 82, 2139 (2003).
28. M.E.T. Molares, V. Buschmann, D. Dobrev, R. Neumann, R. Scholz, I.U. Schuchert, and J. Vetter, *Adv. Mater.* 13, 62 (2001).
29. C.F. Monson and A.T. Woolley, *Nano Lett.* 3, 359 (2003).
30. H. Choi and S.H. Park, *J. Am. Chem. Soc.* 126, 6248 (2004).
31. Z. Liu and Y. Bando, *Adv. Mater.* 15, 303 (2003).
32. Z. Liu and Y. Bando, *Chem. Phys. Lett.* 378, 85 (2003).
33. Z. Liu, Y. Yang, J. Liang, Z. Hu, S. Li, S. Peng, and Y. Qian, *J. Phys. Chem. B* 107, 12658 (2003).
34. M.Y. Yen, C.W. Chiu, C.H. Hsia, F.R. Chen, J.J. Kai, C.Y. Lee, and H.T. Chiu, *Adv. Mater.* 15, 235 (2003).
35. Q. Lu, F. Gao, and D. Zhao, *Nano Lett.* 2, 725 (2002).
36. W. Steinhogl, G. Schindler, G. Steinlesberger, and M. Engelhardt, *Phys. Rev. B* 66, 075414 (2002).
37. I. Lisiecki, H. Sack-Kongehl, K. Weiss, J. Urban, and M.P. Pileni, *Langmuir* 16, 8807 (2000).
38. S.S. Oh, D.H. Kim, M.W. Moon, A. Vaziri, M. Kim, E. Yoon, K.H. Oh, and J.W. Hutchinson, *Adv. Mater.* 9999, 1 (2008).
39. J.L. Lin, D.Y. Petrovykh, A. Kirakosian, H. Rauscher, F.J. Himpsel, and P.A. Dowben, *Appl. Phys. Lett.* 78, 829 (2001).
40. R. Sanz, M. Vazquez, K.R. Pirota, and M. Hernandez-Velez, *J. Appl. Phys.* 101, 114325 (2007).
41. F. Tian, J. Zhu, and D. Wei, *J. Phys. Chem. C* 111, 6994 (2007).
42. F. Tian, J. Zhu, D. Wei, and Y.T. Shen, *J. Phys. Chem. B* 109, 14852 (2005).
43. X.W. Wang, G.T. Fei, X.J. Xu, Z. Jin, and L.D. Zhang, *J. Phys. Chem. B* 109, 24326 (2005).
44. H. Pan, H. Sun, C. Poh, Y. Feng, and J. Lin, *Nanotechnology* 16, 1559 (2005).
45. A.K. Bentley, A.B. Ellis, G.C. Lisensky, and W.C. Crone, *Nanotechnology* 16, 2193 (2005).
46. M. Knez, A.M. Bittner, F. Boes, C. Wege, H. Jeske, E. Mai, and K. Kern, *Nano Lett.* 3, 1079 (2003).
47. K. Nielsch, R.B. Wehrspohn, J. Barthel, J. Kirschner, U. Gosele, S.F. Fischer, and H. Kronmuller, *Appl. Phys. Lett.* 79, 1360 (2001).

48. A.J. Yin, J. Li, W. Jian, A.J. Bennett, and J.M. Xu, *Appl. Phys. Lett.* 79, 1039 (2001).
49. X.Y. Yuan, G.S. Wu, T. Xie, Y. Lin, and L.D. Zhang, *Nanotechnology* 15, 59 (2004).
50. M.Z. Atashbar, D. Banerji, S. Singamaneni, and V. Bliznyuk, *Nanotechnology* 15, 374 (2004).
51. M.P. Zach, K.H. Ng, and R.M. Penner, *Science* 290, 2120 (2000).
52. S. Vaddiraju, H. Chandrasekaran, and M.K. Sunkara, *J. Am. Chem. Soc.* 125, 10792 (2003).
53. X.W. Wang, G.T. Fei, K. Zheng, Z. Jin, and L.D. Zhang, *Appl. Phys. Lett.* 88, 173114 (2006).
54. S. Kar, T. Ghoshal, and S. Chaudhuri, *Chem. Phys. Lett.* 419, 174 (2006).
55. Y. Tong, M. Shao, G. Qian, and Y. Ni, *Nanotechnology* 16, 2512 (2005).
56. J.G. Wang, M.L. Tian, N. Kumar, and T.E. Mallouk, *Nano Lett.* 5, 1247 (2005).
57. E.A. Olson, M.Y. Efremov, M. Zhang, Z. Zhang, and L.H. Allen, *J. Appl. Phys.* 97, 034304 (2005).
58. Y. Wu and P. Yang, *Adv. Mater.* 13, 520 (2001).

9

Oxide Nanowires

9.1 Introduction

Metal oxides are an important class of materials that find applications in a wide variety of fields. For example, ZnO is a direct, wide band gap semiconductor with applications in ultraviolet (UV) and near UV photonics. Many metal oxides find applications either as catalyst supports or active catalyst materials and conducting channels in gas sensors. Several metal oxides also find direct applications in electrochemical energy devices such as lithium ion batteries, electrode materials, and solar cells as will be discussed later in Chapter 16. Finally, the metal oxides are also used as structural materials and composites. Almost all of the above applications can benefit from the availability of metal oxides in the form of nanowires due to various reasons: well-defined crystallinity and surface, reactivity, electronic properties, and high surface area. In some cases, metal oxide nanowires with diameters less than few nanometers are sought for understanding the quantum confinement effects. Even though, nanowires with radius larger than the corresponding Bohr radius may not show distinct properties compared to bulk, they are still promising due to the following reasons: fast charge transport, low thresholds for percolation, high surface area, and well-defined crystal surfaces.

The above-mentioned applications are diverse and require metal oxide nanowires in different formats ranging from epitaxial nanowire arrays, vertical nanowire arrays on a variety of conducting substrates, networked thin films, highly dense thick films, and powders. For epitaxial nanowire arrays, there should be a minimal lattice mismatch between the nucleating metal oxide crystal and the underlying single-crystal substrates. For vertical nanowire arrays, high nucleation densities are typically necessary. A more detailed discussion will be provided under different categories of the metal oxide nanowire growth schemes. The discussion on the synthesis of different metal oxide nanowires is divided into two main categories: the first one refers to metal oxides in which the corresponding metals melt at the temperatures used in the synthesis procedures and they are referred as "low-melting-point metals." The second category refers to metal oxides in which the corresponding metals do not melt at the temperatures used in the respective synthesis procedures and they are referred as "high-melting-point metals." Examples of low-melting point metal oxides include Ga_2O_3, In_2O_3, Al_2O_3, MgO, SnO_2, and ZnO whereas examples of high-melting point metal oxides include WO_3, Ta_2O_5, NiO, and MoO_3.

Many synthesis methods yield pure phase and binary oxide nanowires more easily than oxides with different compositions. So, there is enough interest in understanding whether one can obtain doped and alloy oxide nanowires either during synthesis or using a postsynthesis method. In addition, the synthesis of oxide nanowires is typically easier than the corresponding nitrides, sulfides, etc. So, a section is devoted to discuss the possibility of converting oxide nanowires into other compound nanowires through direct reactions using tungsten oxide–nitride system as an example.

9.2 Synthesis Methodologies

In the case of oxide nanowires, there are five important vapor-phase methodologies used for their synthesis: (a) foreign metal catalyst assisted; (b) direct oxidation of large, molten metal droplets; (c) self-catalyzed; (d) direct chemical vapor transport or chemical vapor deposition (CVD); and (e) thermal/plasma oxidation of metal foils. In each of these synthesis methodologies, a general description of the process followed by experimental conditions and growth mechanisms will be discussed in detail.

9.2.1 Catalyst-Assisted Synthesis

The CVD of metal oxides when performed on substrates coated with a thin film or layer of catalyst particles can lead to metal oxide nanowire growth with catalyst particles at their tips. However, the metal containing gas or liquid precursors are not readily available for deposition of many metal oxides. So, it is easier and inexpensive to start with either solid metal or metal oxide powder sources. The catalyst is typically deposited onto substrates as thin films (<100 nm thickness) either using sputtering or evaporation. In some cases, dispersions of catalyst particles can be synthesized or purchased separately and spin-coated onto the substrates.

Many of the catalyst-assisted synthesis techniques for metal oxide nanowires use thermal CVD in a quartz tube reactor as shown in the schematic in Figure 9.1. This setup includes a horizontal tube heated with a two-zone furnace with one zone for evaporation of solid metal sources and another zone for growth on substrates coated with the catalyst layer. Depending upon the feed source and the type of metal, the distance between zones, d, is important. The metal sources are typically either metal oxide or metal powders. It is possible to reduce the required source temperature for zone 1

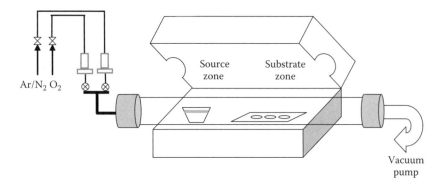

FIGURE 9.1
A schematic showing a typical horizontal quartz tube reactor used for the catalyst-assisted synthesis of nanowires.

by using either carbothermal reduction with carbon or using reaction with halides with source to generate metal containing vapors. The gas sources are typically oxygen, air, or water vapor diluted in a carrier gas. The temperatures for sources and substrates can be anywhere from 400°C to 1200°C depending upon the metal oxides and on the catalyst metal used. The substrate temperature in zone 2 needs to be higher than the melting points of both the catalyst metal used and the eutectic alloy between the source metal and the catalyst metal.

The metal oxide nanowire growth when using catalyst metals occurs according to the typical vapor–liquid–solid (VLS) mechanism similar to that described in earlier chapters. However, it is not completely clear on the composition of the molten catalyst droplet at the tip of the growing metal oxide nanowire. It is expected that the composition will include both the catalyst metal and the metal being deposited. Typically, it has been observed that in the first step, there would always be a metal/catalyst alloy particle formation followed by dissolution of oxygen species into these droplets. Similar to other VLS schemes, the CVD of metal oxides must be performed under conditions where the sticking coefficient of both metal and oxygen species is highly selective on the catalyst metal droplet surface (typically, such conditions correspond to slow growth rates for blanket deposition). Otherwise, the metal oxide films will deposit ubiquitously all over the substrate.

Majority of studies on metal oxide nanowire synthesis using catalysts have been performed using substrates such as quartz, sapphire, and silicon in which nanowires are typically in a thin film format, i.e., nanowires in random directions. In addition, there have been a few studies on obtaining both vertical arrays and epitaxial array growth for metal oxides using catalysts [1–8]. Many of these studies were focused on the synthesis of zinc oxide (ZnO) nanowires due to interest in ZnO as an optoelectronic material. One example involves the use of carbothermal reduction of zinc oxide powders at temperatures ranging from 800°C to 1000°C using argon carrier gas with the silicon substrate kept at a distance ranging from 0.5 to 10 cm from the source [1]. At the synthesis temperature, carbothermal reduction of zinc oxide powders yield zinc vapor according to the following reactions:

Carbothermal reduction (source): ZnO (s) + C (s) → Zn (v) + CO (g)

Catalyst alloy formation: Zn (v) + Au (s) → Au–Zn (l)

Nanowire growth: Au–Zn (l) + Zn(v) + $1/2O_2$(v) → Au–Zn (l) + ZnO (s)

In the initial stage of temperature ramping, the gold catalyst layer breaks down into gold clusters and alloys with zinc through reactions with zinc vapor. Subsequent supply of zinc and oxygen vapors over the molten alloy (Au + Zn) droplets is expected to yield ZnO nanowire growth. For example, this approach on Si substrates coated with a 5 nm thick gold catalyst layer produced randomly oriented nanowires with diameters ranging from 80 to 120 nm and lengths over 10–20 μm as shown in Figure 9.2a. The use of thinner catalyst layer (3 nm) resulted in thinner ZnO nanowires with diameters in the range of 40–70 nm. In order to synthesize epitaxial metal oxide nanowire arrays using a similar kind of catalyst-assisted synthesis technique, the substrate selection plays the most important role. The carbothermal reduction of zinc oxide onto gold coated, *a*-plane sapphire substrate resulted in the synthesis of epitaxial growth of zinc oxide nanowire arrays [2–3]. In the case of *a*-plane sapphire substrate, the minimal lattice mismatch between the *c*-plane ZnO (0001) and the *a*-plane sapphire (110) resulted in the synthesis of epitaxial nanowire arrays. As shown in Figure 9.2b, all the zinc oxide nanowires are epitaxially oriented on the *a*-plane sapphire substrate with diameters in the range of 20–150 nm and lengths over several microns. In another study, a similar kind of carbothermal reduction of the zinc oxide resulted in epitaxial growth of zinc oxide nanowire arrays on gold catalyst patterned silicon carbide substrate [5]. In these experiments, the substrate was patterned

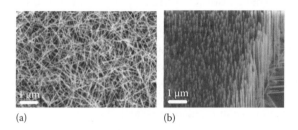

(a)   (b)

FIGURE 9.2
Catalyst-assisted synthesis of ZnO nanowires using carbothermal reduction for source vapors: (a) SEM image of randomly oriented, ZnO nanowire films synthesized on top of a gold-coated silicon substrate; and (b) SEM image of epitaxial ZnO nanowire arrays on gold-coated *a*-plane, sapphire substrate. (Reproduced from Huang, M.H. et al., *Adv. Mater.*, 13, 113, 2001. With permission.)

FIGURE 9.3
SEM image showing epitaxial ZnO nanowire arrays on gold-coated silicon carbide (SiC) substrate grown using a carbothermal reduction route. (Reproduced from Ng, H.T. et al., *Nano Lett.*, 4, 1247, 2004. With permission.)

with gold catalyst pads (7 μm×7 μm) with a diameter of 200 nm and thickness of 1.5–2 nm. As shown in Figure 9.3, the zinc oxide nanowire arrays with diameters of about 40 nm resulted in each region coated with gold with a growth rate of 1 μm/h. The epitaxial nanowire growth was explained based on the minimal lattice mismatch of 5.5% between the c-plane (0001) zinc oxide and the SiC(0001) plane of the substrate [5–6]. Also, other metal vapor phase sources like metal oxide powders (zinc oxide), metal halides (zinc chloride), and metalorganic chemical vapor deposition (MOCVD) precursors (diethyl zinc) have been employed for the nanowire synthesis [4,30]. The latter two sources can produce the metal vapors at relatively lower synthesis temperatures, thereby allowing the use of substrates that do not withstand high temperatures.

The synthesis of epitaxial nanowire arrays of other oxides such as indium and tin oxides has also been demonstrated. The experiments for epitaxial nanowire arrays of tin oxide were conducted using carbothermal reduction and single-crystal, sapphire substrates. The experiments were conducted using argon as a carrier gas for metal vapors resulting from carbothermal reduction of SnO_2 powder source mixed with graphite at 1:1 ratio kept at a temperature of 1000°C. Gold-coated a-plane sapphire substrates (T_{sub}=840°C) were placed at a distance of about few centimeters from the source [7]. The experiments for a duration of 2 h resulted in nanowire arrays as shown in Figure 9.4a. The resulting nanowire arrays were at an angle of 45° with the substrate, indicating heteroepitaxial nanowire growth [7]. This is because the nanowires were grown along <100> direction epitaxially on a-plane of

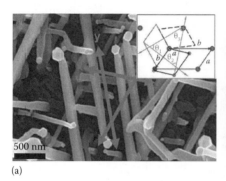 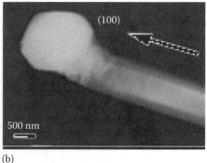

(a)   (b)

FIGURE 9.4

(a) SEM image of epitaxial, tin oxide nanowire arrays grown on *a*-plane, sapphire substrate. Inset showing overlay of unit cell of (110) plane of sapphire with that of (110) plane of rutile SnO_2 (atoms in the red represent oxygen of sapphire and gray atoms represent tin). (b) TEM image of a single nanowire with catalytic alloyed particle at the tip of the nanowire. (Reproduced from Nguyen, P. et al., *Nano Lett.*, 3, 925, 2003. With permission.)

the sapphire substrate. The unit cell of (110) SnO_2, represented by the gray Sn atoms with lattice constants *a* and *b*, is found to match reasonably well along three different directions (also the directions of (100) plane with respect to (110)) with that of the *a*-sapphire (lattice constants *a* and *b*) and found to agree with the experimental values. Lattice mismatches between *ba*, *ca*, and *aa* are about 6.03%, 6.03%, and 8.92%, respectively, with $\theta_1 = \theta_2 = 64.56°$ and $\theta_3 = 50.88°$ and agree well with the experimentally observed values. Due to mirror images of the unit cell, nanowires can be grown in opposite directions along each *x*-, *y*-, and *z*-direction. The nanowire growth along the *x*-direction is not as structurally uniform, most likely due to the larger lattice mismatch along this particular direction. On the (100) sapphire (*m* face), however, preferred lattice matches with SnO_2 (110) are found only along two major directions that are orthogonal to each other. Lattice mismatches between *2yy* and *xy* are 1.89% and 3.14%, comparatively lower than on *a*-sapphire. This evidently explains the different directional growth phenomena and the more favorable growth leading to higher nanowire density in the latter. The angle between (110) and (100) of SnO_2 is calculated to be exactly 45°, which coincides well with the nanowires projected angles. As shown in Figure 9.4b, the TEM image of a single tin oxide nanowire shows catalyst particle at the tip of the nanowire. The synthesized nanowires are heteroepitaxially oriented but only on substrate surfaces that exhibit slight lattice mismatch with the metal oxide. A similar kind of carbothermal reduction procedure of indium oxide resulted in the synthesis of epitaxial nanowire arrays on *a*-plane sapphire substrate. As shown in Figure 9.5, the synthesized epitaxial nanowire arrays are tapered with catalyst particles at the tip of the nanowires. The synthesized nanowires are tapered with an aspect ratio of 4, where the (100) plane of indium oxide crystal is in heteroepitaxy with the underlying (11–20)

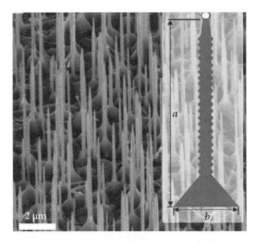

FIGURE 9.5
SEM image of epitaxial indium oxide nanowire arrays grown on gold catalyst patterned *a*-plane, sapphire substrate using a carbothermal reduction procedure. (Reproduced from Nguyen, P. et al., *Nano Lett.*, 4, 651, 2004. With permission.)

plane of the sapphire substrate with a minimal lattice mismatch. The tapered indium oxide nanowires resemble a pillar having nanothreads cladding it along the axial direction of the nanowire [8].

As mentioned earlier, the source generation in metal oxide nanowire growth can take many forms besides carbothermal reduction, such as direct thermal evaporation of metal powder and laser ablation [9–12]. A metal source temperature of 900°C is typically employed in direct metal evaporation of Ga, In, and Sn powders. Due to the difference in the vapor pressures of the three metal sources (In, Ga, and Sn) at the source temperature, growth rates differ for gallium, indium, and tin oxide nanowires respectively [9]. In the case of cadmium oxide (CdO), randomly oriented nanowires have been synthesized at a source (CdO/C mixture) temperature of 500°C, substrate temperature of 400°C, and Ar/O_2 (250/0.32 sccm) as the carrier gas [10]. In the case of zinc oxide, randomly oriented nanowires have also been synthesized using an oxide catalyst [11]. Also, indium-doped tin oxide nanowire arrays were synthesized using thermal evaporation of SnO/In powder mixtures onto gold-coated Yttria-stabilized zirconia (YSZ) substrates [13–14]. The process parameters employed for the nanowire synthesis were source (SnO/In mixture) temperature of 900°C, 500 sccm of nitrogen (N_2), and experimental duration of 2 h. In the first step, an indium tin oxide (ITO) buffer layer with (100) orientation was deposited using pulsed laser deposition (PLD) for lattice-matching with the substrate. In the next step, thermal evaporation of powder mixtures onto the buffer layer coated substrates resulted in epitaxial growth with (100) growth direction. Experiments performed at similar process conditions without the deposition of the buffer layer on top of YSZ substrates yielded a web-like nanowire morphology. In addition, the synthesis of ternary metal oxide nanowires has been claimed using catalyst-assisted synthesis procedure similar to that described above [15–16].

9.2.2 Direct Oxidation Schemes Using Low-Melting Metals

The synthesis procedures involving direct oxidation of metals are gaining importance compared to catalyst-assisted schemes mainly because of two reasons: (1) possible contamination with foreign catalyst metals in the catalyst-assisted schemes and (2) the costs associated with the respective catalyst metals and corresponding processes. In this section, two types of direct oxidation schemes for oxides of low-melting point metals are discussed: (a) direct oxidation of large (micron scale) molten metal droplets to produce the respective metal oxide nanowires; and (b) chemical/reactive vapor transport of low-melting metal and oxygen vapors onto substrates kept at temperatures higher than the melting point of the metal.

9.2.2.1 Direct Oxidation of Molten Metal Clusters

The solubilities of oxygen and the respective oxides in the low-melting point metal melts are expected to be negligible. So, any amount of dissolution of oxygen species during direct oxidation of large, micron-scale molten metal droplets leads to spontaneous nucleation of nanometer scale, metal oxide nuclei from molten metal. Further growth of metal oxide nuclei via basal attachment leads them to one-dimensional structures. High density of nucleation followed by growth via basal attachment (from dissolved oxygen) from large, molten metal clusters results in flowery morphology in which a high density of nanowires emanates from one point. After nucleation stage, further growth of metal oxide nuclei can also occur laterally in addition to basal growth. Such growth behavior leads to polycrystalline crust formation on large molten metal clusters. It has been found that the presence of reducing species such as hydrogen or chlorine reduces the lateral propagation of metal oxide nuclei. This is schematically illustrated in Figure 9.6 [17].

The direct oxidation scheme concept for the nanowire growth can be understood with the following example—direct oxidation of molten gallium droplets using microwave plasma containing oxygen and hydrogen. In these experiments, Ga-film-coated quartz substrates were exposed to microwave plasma at 700 W power, 40 Torr pressure with a substrate temperature of 550°C, and gas flow rates of 8 sccm of O_2 in 100 sccm of H_2. Quartz substrates were coated with thick Ga film by spreading molten Ga with the help of a hot plate. In the initial step, the gallium-coated substrates were exposed to pure hydrogen plasma for the formation of large spherical metal droplets (micron scale) over the substrate. In the next step, the hydrogen and oxygen plasma exposure for about 30 min resulted in complete oxidation of micron-sized droplets. The resulting morphologies as shown in Figure 9.7a and b indicate

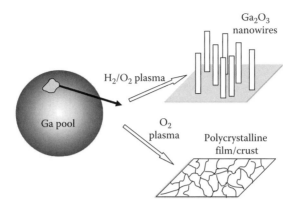

FIGURE 9.6
A schematic illustrating the effect of hydrogen or other reducing gas-phase precursors during direct oxidation schemes for bulk nucleation and growth of oxide nanowires from their respective molten metal droplets using Ga as an example.

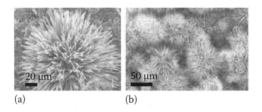

(a)   (b)

FIGURE 9.7
SEM images showing a high density of gallium oxide nanowires growing out of micron-scale Ga droplets. (a) High-magnification SEM image; and (b) Low-magnification image showing flowery growth from several Ga droplets. (Reproduced from Sharma, S. and Sunkara, M.K., *J. Am. Chem. Soc.*, 124, 12288, 2002. With permission.)

high densities of nanowires emanating from each droplet. As described earlier, the flowery type morphologies indicate high density of nucleation from micron-size, molten metal droplets followed by growth via basal attachment. During the oxidation of gallium droplets, it was observed that there was no weight loss of the gallium before and after oxidation. This led to the conclusion that there were no Ga-containing species in the gas phase at these temperatures confirming that the growth primarily occurred via basal attachment, i.e., at the molten metal–nanostructure interface. Due to the high reactivity of these low-melting metals, only few substrates like sapphire, pyrolytic boron nitride, and highly ordered pyrolytic graphite (HOPG) can be used without the use of any oxide buffer layers. In all other cases, substrates coated with an oxide buffer layer have to be used to support molten metal droplets for direct oxidation experiments.

Several experiments performed using a range of process conditions of 600–1200 W, 30–60 Torr, and 0.6 sccm O_2 in 100 sccm of H_2 led to several different

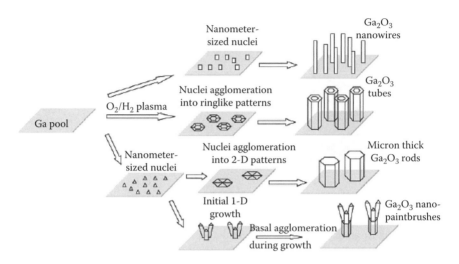

FIGURE 9.8

A model of nuclei dynamics on molten metal surfaces during initial stages for explaining various one-dimensional morphologies observed during bulk nucleation and growth from large molten metal droplets. (Reproduced from Sharma, S. and Sunkara, M.K., *J. Am. Chem. Soc.*, 124, 12288, 2002. With permission.)

and interesting morphologies. The resulting morphologies showed thicker rods, hollow tubular, and paintbrush-like nanostructures. The observation of such distinct morphologies is a likely result of an interesting phenomenon of self-assembly process. As explained above, high density of nanometer-scale metal oxide nuclei occur on top of molten metal surface and lead to one-dimensional structures due to growth via basal attachment. The growth of different kind of morphologies can be explained based on the initial stages of nucleation, where the nanometer-scale metal oxide nuclei assemble to form agglomerates with different, two-dimensional morphologies as illustrated in the schematic in Figure 9.8. Further growth of such self-assembled structures can lead to differences in the morphologies such as paintbrush-like, hollow tubular, and well faceted, thick rods as seen in Figure 9.9. As shown in Figure 9.9d, the SEM image of the backside of the tubular morphology indicates the agglomerate pattern that led to spiraling tubular morphology.

The direct oxidation schemes were successfully employed for the synthesis of metal oxide nanowires of different metals such as Sn, Al, Zn, Bi, and In. In addition, direct thermal heating of metal powders and carbothermal reduction of different low-melting metal oxide powders like gallium oxide and indium oxide powders were also employed for direct oxidation schemes [18–21]. In the carbothermal reduction of gallium oxide powders, a more direct evidence of flowery kind of nanowire morphology growth from molten gallium pools was observed [18]. In the case of direct thermal heating, metal powders like gallium, magnesium, and zinc were heated in oxygen gas phase environment at temperatures above their melting points, resulting in oxide

FIGURE 9.9

SEM images showing paintbrush-like morphology for one-dimensional, gallium oxide structures using direct oxidation of Ga droplets: (a) Cluster of gallium oxide paintbrushes; (b) an individual gallium oxide paintbrush-like morphology; (c) SEM image showing gallium oxide nanotubes synthesized with direct oxidation of large Ga droplets with oxygen plasma. (d) SEM image showing the backside of the tube showing the nuclei agglomeration pattern in a spiral fashion that resulted in the complicated tubular morphology shown in (c). (Reproduced from Sharma, S. and Sunkara, M.K., *J. Am. Chem. Soc.*, 124, 12288, 2002. With permission.)

nanowires [19]. In all of the direct oxidation schemes, the addition of chlorine (by placing KCl or NaCl in the reactor) also helped with the production of one-dimensional structures, reducing the diameter of resulting nanowires and the required synthesis temperature. The use of halides and hydrogen helps in preventing the lateral growth of metal oxide nuclei during their growth on molten metal droplets, thus reducing the diameter of the resulting nanowires. In the absence of hydrogen or halides in the gas phase, the direct oxidation leads to crust formation of molten metal clusters rather than the growth of one-dimensional structures. But, at high-enough synthesis temperatures, the direct oxidation can also lead to one-dimensional structures from molten metal clusters but with much larger diameters.

9.2.2.2 Direct Chemical/Reactive Vapor Deposition of Low-Melting Metal Oxides

The direct oxidation of low-melting metals can also be carried out using either chemical or reactive vapor transport of metal vapor onto substrates in the presence of oxygen containing gas phase. Different synthesis procedures like thermal evaporation, carbothermal reduction, and pulsed laser ablation of targets can be employed for the reactive vapor transport of the metal containing vapors onto substrates containing no foreign catalyst layers [22–28].

As there is no foreign molten metal catalyst used during synthesis, it is common to misrepresent this as a "vapor–solid" scheme. In contrast, these schemes can still be categorized as VLS methods because the melting points of the respective metals are below the synthesis temperatures used. These methods can be referred as "self-catalytic" schemes because the metal oxide nanowire growth is led by the corresponding metal droplet and not by a foreign metal cluster. However, thermal evaporation of metal oxide powders onto catalyst-coated substrates can also be categorized under these self-catalytic schemes because of growth of an oxide buffer layer on top of the catalyst-coated substrates [25–26]. The overall mechanism of self-catalyzed growth shown in Figure 9.10 is as follows: the metal or metal oxide or metal organic vapors react on the substrate to form metal oxide crystals; metal droplets form on these crystals and selective growth at the molten metal droplet and the underlying crystal occurs through "liquid-phase epitaxy" with exposure to metal and oxygen containing gas-phase species. Such selective growth underneath the molten metal droplets leads to metal oxide nanowire growth with molten metal droplets at their tips.

Growth of tin oxide nanowires using a reactive vapor transport–assisted direct oxidation in a self-catalytic fashion is discussed in the following. The process involves evaporation of tin from a heater cup onto quartz substrates as shown in the schematic in Figure 9.11a. The tin-containing cup was heated using a resistive heater to about 1400°C. The tin metal vapor reacted with oxygen on a quartz substrate maintained at temperatures around 850°C in the presence of gas mixtures of 95% oxygen and 5% hydrogen [29]. The above experiment yielded randomly oriented, long (>50 μm) nanowires with diameters ranging from 15 to 50 nm as seen in Figure 9.11b. In some experiments, tin oxide nanowires with molten metal tin droplets at the tip of the

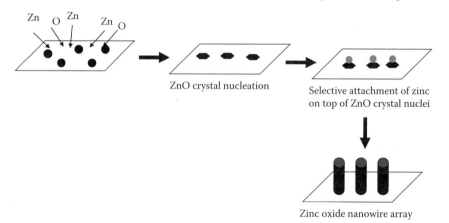

ZnO crystal nucleation

Selective attachment of zinc
on top of ZnO crystal nuclei

Zinc oxide nanowire array

FIGURE 9.10

A schematic representing the nucleation and growth mechanism of metal oxide nanowires like zinc oxide and tin oxide in an array fashion using self-catalysis schemes. (Reproduced from Ye, Z.Z. et al., *Solid State Commun.*, 141, 464, 2007. With permission.)

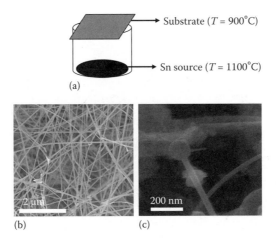

FIGURE 9.11

(a) A schematic of the experimental setup used for reactive vapor transport experiments involving tin in oxygen environment. (b) SEM image of the resulting tin oxide nanowires synthesized. (c) SEM image of tin oxide nanowires showing the presence of spherical tin cluster at the tip of the nanowire.

nanowires were observed (Figure 9.11c). Alternatively, the carbothermal reduction approach using a mixture of tin oxide and graphite powders also yielded randomly oriented tin oxide nanowires based on the source–substrate distance [22].

The self-catalysis schemes, unlike bulk nucleation and growth from large molten metal droplets, can lead to both vertical arrays on nonsingle-crystal substrates and epitaxial arrays on single-crystal templates. In the case of single-crystal substrates, the metal oxide nuclei can occur in epitaxy with the underlying substrate and further growth via liquid-phase epitaxy through molten metal droplet will lead to nanowires in epitaxial relationship with the substrates. In the case of noncrystalline substrates, the conditions leading to high nucleation densities in the initial stages can lead to vertical array growth. For example, ZnO nanowire array growth has been demonstrated by direct, thermal CVD using metalorganic precursors [30,31]. In this case, diethyl zinc as zinc precursor and an oxygen precursor (N_2O or O_2 gas) were used in a low-pressure MOCVD system. A buffer layer of ZnO was grown at a low deposition temperature of 400°C and 60 Torr pressure on Si substrate to protect it from Zn droplets at higher synthesis temperatures used for nanowire growth. Later, the MOCVD of ZnO onto ZnO buffer layer on Si substrate at a higher substrate temperature of 650°C resulted in ZnO nanowire arrays as shown in Figure 9.12 [31]. The SEM images of resulting ZnO nanowires did not show the presence of Zn droplets at their tips [31].

A similar observation has been reported even with synthesis of several kinds of low-melting metal oxides (gallium oxide, indium oxide) and nitrides (gallium nitride, indium nitride). For example, the self-catalyzed growth

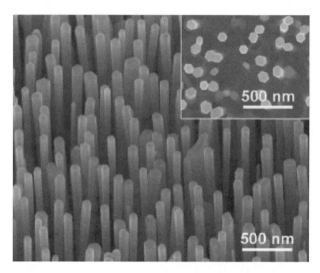

FIGURE 9.12
SEM image of ZnO nanowire arrays synthesized using the reactive vapor transport approach. (Reproduced from Ye, Z.Z. et al., *Solid State Commun.*, 141, 464, 2007. With permission.)

of GaN nanowires in *a*-plane orientation also did not show the presence of Ga droplets at their tips [32]. In all the nanowire synthesis procedures involving low-melting metals, it is possible to lose the molten metal droplet through vapor pressure and flow onto the surface due to the high synthesis temperatures used within the timescales involved with experimental shutdown.

9.2.3 Chemical Vapor Transport or Deposition of High-Melting Metal Oxides

In the case of high-melting point metals (for example, tungsten, tantalum, and molybdenum), their melting points are always much higher than the synthesis temperatures used in any procedure. For such metals, the respective oxide vapors can easily be produced through oxidation of the corresponding metal at temperatures much lower than their melting points of metals. The transport of the respective metal oxide vapors onto the substrates can be used to create metal oxide films similar to any CVD scheme. However, it has been shown that under particular circumstances with low oxygen partial pressure, such chemical vapor transport procedures, resulted in the synthesis of respective oxide nanowires without the aid of any external catalyst. Such a scheme can be appropriately described as "vapor–solid" since the respective metal is not molten at the synthesis temperature. The underlying mechanism leading to one-dimensional growth in such "vapor–solid" schemes still needs to be understood. In a typical chemical vapor transport scheme, the oxide vapors from a source are transported on to substrates kept

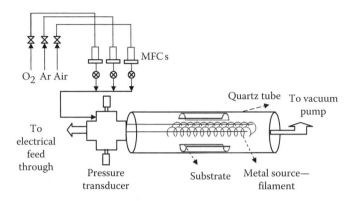

FIGURE 9.13
A schematic of the hot-wire CVD reactor setup used for chemical vapor transport experiments using a variety of high-melting point metals as filaments. (Reproduced from Thangala, J. et al., *Small*, 3, 890, 2007. With permission.)

at a temperature lower than that of the source. In terms of metal sources, one can use a number of materials that include metal powders, metal oxide powders, carbon mixed with oxide powders, metal organic precursors, and hot-filaments to produce the necessary metal-containing precursors and transport them onto substrates kept at a distance from the source.

All of the above "vapor–solid" observations can be better understood through the hot-wire-assisted CVD-based experiments for tungsten oxide nanowires. The hot wire (or filament) CVD experiments were conducted using oxygen flow over tungsten filaments for the synthesis of tungsten oxide nanowires [33,34]. As shown in Figure 9.13, the reactor consists of a 2 in. diameter quartz tube with the metal filaments wound on top of two ceramic tubes, which are connected to an electrical feedthrough. The filament is resistively heated by flowing electric current and the temperature is controlled by controlling the applied power to the filament using a variac power supply. The quartz tube is placed inside a furnace oven. In addition to the oven, the radiation from the filaments can heat the substrates to high temperatures depending upon their distance from the source. The substrates can be easily heated up to temperatures around 550°C in 2 in. quartz tube without the use of external heating via furnace. Experiments using tungsten filaments and oxygen flow at low partial pressures in the range of 0.05 to 0.3 Torr, filament temperature around 1600°C and substrate temperatures around 400°C to 800°C resulted in the synthesis of tungsten oxide nanowires. The experiments conducted using substrate temperatures lower than 550°C resulted in high nucleation densities and led to vertically oriented nanowire arrays (see Figure 9.14a and b). The vertical nanowire array growth was observed on both amorphous and crystalline substrates.

The experiments conducted using higher substrate temperatures of 750°C with furnace heating resulted in lower nucleation densities, leading to thin

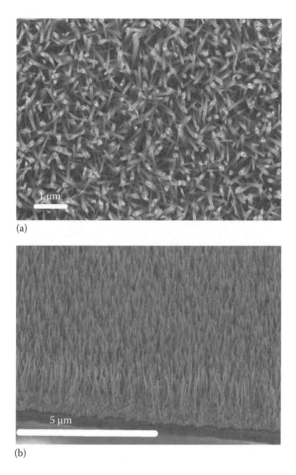

FIGURE 9.14
(a) SEM image showing the top view of tungsten oxide nanowire arrays synthesized on amorphous quartz substrates. (b) SEM image showing the side view of the tungsten oxide nanowire arrays.

films containing randomly oriented, interconnected nanowires (termed as mats), as seen in Figure 9.15. Nanowires of other transition metal oxides (tantalum oxide, nickel oxide, molybdenum oxide) were also synthesized by using the respective metals as filaments. The SEM image for molybdenum oxide nanowire arrays synthesized using Mo filaments is shown in Figure 9.16. For molybdenum oxide nanowires, the process parameters include filament temperature of 750°C, substrate temperature of around 350°C–400°C, 10 sccm of O_2, and experimental duration of 10–20 min. The filament temperature, substrate temperature, and oxygen partial pressure play an important role in determining the density of nucleation and the resulting metal oxide characteristics (nanowires versus nanoparticles) as described below.

FIGURE 9.15
SEM image showing the as-synthesized, interconnected tungsten oxide nanowires using high
substrate temperatures (>700°C) in the hot-wire CVD process.

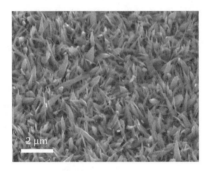

FIGURE 9.16
SEM image showing molybdenum oxide nanowires synthesized using hot-wire-assisted CVD
process.

Thermodynamically, the nucleation density can be modeled as that proportional to the gas-phase supersaturation of solute of interest. In a simplification, the gas-phase composition near filament can be assumed to be
"near" equilibrium due to the high filament temperatures used. Assuming
that there is no gas-phase recombination, the gas-phase composition at the
substrate is same as that at the filament. As the substrate temperature is
always much lower than that of filaments, the gas-phase composition near
the substrate in this technique will be much higher than the expected equilibrium composition at the substrate temperature. Supersaturations for
various tungsten oxide species such as WO_2, W_3O_8, WO, and WO_3 can be
computed using thermodynamic data for each of the species taken from
JANAF thermodynamic tables. Based on the above information, the critical nuclei radius is estimated for metal oxide nuclei condensation from the
vapor phase as the following:

$$R^* = 2\sigma\Omega/\Delta G_v$$

$$\Delta G_v = RT\ln(P/P^*) \qquad (9.1)$$

In the present case of metal filament acting as the source,

$$\Delta G_v = RT\ln(P_f^*/P_s^*) \qquad (9.2)$$

$$\text{Gas-phase supersaturation} = \text{constant} * RT\ln(P_f^*/P_s^*) \qquad (9.3)$$

$$\text{Nanowire density} = \text{constant} * RT\ln(P_f^*/P_s^*) \qquad (9.4)$$

The supersaturation values are estimated as a function of substrate temperature as shown in Figure 9.17a and b. In general, the trends shown in Figure 9.17a for gas-phase supersaturations are consistent with that of the experimental observations shown in Figure 9.17b, i.e., nucleation density of nanowires is higher at lower substrate temperatures [33]. The supersaturation for WO_2 is much higher than that of WO_3 species in the gas phase (Figure 9.17b). So, the WO_2 condenses readily to form WO_2 clusters in the initial stages, which further grow in to tungsten oxide nanowires. It is also likely that these suboxide clusters play a key role in the one-dimensional growth process.

So, the nucleation and growth mechanism for the "vapor–solid" schemes can be explained as that of a "vapor–solid–solid" type mechanism involving suboxide clusters, leading the growth of nanowires. Several steps in this mechanism are shown explicitly in the schematic in Figure 9.18. The main aspects of this mechanism using tungsten oxide (WO_3) system are as follows: (1) In the nucleation step, high gas-phase supersaturation of WO_2 leads to condensation of WO_2 solid clusters onto the substrate. (2) In the second step, selective oxidation at the cluster–substrate interface leads to the precipitation of the WO_{3-x} crystal with WO_2 cluster at the tip. Finally, the enhanced adsorption of WO_2/WO_3 species on the tip relative to the bulk surface followed by further oxidation at the cluster–crystal interface might lead to one-dimensional growth of WO_{3-x} nanowire. The enhanced adsorption at the tip is also due to the competition between the net condensation reaction and the desorption/oxidation reactions on WO_2 cluster in comparison to the WO_{3-x} nanowire and determine the relative rates of axial versus radial growth. In essence, the WO_2 cluster nucleation followed by enhanced precipitation via the amorphous WO_{2+y} cluster (with $2 + y$ less than $3 - x$) at the tip could lead to one-dimensional growth in the so-called vapor–solid schemes involving high melting-point metal oxides [33]. However, there are no direct observations of such clusters at the tips of the resulting nanowires at the end of the experiment. It is likely that such clusters may get oxidized quickly during

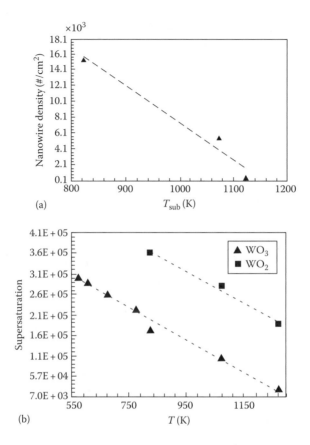

(a)

(b)

FIGURE 9.17
(a) A plot of nanowire density as a function of substrate temperature. (b) A plot of the estimated, theoretical supersaturation for both WO_2 and WO_3 species as a function of the substrate temperature. (Reproduced from Thangala, J. et al., *Small*, 3, 890, 2007. With permission.)

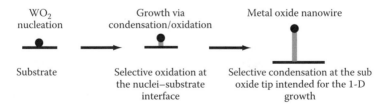

FIGURE 9.18
A schematic illustrating possible steps responsible for one-dimensional growth of transition metal oxide nanowires in the chemical vapor transport approach. (Reproduced from Thangala, J. et al., *Small*, 3, 890, 2007. With permission.)

the shutdown of the experiment. A similar kind of chemical vapor transport experiment using tungsten filaments onto different substrates resulted in nanowires, nanobelts, nanobundles, and other morphologies [35–38].

In addition to binary oxide nanowires, it is also possible to synthesize tungstate nanowires such as $MgWO_4$ using the hot filament–assisted technique [39]. The alloy nanowires resulted under similar as that described earlier for tungsten oxide nanowire arrays. In these experiments, the Mg powders were placed underneath the filament wire, with substrates placed few centimeters from the powder source.

The chemical vapor transport of metal oxide vapors using thermal evaporation using powder sources performed onto gold-coated silicon substrates also result in the synthesis of nanowires [40,41]. The resulting nanowires do not exhibit gold catalyst clusters at their tips, indicating that the gold catalyst does not play a role. In fact, the observations of nanowires on regions not coated with gold catalyst layer also suggest that the underlying growth mechanism is similar to the one discussed in this section. Similar chemical vapor transport experiments using powder sources using bare substrates without any foreign metal catalysts also resulted in respective metal oxide nanowires in the case of tungsten and molybdenum [42–47]. At elevated temperatures, direct reaction of tungsten powder in presence of water vapor resulted in the synthesis of tungsten oxide nanoneedles [48]. In this experiment, the one-dimensional growth probably resulted with chemical vapor transport of tungsten oxide vapors rather than direct reaction of oxygen with tungsten metal. Other types of synthesis methodologies using potassium and boron source powders on top of the tungsten metal plates have also resulted in the synthesis of tungsten oxide nanowires [49–51]. The tungsten oxide nanowire growth was also observed during the heating of tungsten wire covered with boron oxide powders at higher synthesis temperatures [51]. In all these synthesis procedures, the potassium and boron oxides was suspected to provide the molten phase clusters during synthesis. The lack of observations of any spherical clusters containing either potassium or boron at the nanowire tips suggests that these experiments are also similar to those described in this section.

9.2.4 Plasma and Thermal Oxidation of Foils

Another concept that is interesting for synthesizing metal oxide nanowire arrays involves direct oxidation of metal foils. Here, the metal foils are oxidized either thermally or using highly dissociated oxygen plasmas. The underlying nucleation and growth modes in these cases are different from that of "tip-led" growth in catalyst-assisted and self-catalyzed schemes but are somewhat similar to that of bulk nucleation and growth from low-melting metal melts. The main difference is however that the nanowire growth occurs here with direct oxidation of solid metal foils and at temperatures lower than the melting temperatures of the respective metals.

FIGURE 9.19

SEM images of iron oxide nanowires synthesized through plasma oxidation of an iron foil: (a) side view of an iron oxide nanowire array on top of the iron foil substrate; and (b) inset shows SEM image of a single tapered iron oxide nanowire. (Reproduced from Cvelbar, U. et al., *Small*, 4, 1610, 2008. With permission.)

This scheme is illustrated with plasma oxidation of iron foils using highly disassociated oxygen plasma created using an inductively coupled radio-frequency (RF) plasma. These oxidation experiments were conducted in a vacuum chamber equipped with an inductively coupled RF generator with a maximum power of 5 KW, reactor pressure of 2 Pa, and experimental duration of about 2 min [52]. The resulting nanowire arrays are shown in Figure 9.19a and b. Under these experimental conditions, the typical O ion densities, as measured by a catalytic probe, were in the range of 1×10^{16} to 6×10^{16} m^{-3} and the neutral O atom densities were in the range of 1.6×10^{21} to 2.3×10^{21} m^{-3}. The sample was heated with exposure to oxygen radicals in the plasma due to heat resulting from the exothermic reactions of oxygen radical recombination and metal oxidation. The temperature profile of the substrate during oxidation process determines both density and one-dimensional morphology of the resulting nanostructures. As shown in Figure 9.20, the initial temperature increase during the oxidation process determined the nucleation density of the nanowires. The Fe–O phase diagram suggests that a temperature of 580°C is needed for the formation of α-Fe$_2$O$_3$ phase. The experimental data shows a decrease in nanowire density when the initial synthesis temperature was below 570°C. The data further suggests that the substrate temperature beyond initial stages of nucleation must be maintained at temperatures lower than 580°C to achieve one-dimensional growth of nanowire. Again, high substrate temperatures of about 680°C resulted in nanobelt synthesis and substrate temperatures around 780°C resulted in thin film formation with no one-dimensional structures. When the substrate temperature was kept lower than 580°C after initial stages of nucleation, nanowire arrays with high densities can be obtained.

Thus, based on the above experimental observations, the underlying nucleation and growth mechanisms involved with one-dimensional growth

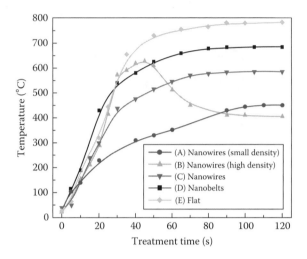

FIGURE 9.20
A plot showing the substrate temperature profiles during various foil oxidation experiments and their correlation to the resulting one-dimensional morphologies. (Reproduced from Cvelbar, U. et al., *Small*, 4, 1610, 2008. With permission.)

are schematically represented in Figure 9.21. In the initial stages, the oxygen radicals can diffuse in the top layers of metal foils, thus supersaturating a thin surface layer of metal foil. As the substrate temperature undergoes a jump beyond metal–metal oxide transition temperature, the supersaturation

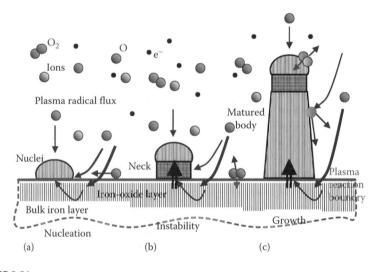

FIGURE 9.21
A schematic illustration of nucleation and growth for the metal oxide nanowires from direct oxidation schemes of bulk foils when exposed to highly dissociated oxygen plasma. Three main stages of nanowire growth are (a) nucleation, (b) instability, and (c) growth. (Reproduced from Cvelbar, U. et al., *Small*, 4, 1610, 2008. With permission.)

of dissolved oxygen will lead to spontaneous nucleation of metal oxide nuclei over the metal foil surface. Subsequent growth of the nuclei will depend heavily upon the substrate temperature. If the substrate is kept lower than the phase transition temperature, then the growth will primarily take place at the metal oxide and metal interface. Such basal growth will lead to one-dimensional growth. Elevated temperatures can make the metal atoms mobile and increase the lateral growth of nanowires. Such lateral growth can lead to either belt-like shape or to tapered shape depending upon the kinetics. At very high substrate temperatures beyond initial stages, high mobility of metal atoms lead to three-dimensional growth of nuclei, thus leading to polycrystalline oxide film growth.

The above scheme also applies to the synthesis of corresponding oxide nanowires for all high-melting-point metals. The concept has been demonstrated for the synthesis of niobium pentoxide nanowire arrays and vanadium pentoxide nanowire arrays using direct oxidation of respective metal foils in an inductively coupled plasma [53]. A similar kind of metal oxide nanowire array growth using direct oxidation of metal foils can be carried out in atmospheric microwave plasma reactors. For example, atmospheric plasmas with some control over substrate temperature could be used for producing metal oxide nanowire arrays directly on the metal foils. Such oxidation schemes using atmospheric plasmas can be simple and scalable. For example, Figure 9.22 shows oxidation of copper, iron, and niobium foils using atmospheric plasma, producing high densities of corresponding metal oxide nanowire arrays.

Thermal oxidation of metal foils is also possible for producing metal oxide nanowire arrays. For example, thermal oxidation of niobium foils (2.5×2.5 mm) inside a horizontal quartz tube furnace reactor at a temperature of 900°C and total pressure of 1 Torr resulted in the synthesis of vertical nanowire arrays on top of the foil [54]. Also, direct thermal oxidation of iron foils have produced both nanowire and nanobelt morphology based on the substrate temperature and the type of gas mixtures (pure O_2 versus mixture of O_2 with inert gases) [55]. The experimental durations for thermal oxidation

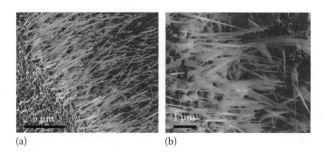

(a)   (b)

FIGURE 9.22
Metal oxide nanowire arrays synthesized by direct oxidation of metal foils using a microwave atmospheric plasma jet reactor. (a) SEM image of the resulting copper oxide nanowire arrays on copper foils. (b) SEM image of the resulting iron oxide nanowire arrays on iron foils.

schemes are in the order of several hours whereas the durations used for plasma oxidation schemes are in the order of a few minutes. Because of such large-scale, time-scale difference, the resulting interfacial layers between nanowire arrays and the underlying metal substrate can be very different in terms of both composition and thickness. Similar experiments using low-melting metal foils will result in the surface melting of foils due to oxidation reaction irrespective of the bulk foil temperature. So, the resulting oxide nanostructure growth will be similar to that of bulk nucleation and growth from low-melting metals. For example, direct oxidation of zinc metal blocks was performed at different temperatures ranging from 200°C to 450°C in presence of air and oxygen gases. The direct oxidation experiments at temperatures of 350°C and 400°C resulted in the synthesis of zinc oxide nanowire arrays on top of the zinc metal plates [56,57]. At the synthesis temperatures, the surface of the metal plates melts, resulting in molten zinc metal formation. The formation of nanowires follows that of bulk nucleation and growth from large molten metal droplets as described earlier for direct oxidation of low-melting metals.

9.3 Directed Growth and Morphological Control

9.3.1 Branched Nanowire Structures

Nucleation and growth of a new nanowire on an existing nanowire lead to branching of the existing nanowires. In the case of catalyst-cluster-assisted techniques, such new nanowires can be initiated by selective deposition of foreign metal catalyst clusters on the existing nanowires. For example, branched indium oxide, gallium oxide, and tin oxide nanostructures were synthesized using catalyst particles deposited on top of the synthesized metal oxide nanowires [58,59]. In both these material systems, the nanowire growth was carried out through direct thermal evaporation of the respective metal powders onto gold-coated substrates. In this case, the size and density of new branches are determined by the catalyst clusters deposited. In noncatalyst-assisted synthesis procedures, branching can only be possible through secondary nucleation events onto the existing nanowire. For example, such branched growth through new nucleation events during CVD experiments was achieved by shutting the reactor on/off or by pulsing the substrate temperature or by pulsing the gas flow into the chamber. As shown in Figure 9.23a and b, the intermittent shutdown of the reactor and re-flowing of the gases resulted in secondary nucleation and led to branching of nanowires [33]. The synthesis of branched nanostructures was carried out by first performing the experiments at the nanowire array synthesis conditions explained in Section 9.2.3. After the shutdown of the reactor, nanowire growth was performed for the second time (same partial pressure of oxygen,

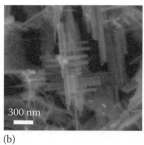

(a)   (b)

FIGURE 9.23
SEM image showing highly branched tungsten oxide nanowires synthesized using a hot-wire-assisted CVD process. (Reproduced from Thangala, J. et al., *Thin Solid Films*, 517, 3600, 2009. With permission.)

filament temperature, and substrate temperature) without taking out the original substrates. During the second set of experiments, renucleation of several suboxide clusters on top of the already synthesized nanowire resulted in the branching of nanowires. A similar kind of intermittent on–off experiments without taking out the substrate resulted in several levels of branching. In the case of tungsten oxide, continuous thermal evaporation also led to new nucleation events on the existing nanowire structures, resulting in the synthesis of three-dimensional, branched nanowires similar to that shown in Figure 9.23a and b [60]. Similar kinds of three-dimensional tungsten oxide tree structures were synthesized by heating of tungsten foil to high temperatures [61].

In the case of low-melting metals, the formation of the respective metal clusters on the metal oxide nanowires can lead to the formation of branches in a similar fashion to that of foreign metal catalysts. In this case, the size and density of resulting branches depends heavily on the process parameters and the interfacial properties between the metal and metal oxide. In a reactive vapor transport approach, nucleation of new molten metal clusters can be initiated using fluctuations of process conditions during the experiment, i.e., source temperature fluctuation. A similar kind of intermittent shutdown experiments performed during the growth of tin oxide nanowires using reactive vapor transport approach resulted in branched nanowires as shown in Figure 9.24a. In the first step, randomly oriented tin oxide nanowire films are synthesized at the experimental conditions described in Section 9.2.2.2. In the second step, experiments performed at the nanowire synthesis conditions onto the already synthesized nanowires resulted in secondary nucleation on top of the nanowires resulting in branching of the nanowires. As shown in Figure 9.24b, intermittent on–off experiments at higher substrate temperatures of 950°C resulted in branched nanowire growth for gallium oxide. Direct carbothermal reduction of a mixture of gallium/indium oxide powders resulted in the growth of heterostructured nanowires [62]. In the first step, carbothermal reduction of metal oxide powders resulted in the formation of molten gallium clusters on top of the substrates placed in

(a)   (b)

FIGURE 9.24
(a) SEM image showing highly branched tin oxide nanowires synthesized using reactive vapor transport approach. (b) SEM image showing highly branched gallium oxide nanowires synthesized using a reactive vapor transport approach.

the downstream side of the chamber. This step is followed by the indium oxide nanobelt growth through the formation of a Ga–In–O eutectic alloy followed by precipitation of InO crystal with gallium led 1-D crystal growth. This step is followed by selective adsorption of gallium oxide vapors on the side facets of the nanobelt, leading to branched gallium oxide nanowires [62]. Direct thermal oxidation of copper foils resulted in the synthesis of branched copper oxide nanostructures [63]. In several studies, a similar kind of unintentional branching can be observed without any high degree of uniformity because of renucleation onto some growing nanowires. In reactive vapor transport schemes involving low-melting-point metals, the low-melting metal droplet formation depends heavily on the process conditions and the interfacial wetting characteristics. If the conditions favor nonwetting, then the diameter of the resulting branch will be smaller and gets bigger in size with further growth. Also, the density of droplets and uniformity of sizes depend upon a number of factors such as adatom mobility and metal vapor flux. So, the self-catalyzed branching of nanowires both for low-melting and high-melting metals offer a wide variety in terms of density, diameters, and tapering of branching, etc. compared to that using foreign metal cluster mediated branching.

9.3.2 Networking of Nanowires

In addition to branching, one can achieve networking of nanowires in the case of low-melting metal oxides by directly oxidizing small clusters of low-melting metals supported on substrates. The density of nucleation from such small clusters (<100 nm size) tends to be few in number and the resulting nanowire growth tends to be closer and parallel to the substrate. Subsequent growth of nanowires from such small clusters can lead to two-dimensional networking of nanowires in a web-like form (nanowebs). This is illustrated here with the synthesis of Ga_2O_3 nanowire network [64]. First, the quartz

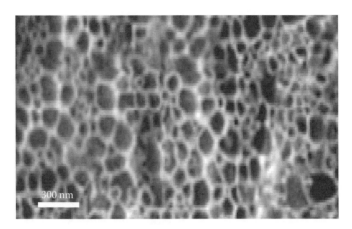

FIGURE 9.25
SEM image showing 3D networking of gallium oxide nanowires ("nanowebs") synthesized using direct oxidation of tiny (<100 nm) metal droplets. (Reproduced from Graham, U.M. et al., *Adv. Funct. Mater.*, 13, 576, 2003. With permission.)

substrate coated with Ga film is exposed to hydrogen plasma to create a high density of small Ga droplets. The typical experimental conditions are as follows: microwave power of 500–1200 W, total pressure of 20–60 Torr, and 2–30 min of hydrogen plasma treatment. Later, the substrate is exposed to a hydrogen/oxygen plasma mixture for about 15 min to 3 h. This process yields a 2D network of nanowires in the form of nanowebs as shown in Figure 9.25 with nanowire diameters of 5–150 nm and lengths of about 0.5 μm.

The decreased surface tension with exposure to oxygen containing gas phase can flatten out the small size Ga droplets easily compared to micron-sized Ga droplets. So, the nucleation and growth of nanowires from such disk-shaped droplets lead the nanowire growth parallel to the substrate. The nanowires growing from neighboring Ga droplets can collide with each other and join during growth, thus forming a nanowire network. These steps are illustrated in the schematic shown in Figure 9.26.

9.3.3 Nanobelts

Nanobelts are another type of one-dimensional materials with nanometer-scale thickness and micron-scale width with lengths extending to tens of microns. Even though, the origin of such morphological evolution during nanowire synthesis experiments is not entirely clear, such nanobelt morphology has been observed for several metal oxides such as tin oxide, indium oxide, gallium oxide and zinc oxide. In many cases, such morphological observations were made directly in the experiments leading to nanowires but with slightly different process conditions (for example, higher substrate temperature and pressures). The nanobelt-like morphology was first observed during the silicon whisker growth reported using gold catalyst clusters [65].

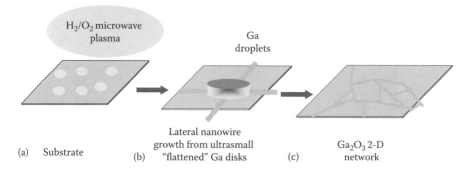

FIGURE 9.26

A schematic illustrating various stages involved with the formation of highly networked nano-wires (nanowebs): (a) formation of smaller gallium droplets; (b) flattening of Ga droplets with exposure to oxygen containing atmosphere; (c) nucleation and growth of nanowires parallel to the substrate from the diskette-shaped droplets; (d) joining of nanowires upon physical impingement during growth to form the network.

The synthesis approaches for nanobelts are similar to those used for nanowires. For example, the thermal CVD in a horizontal quartz tube furnace reactor using metal powders as the metal source and quartz substrates in the center of the quartz tube resulted in the synthesis of nanobelts at ambient pressure and under argon flow. In case of gallium oxide, a synthesis temperature of about 950°C and 400 sccm of oxygen flow resulted in nanowires whereas raising the temperature to 1050°C produced nanobelt morphologies. As shown in Figure 9.27a, the widths of the nanobelts were in the range of 100 nm to 1 μm with a thickness of about 50 nm [29]. As shown in Figure 9.27b, electron diffraction analysis on the nanobelt indicates that the growth direction of the nanobelt is [111] and the phase is identified as monoclinc β-Ga$_2$O$_3$. Similar results were obtained in the case of tin oxide, i.e., the nanowires appeared at about 900°C and belt-like morphologies appeared at 1200°C as seen in Figure 9.28. The growth direction of both nanowire and nanobelts in

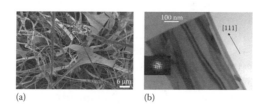

FIGURE 9.27

(a) SEM images showing several gallium oxide nanobelts synthesized using a reactive vapor transport approach. (b) High-resolution electron microscopy (HREM) image of a gallium oxide nanobelt indicative of a monoclinic β-Ga$_2$O$_3$ phase with [111] growth direction. (Reproduced from Rao, R. et al., *J. Electron. Mater.*, 35, 941, 2006. With permission.)

FIGURE 9.28
SEM images showing tin oxide nanobelts synthesized using a reactive vapor transport approach.

the case above is the same. So, the transition between nanowire to belt-like morphology is dictated by the synthesis temperature. This is because the vapor pressure of gallium at the nanowire synthesis conditions is about an order of magnitude less than that at the nanobelt synthesis temperature. Higher Ga vapor pressures lead to enhanced homoepitaxial growth of nanowire. The nanowires grown in the above experiments have rectangular cross-section. As explained in detail in Chapter 10 with GaN, the homoepitaxial growth onto nanowires with rectangular cross-sections can lead to belt-like morphology with preferential growth of one of the polar surfaces of metal oxide systems (metal-terminated surface).

In the case of both gallium oxide and indium oxide nanobelts, vapor transport of the respective metal vapors onto gold-coated silicon substrates placed few centimeters from the source resulted in both nanowires and nanobelts [12,66]. The resulting nanobelts were ten to several hundred nanometers in width and several microns long. For gallium oxide nanobelt synthesis, the process parameters include source temperature of 750°C, experimental duration of 75 min, and 20 sccm of Ar gas [12]. For indium oxide nanobelt synthesis, the process parameters include source temperature of 900°C, experimental duration of 1 h, and 120 sccm of N_2 gas [66]. Even though several such experiments were reported on the use of catalyst layer for nanobelt synthesis, the role of catalyst particle on the nanobelt synthesis is clearly not directly related. Direct thermal evaporation of metal oxide powders onto alumina plates resulted in the synthesis of nanobelts-like zinc oxide, tin oxide, indium oxide, and cadmium oxide [67]. For zinc oxide synthesis, the process parameters employed were source temperature of 1400°C, experimental duration of 2 h, and a chamber pressure of 300 Torr. Electron microscopy characterization of zinc oxide nanobelts indicated single crystallinity and the nanobelts grown along [0001] showed the presence of no defects. The zinc oxide nanobelts grown along [01–10] growth direction showed the presence of one stacking fault parallel to the growth direction of the nanobelt.

9.3.4 Tubular Nanostructures

The tubular structures with hollow core are another morphological form for one-dimensional materials. In the processing schemes described above and later in this section, several metal oxide materials systems (both high-melting and low-melting point systems) exhibited one-dimensional morphologies with hollow core. The synthesis routes and possible pathways for such tubular morphologies for both systems are discussed here.

FIGURE 9.29
SEM image showing molybdenum oxide nanotubes synthesized using a hot-wire-assisted CVD process. (Reproduced from Thangala, J. et al., *Thin Solid Films*, 517, 3600, 2009. With permission.)

9.3.4.1 High-Melting Metal Oxides

In the hot-wire-assisted CVD procedure described in Section 9.2.3, the use of molybdenum as the filament at 1200°C, quartz substrate at 550°C, and oxygen at a partial pressure of 0.54 Torr results in tubular morphology as shown in Figure 9.29 [39]. Also, similar tubular morphologies for MoO₃ and WO₃ were reported using direct oxidation of a Mo foil and W foil for deposition onto Ta substrate [68,69]. In the case of tungsten trioxide, tubular structure formation was also observed during the heating of tungsten loop in the presence of water vapor and oxygen to temperatures around 1300°C [70]. There are two possible pathways for explaining the formation of such hollow morphologies. In the first possibility, the clustering and sintered growth of metal oxide clusters in the initial stages could lead to hollow core as illustrated in Figure 9.30. In the second possibility (Figure 9.31), the suboxide clusters themselves could lead the growth of tubular morphologies in a similar fashion as that of carbon tubular growth, i.e., outer oxidation of suboxide clusters to induce wall formation and further growth. Different material systems like vanadium oxide, iron oxide, and titania were synthesized in the form of tubular structures using alumina as templates [71–73].

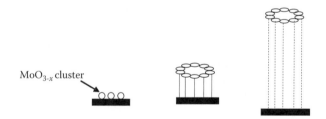

MoO_{3-x} cluster

FIGURE 9.30
A schematic illustrating another possible route to explain the growth of tubular structures of transition metal oxides, which involves agglomeration of oxide nuclei in the initial stages. (Reproduced from Thangala, J. et al., *Thin Solid Films*, 517, 3600, 2009. With permission.)

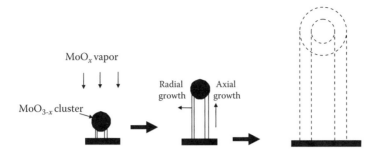

FIGURE 9.31
A schematic illustrating the possible steps involved with the growth of nanotubular structures of transition metal oxides, which involves outer oxidation of suboxide clusters. (Reproduced from Thangala, J. et al., *Thin Solid Films*, 517, 3600, 2009. With permission.)

(a)   (b)

FIGURE 9.32
(a) SEM image showing gallium oxide nanotubes synthesized with direct oxidation of large Ga droplets with oxygen plasma. (b) SEM image showing the backside of the tube showing the nuclei agglomeration pattern in a spiral fashion that resulted in the complicated tubular morphology evident from shown in (a). (Reproduced from Sharma, S. and Sunkara, M.K., *J. Am. Chem. Soc.*, 124, 12288, 2002. With permission.)

9.3.4.2 Low-Melting Metal Oxides

As explained in Section 9.2.2.1, the direct oxidation of low-melting metal melts using hydrogen and oxygen plasmas result in tubular morphologies in the case of Ga_2O_3 as shown in Figure 9.32. Referring to the schematic in Figure 9.9, the tubular morphology can be explained to be the consequence of the self-assembly of nuclei on molten metal surface into hollow patterns in the initial stages [17]. This is similar to the argument of sintered growth of nanowires on solid substrates in the case of high-melting metal oxides such as molybdenum oxide. In the case with low-melting metal oxides, the argument of aggregation of metal oxide nuclei on molten metal surfaces into hollow patterns is much more plausible. The aggregation of nuclei seems to occur in hexagonally shaped but seems to spiral leading to tube-inside-tube morphologies. The SEM images in Figure 9.32b show the pattern at the backside (at the interface with molten metal).

Also, carbothermal reduction/evaporation of a mixture of magnesium oxide, gallium oxide, and carbon powder produces magnesium oxide nanotubes [74]. The thermal evaporation of a mixture of indium/indium oxide powders yield indium oxide tubes filled with indium [75]. The underlying mechanism leading to tubular morphologies in the above two cases must be different from the case of tubular morphology evolution on molten metal surface since the resulting tubules are filled with the low-melting metal.

9.4 Oxygen Vacancies, Doping, and Phase Transformation

Metal oxide nanowires are typically synthesized under oxygen-lean conditions and thus are prone for oxygen deficiency. So, it is possible that the growth mechanism involved for nanowires itself may involve oxygen vacancies. In addition, it is important to understand the presence of oxygen vacancies because they will influence the properties and play an important role in the doping and alloying process of these metal oxide nanowires during both in situ and post synthesis methods. Oxide nanowires are much easier to synthesize than other compounds such as sulfides or nitrides. So, there is interest in transforming oxide nanomaterials to other compounds through postsynthesis phase transformation. In this section, the nature of oxygen vacancies in metal oxide nanowire systems and basic aspects about doping into nanowires are discussed. Finally, the transformation of oxide nanowires into other compounds is discussed using tungsten oxide as an example material system.

9.4.1 Oxygen Vacancies

The synthesis of high-melting point metal oxides occurs under oxygen-lean conditions and the resulting nanowires are always of oxygen-deficient phases. Tungsten oxide is a good example of this where the as-synthesized tungsten oxide exhibits oxygen deficient phases ($WO_{2.7}$–$WO_{2.9}$) instead of the stoichiometric phase of WO_3. In addition, the oxygen-deficient phases themselves exhibit oxygen vacancies. In all of the high-melting point metal oxide systems such as tungsten oxide, molybdenum oxide, and iron oxide, the oxygen vacancies seem to occur as oxygen vacancy planes and in some cases, these vacancy planes seem to be ordered forming superlattice structures.

The oxygen vacancy plane ordering can be illustrated by considering iron oxide nanowires as the example material system. The iron oxide nanowires synthesized using direct plasma oxidation of iron foils were characterized using transmission electron microscopy [76]. The diffraction patterns (Figure 9.33) from nanowires and nanobelts feature a closely spaced spot pattern within the diffraction pattern representing α-Fe_2O_3, indicating a

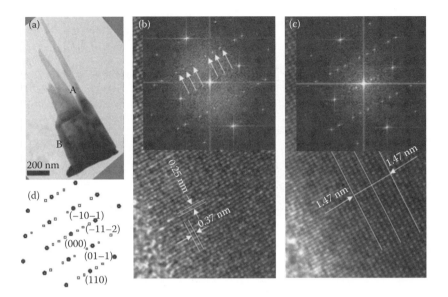

FIGURE 9.33
(a) A TEM image showing straight α-Fe$_2$O$_3$ nanowires partly merged to form a nanobelt.
(b) A HREM image taken in region A and the inset fast Fourier transform (FFT) image shows
extra maxima (indicated by arrows). (c) HREM image taken from region B presenting clear
ordering features with a ordering distance of 1.47 nm; the inset FFT image shows that this
ordering is the same as the one in region A. (d) The indexed schematic drawing of FFT image,
where ordering plane in α-Fe$_2$O$_3$ is 1/4(1–12). (Reproduced from Chen, Z. et al., *Chem. Mater.*,
20, 3224, 2008. With permission.)

long-range ordering of a superstructure. Two such patterns involving different planes are identified and the growth direction of the nanowire for both
cases is found to be along the [110] direction. The HREM images indicate
that the fringe distances of 0.37 and 0.25 nm correspond to the planar spac-
ings of α-Fe$_2$O$_3$ of (–11–2) and (110) planes as shown in Figure 9.33 [77]. The
vacancy planes are along (1–12), which are parallel to the growth direction
but are ordered at every fourth plane. Similarly, different iron oxide nanowires from the same sample also show vacancy planes to be along (3–30)
with ordering over every tenth plane. The ordering distances in both cases
happens to be about 1.45 nm as shown in Figure 9.34. This common ordering
distance is thought to arise from strain due to lattice mismatch with iron
substrate during growth. The epitaxial relationship between the 1/2(3–30)
Fe$_2$O$_3$||(12–3)Fe and (1–12)Fe$_2$O$_3$||(001)Fe could explain the observed ordering
distance from the lattice mismatch considerations at synthesis temperature.
At reaction temperatures, the dislocation distances are estimated as 1.405
and 1.329 nm for (3–30) and (1–12) planes with discrepancies of about 3.8%
and 10% respectively [76]. The oxygen vacancy plane ordering is explained
based on the interfacial stress created by the lattice mismatch between the
iron oxide with higher lattice constants in comparison to iron with lower

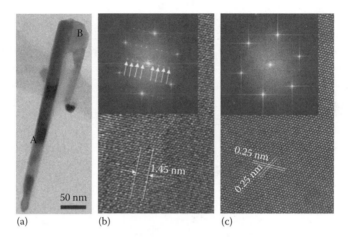

FIGURE 9.34
(a) The BF TEM image with analysis spots (marked) showing the straight α-Fe$_2$O$_3$ nanowire with a tip and merger area with nanowire of different orientation. (b) HREM image taken from the A region showing the ordering distance of about 1.45 nm, and corresponding the fast-Fourier transform (FFT) image on the bottom illustrating that ordering occurred by 1/10(3–30). (c) HREM image taken from position B revealing few ordering features. (Reproduced from Chen, Z. et al., *Chem. Mater.*, 20, 3224, 2008. With permission.)

lattice constants. The above hypothesis probably indicates that the oxygen vacancy plane ordering is more a process-dependent than the material system itself. A similar kind of oxygen vacancy planes were observed in other high-melting point metal oxides like tungsten oxide and niobium pentoxide nanowires [33,76]. However, both the nature of the oxygen vacancy planes and the ordering are dependent upon the process and the material system.

The understanding about the nature of oxygen vacancies within low-melting metal oxides such as ZnO and SnO$_2$ is limited. The random occurrence of oxygen vacancies throughout metal oxide nanowires is very likely, but it is difficult to visualize and quantify such oxygen vacancies compared to oxygen vacancy planes.

9.4.2 Doping and Alloying

Doping of metal oxide nanowires (at concentrations less than 1 at%) is important for modification of electrical and optical properties, which can be achieved by either in situ during synthesis or in a postprocess treatment. As the nanowire growth occurs through either molten metal or solid or suboxide clusters, the mechanism of dopant incorporation is difficult to understand and control. Attempts to obtain doped nanowires have mostly resulted in alloy compositions greater than 1 at%. For example, zinc oxide nanowires have been doped or alloyed with elements such as S, Sc, N, Mg, Bi, Ga, In, Sn, Cu, As, and Mn either through in situ or postprocess synthesis procedures [77–83]. Results include ZnO nanowires with 4 at% S, 5.78% Sc, and 1 at% Mg wherein the respective alloying elemental powders are mixed with the source metal during synthesis. In the case of Sc incorporation, the carbothermal reduction of a mixture of Sc_2O_3, ZnO, and C powders resulted in the synthesis of Sc-doped zinc oxide nanowires through self-catalysis. In the case of gallium oxide, the presence of manganese (Mn) resulted in alloy nanowires with 20 at% composition for Ga–Mn–O alloy nanowires. Similarly, the antimony doping of tin oxide nanowires was achieved by evaporation of a mixture of tin and antimony onto gold catalyst-coated silicon substrate.

9.4.3 Phase Transformation of Metal Oxide Nanowires

As mentioned earlier, for most metals, oxide nanowires have been found to be easier to produce than, say, nitrides or sulfides of the same system. Then, the question is whether one can convert the oxide nanowires into other compounds without losing the one-dimensional morphology and crystallinity. Nanowire morphology offers an interesting platform for doing such phase transformations compared to thin films because the nanowire diameters can be much less than or on the order of the diffusion lengths for the reacting species. Therefore, one can easily imagine transforming the entire nanowire from one composition to another, but, to achieve this, it is important to understand the factors that control the nucleation and the overall crystallinity within nanowires during phase transformation. In this section, the discussion uses nitridation of tungsten oxide nanowires as a model system for phase transformation of metal oxide nanowires.

The nitridation experiments using tungsten oxide nanowires were conducted using reaction with NH_3 (50 sccm/0.255 Torr) at a temperature of around 750°C. The nitridation process was complete in 1 h and the XRD of nanowire array films confirmed the resulting pure tungsten nitride phase with predominant peak in [200] direction indicating preferred orientation

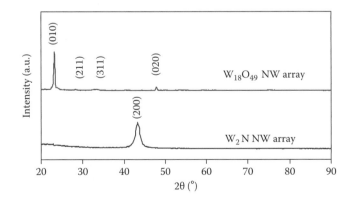

FIGURE 9.35

XRD spectra indicative of the presence of only W_2N phase after the phase transformation process. (Reproduced from Thangala, J. et al., *Cryst. Growth Des.*, 9, 3177, 2009. With permission.)

of the nanowires (see Figure 9.35) [84]. The process of nitridation was understood through HREM analysis for nanowires of different sizes and different processing timescales. The analysis showed that the thinner wires (less than 10 nm) to be single crystalline while thicker wires to be highly oriented but polycrystalline [84]. The diffraction patterns of tungsten oxide nanowires before nitridation show the presence of streak lines along the growth direction of the nanowire, indicating the presence of oxygen vacancy planes parallel to the growth direction. The HREM analysis and the FFT pattern of the partially nitrided nanowires show that the nitride nuclei are in epitaxial relationship with the oxide matrix [84]. The transformation of $W_{18}O_{49}$ to W_2N follows the orientation relationship: $[110]_{W_2N}||[001]_{W_{18}O_{49}}$ and $(002)_{W_2N}||(010)_{W_{18}O_{49}}$. However, there exists a large lattice mismatch between the nitride and oxide phases, which causes a large strain. In this case of heteroepitaxial nucleation, such lattice mismatch induced strain energy contributes to the energy of nuclei formation as

$$\Delta G = -V\left(\Delta G_v + \Delta G_s\right) + A\sigma \qquad (9.5)$$

Hence, the critical nuclei radius (r^*) in the present case is given by

$$r^* = 2\gamma\Omega/\left(\Delta G_v + \Delta G_s\right) \qquad (9.6)$$

where
  γ is the surface energy
  Ω is the molar volume
  ΔG_v is the volume free energy term
  ΔG_s is the strain energy contribution

In the present case, the strain energy based on the mismatch between the tungsten nitride phase with tungsten oxide is computed using

$$\Delta G_s = Y/(1 - v) * \left(\varepsilon^2\right) \qquad (9.7)$$

where
  Y is Young's modulus of the tungsten nitride (W_2N)
  v is Poisson's ratio
  ε is the lattice mismatch between the nitride and the oxide phase

  In the present case of strain-induced epitaxial nucleation, the critical nuclei size is estimated based on only the strain energy contribution resulting from the lattice mismatch between the nitride and the oxide phase. The strain energy contribution is estimated based on Young's modulus, Poisson's ratio, and thermal expansion coefficient for tantalum nitride. The lattice mismatch is computed at different temperatures based on the thermal expansion coefficient difference of the two phases. At the experimental nitridation temperatures (1023 K), the lattice mismatch is measured to be about 6.7% and the grain size to be about 6 nm. Simple sensitivity analysis based on variations in the surface energy, Young's modulus, and Poisson's ratio values showed that there is an increase in the grain size with increase in the nitridation temperature. Nevertheless, such analysis predicts the nuclei size to be below 10 nm at the nitridation temperatures.

The phase transformation of transition metal oxide to nitride nanowires can be understood with epitaxial nucleation followed by growth within the oxide nanowire [84]. In thinner nanowires with sizes below the nuclei diameter, only one nucleation event may occur in the nanowire along the cross section. At the same time, multiple nucleation sites could form along the longitudinal direction of the nanowire. However, as the nucleation of W_2N has a preferred orientation parallel to the growth direction of the nanowire, the nucleated grains in this direction could easily be aligned with each other to become a single-crystal nanowire. In thicker nanowires, multiple nucleation events can occur across the cross-section of the nanowire while being oriented with respect to the longitudinal axis. Some of the grains can be rotated with respect to each other. Hence, it can be difficult for all these grains to attach with each other to become one single crystal even when majority of the grains are oriented along both radial and axial directions of the nanowire as seen in Figure 9.36 [84].

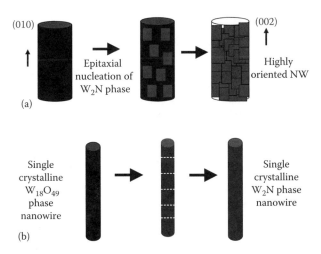

FIGURE 9.36
A schematic illustrating the phase transformation of metal oxide nanowires during nitridation depending upon the size. (a) Nanowires with diameters greater than twice the critical nuclei size exhibit multiple nucleation along both longitudinal and radial directions leading to nanowires with highly oriented grains. (b) Nanowires with diameters less than twice the critical nuclei size exhibit nucleation only along longitudinal axis. (Reproduced from Thangala, J. et al., *Cryst. Growth Des.*, 9, 3177, 2009. With permission.)

The observed orientation of nitride nuclei with respect to the longitudinal axis of the oxide nanowires is a direct result of preferred nitridation along the oxygen vacancy planes in the initial stages. The nitridation experiments using fully oxidized tungsten trioxide nanowires also result in oxygen-deficient nitride phases in the initial stages of the phase transformation process. The observations of streak lines indicate that there is no long-range ordering of resulting oxygen vacancy planes. So, the role of oxygen vacancy planes in the high-melting point metal oxides can help with oriented nucleation for highly crystalline transformations of oxide nanowires into other compositions. In the case of low-melting point metal oxides, the gas-phase reactions result in the creation of molten metal droplets, which further cause selective pitting of the nanowires. So, the phase transformation of low-melting point metal oxide nanowires using gas–solid reactions can be very difficult and must be performed at very low process temperatures to avoid the formation of molten metal clusters during the process. In the case of zinc oxide, phase transformation using gas–solid reaction in presence of hydrogen sulfide (H_2S) gas resulted in the synthesis of zinc sulfide nanocolumns and tubular structures [85]. The transformation of oxide to sulfide nanocolumns is explained based on surface sulfide layer formation followed by the entire conversion of oxide to sulfide nanocolumns.

References

1. M. H. Huang, Y. Wu, H. Feick, N. Tran, E. Weber, and P. Yang, *Adv. Mater.* 13, 113 (2001).
2. M. H. Huang, S. Mao, H. Feick, H. Yan, Y. Wu, H. Kind, E. Weber, R. Russo, and P. Yang, *Science* 292, 1897 (2001).
3. P. Yang, H. Yan, S. Mao, R. Russo, J. Johnson, R. Saykally, N. Morris, J. Phao, R. He, and H. J. Choi, *Adv. Funct. Mater.* 12, 323 (2002).
4. C. X. Xu, X. W. Sun, Z. L. Dong, M. B. Yu, T. D. My, X. H. Zhang, S. J. Chua, and T. J. White, *Nanotechnology* 15, 839 (2004).
5. H. T. Ng, J. Han, T. Yamada, P. Nguyen, Y. P. Chen, and M. Meyyappan, *Nano Lett.* 4, 1247 (2004).
6. H. T. Ng, J. Li, M. K. Smith, P. Nguyen, A. Cassell, J. Han, and M. Meyyappan, *Science* 300, 1249 (2003).
7. P. Nguyen, H. T. Ng, J. Kong, A. M. Cassell, R. Quinn, J. Li, J. Han, M. McNeil, and M. Meyyappan, *Nano Lett.* 3, 925 (2003).
8. P. Nguyen, H. T. Ng, T. Yamada, M. K. Smith, J. Li, J. Han, and M. Meyyappan, *Nano Lett.* 4, 651 (2004).
9. M. C. Johnson, S. Aloni, D. E. McCready, and E. D. Bourret-Courchesne, *Cryst. Growth Des.* 6, 1936 (2006).
10. T. J. Kuo and M. H. Huang, *J. Phys. Chem. B* 110, 13717 (2006).
11. T. Y. Kim, J. Y. Kim, M. Senthil Kumar, E. K. Suh, and K. S. Nahm, *J. Crystal Growth* 270, 491 (2004).

12. C. L. Kuo and M. H. Huang, *Nanotechnology* 19, 155604 (2008).
13. Q. Wan, E. N. Dattoli, W. Y. Fung, W. Guo, Y. Chen, X. Pan, and W. Lu, *Nano Lett.* 6, 2909 (2006).
14. Q. Wan, M. Wei, D. Zhi, J. L. MacManus-Driscoll, and M. G. Blamire, *Adv. Mater.* 18, 234 (2006).
15. J. Jie, G. Wang, X. Han, J. Fang, Q. Yu, Y. Liao, B. Xu, Q. Wang, and J. G. Hou, *J. Phys. Chem. B* 108, 8249 (2004).
16. M. Lorenz, E. M. Kaidashev, A. Rahm, Th. Nobis, J. Lenzner, G. Wagner, D. Spemann, H. Hochmuth, and M. Grundmann, *Appl. Phys. Lett.* 86, 143113 (2005).
17. S. Sharma and M. K. Sunkara, *J. Am. Chem. Soc.* 124, 12288 (2002).
18. C. Cao, Z. Chen, X. An, and H. Zhu, *J. Phys. Chem. C* 112, 95 (2008).
19. H. Y. Dang, J. Wang, and S. S. Fan, *Nanotechnology* 14, 738 (2003).
20. P. Wu, Q. Li, C. X. Zhao, D. L. Zhang, L. F. Chi, and T. Xiao, *Appl. Surf. Sci.* 255, 3201 (2008).
21. G. Wang, J. Park, X. Kong, P. R. Wilson, Z. Chen, and J. Ahn, *Cryst. Growth Des.* 8, 1940 (2008).
22. S. H. Sun, G. W. Meng, M. G. Zhang, X. H. An, G. S. Wu, and L. D. Zhang, *J. Phys. D: Appl. Phys.* 37, 409 (2004).
23. X. C. Wu, J. M. Hong, Z. J. Han, and Y. R. Tao, *Chem. Phys. Lett.* 373, 28 (2003).
24. Y. Yin, G. Zhang, and Y. Xia, *Adv. Funct. Mater.* 12, 293 (2002).
25. S. C. Lyu, Y. Zhang, C. J. Lee, H. Ruh, and H. J. Lee, *Chem. Mater.* 15, 3294 (2003).
26. C. J. Lee, T. J. Lee, S. C. Lyu, Y. Zhang, H. Ruh, and H. J. Lee, *Appl. Phys. Lett.* 81, 3648 (2002).
27. J. S. Lee, M. Kang, S. Kim, M. S. Lee, and Y. K. Lee, *J. Cryst. Growth* 249, 201 (2003).
28. S. T. Ho, K. C. Chen, H. A. Chen, H. Y. Lin, C. Y. Cheng, and H. N. Lin, *Chem. Mater.* 19, 4083 (2007).
29. R. Rao, H. Chandrasekaran, S. Gubbala, M. K. Sunkara, C. Daraio, S. Jin, and A. M. Rao, *J. Electron. Mater.* 35, 941 (2006).
30. W. I. Park, D. H. Kim, S. W. Jung, and Y. Gyu-Chul, *Appl. Phys. Lett.* 80, 4232 (2002).
31. Z. Z. Ye, J. Y. Huang, W. Z. Xu, J. Zhou, and Z. L. Wang, *Solid State Commun.* 141, 464 (2007).
32. H. Li, A. H. Chin, and M. K. Sunkara, *Adv. Mater.* 18, 216 (2003).
33. J. Thangala, S. Vaddiraju, R. Bogale, R. Thurman, T. Powers, B. Deb, and M. K. Sunkara, *Small* 3, 890 (2007).
34. S. Vaddiraju, H. Chandrasekaran, and M. K. Sunkara, *J. Am. Chem. Soc.* 125, 10792 (2003).
35. L. Chi, N. Xu, S. Deng, J. Chen, and J. She, *Nanotechnology* 17, 5590 (2006).
36. Y. B. Li, Y. Bando, D. Golberg, and K. Kurashima, *Chem. Phys. Lett.* 367, 214 (2003).
37. J. Liu, Y. Zhao, and Z. Zhang, *J. Phys.: Condens. Matter* 15, L453 (2003).
38. A. H. Mahan, P. A. Parilla, K. M. Jones, and A. C. Dhillon, *Chem. Phys. Lett.* 413, 88 (2005).
39. J. Thangala, S. Vaddiraju, S. Malhotra, V. Chakrapani, and M.K. Sunkara, *Thin Solid Films*, 517, 3600 (2009).

40. K. Hong, M. Xie, and H. Wu, *Nanotechnology* 17, 4830 (2006).
41. K. Hong, M. Xie, and H. Wu, *Appl. Phys. Lett.* 90, 173121 (2007).
42. K. Senthil and K. Young, *Nanotechnology* 18, 395604 (2007).
43. J. Zhou, L. Gong, S. Z. Deng, J. Chen, J. C. She, N. S. Xu, R. Yang, and Z. L. Wang, *Appl. Phys. Lett.* 87, 223108 (2005).
44. Y. Baek and K. Yong, *J. Phys. Chem. C* 111, 1213 (2007).
45. K. Hong, M. Xie, and H. Wu, *Nanotechnology* 17, 4830 (2006).
46. J. Zhou, N. S. Xu, S. Z. Deng, J. Chen, and J. C. She, *Chem. Phys. Lett.* 382, 443 (2003).
47. C. C. Liao, F. R. Chen, and J. J. Kai, *Solar Energy Materials & Solar Cells* 90, 1147 (2006).
48. Y. Z. Jin, Y. Q. Zhu, R. L. D. Whitby, N. Yao, R. Ma, P. C. P. Watts, H. W. Kroto, and D. R. M. Walton, *J. Phys. Chem. B* 108, 15572 (2004).
49. K. Hong, W. Yiu, H. Wu, J. Gao, and M. Xie, *Nanotechnology* 16, 1608 (2005).
50. H. Qi, C. Wang, and J. Liu, *Adv. Mater.* 15, 411 (2003).
51. Z. Liu, Y. Bando, and C. Tang, *Chem. Phys. Lett.* 372, 179 (2003).
52. U. Cvelbar, Z. Q. Chen, M. K. Sunkara, and M. Mozetic, *Small* 4, 1610 (2008).
53. M. Mozetic, U. Cvelbar, M. K. Sunkara, and S. Vaddiraju, *Adv. Mater.* 17, 2138 (2005).
54. B. Varghese, S. C. Haur, and C. T. Lim, *J. Phys. Chem. C* 112, 10008 (2008).
55. X. Wen, S. Wang, Y. Ding, Z. L. Wang, and S. Yang, *J. Phys. Chem. B* 109, 215 (2005).
56. S. Ren, Y. F. Bai, J. Chen, S. Z. Deng, N. S. Xu, Q. B. Wu, and S. Yang, *Mater. Lett.* 61, 666 (2007).
57. T. Ghoshal, S. Biswas, S. Kar, A. Dev, S. Chakrabarti, and S. Chaudhuri, *Nanotechnology* 19, 065606 (2008).
58. Q. Wan, E. N. Dattoli, W. Y. Fung, W. Guo, Y. Chen, X. Pan, and W. Lu, *Nano Lett.* 6, 2909 (2006).
59. H. W. Kim and S. H. Shim, *Vacuum* 82, 1395 (2008).
60. J. Zhou, Y. Ding, S. Z. Deng, L. Gong, N. S. Xu, and Z. L. Wang, *Adv. Mater.* 17, 2107 (2005).
61. Y. Q. Zhu, W. Hu, W. Hsu, M. Terrones, N. Grobert, J. P. Hare, H. W. Kroto, D. R. M. Walton, and H. Terrones, *Chem. Phys. Lett.* 309, 327 (1999).
62. L. Xu, Y. Su, S. Li, Y. Chen, Q. Zhou, S. Yin, and Y. Feng, *J. Phys. Chem. B* 111, 760 (2007).
63. M. Kaur, K. P. Muthe, S. K. Despande, S. Choudhury, J. B. Sing, N. Verma, S. K. Gupta, and J. V. Yakhmi, *J. Cryst. Growth* 289, 670 (2006).
64. U. M. Graham, S. Sharma, M. K. Sunkara, and B. H. Davis, *Adv. Funct. Mater.* 13, 576 (2003).
65. R. S. Wagner and W. C. Ellis, *Appl. Phys. Lett.* 4, 89 (1964).
66. T. Gao and T. Wang, *J. Cryst. Growth* 290, 660 (2006).
67. Z. W. Pan, Z. R. Dai, and Z. L. Wang, *Science* 291, 1947 (2001).
68. Y. Li and Y. Bando, *Chem. Phys. Lett.* 364, 484 (2002).
69. Y. Li, Y. Bando, and D. Golberg, *Adv. Mater.* 15, 1294 (2003).
70. Y. Wu, Z. Xi, G. Zhang, J. Yu, and D. Guo, *J. Cryst. Growth* 292, 143 (2006).
71. Y. Wang, K. Takahashi, H. Shang, and G. Cao, *J. Phys. Chem. B* 109, 3085 (2005).
72. H. Imai, Y. Takei, K. Shimizu, M. Matsuda, and H. Hirashima, *J. Mater. Chem.* 9, 2971 (1999).
73. J. Chen, L. Xu, W. Li, and X. Gou, *Adv. Mater.* 17, 582 (2005).

74. J. Zhan, Y. Bando, J. Hu, and D. Golberg, *Inorg. Chem.* 43, 2462 (2004).
75. Y. Li, Y. Bando, and D. Golberg, *Adv. Mater.* 15, 581 (2003).
76. Z. Chen, U. Cvelbar, M. Mozetic, J. He, and M. K. Sunkara, *Chem. Mater.* 20, 3224 (2008).
77. S. Y. Bae, C. W. Na, J. H. Kang, and J. Park, *J. Phys. Chem. B* 109, 2526 (2005).
78. S. Y. Bae, H. W. Seo, and J. Park, *J. Phys. Chem. B* 108, 5206 (2004).
79. C. Xu, J. Chun, D. E. Kim, J. J. Kim, B. Chon, and T. Joo, *Appl. Phys. Lett.* 90, 083113 (2007).
80. W. Lee, M. C. Jeong, S. W. Joo, and J. M. Myoung, *Nanotechnology* 16, 764 (2005).
81. C. X. Xu, X. W. Sun, and B. J. Chen, *Appl. Phys. Lett.* 84, 1540 (2004).
82. R. Liu, A. Pan, H. Fan, F. Wang, Z. Shen, G. Yang, S. Xie, and B. Zou, *J. Phys.: Condens. Matter* 19, 136206 (2007).
83. S. M. Zhou, X. H. Zhang, X. M. Meng, X. Fan, S. K. Wu, and S. T. Lee, *Physica E* 25, 587 (2005).
84. J. Thangala, Z. Q. Chen, A. H. Chin, C. Z. Ning, and M. K. Sunkara, *Cryst. Growth Des.* 9, 3177 (2009).
85. L. Dloczik and R. Konenkamp, *Nano Lett.* 3, 651 (2003).

10

Nitride Nanowires

10.1 Introduction

III-nitride materials (aluminum nitride [AlN], gallium nitride [GaN], and indium nitride [InN]) have a wide range of applications in electronics, optoelectronics, LEDs, lasers, and photoelectrochemical systems. AlN with its very wide band gap of 6.4 eV, high dielectric constant, and thermal conductivity, has potential applications as a deep UV emitter, optical storage media, and as a substrate for high-temperature devices. GaN with a 3.4 eV band gap in the UV region, high thermal, voltage, and chemical stability has applications in high-power, high-speed, and high-temperature electronics, blue lasers, LEDs, and photoelectrochemical hydrogen production. InN with its narrow band gap of 0.7 eV has potential applications in near infrared (NIR) lasers and solar cells. III-nitrides have low sensitivity to ionizing radiation, which allows for potential applications in solar cells in outer space. Apart from the applications of binary species, the most important feature of III-nitrides as a system is that their alloys can exist as solid solutions, which translates to a composition-dependent control over the properties. Hence, it is possible to tune the properties of III-nitrides all the way from InN to AlN by varying the alloy composition. Another important property in this regard is band gap bowing, i.e., the band gap of the alloy does not vary linearly with composition. These alloys can have properties beyond the range of binary III-nitrides, which opens up new applications for ternary and quaternary III-nitride alloys.

10.2 Synthesis of Group III–Nitride Nanowires

III-nitride nanowires provide advantages offered by nanowires in terms of miniaturization and property modulation/enhancement on the nanoscale. They have been synthesized using a variety of techniques such as thermal chemical vapor deposition (CVD), plasma-assisted CVD, metalorganic chemical vapor deposition (MOCVD), molecular beam epitaxy (MBE), pulsed laser deposition (PLD) or laser ablation, and oxide-assisted growth. The nitridation reaction with Group III metal requires the use of ammonia. However, it is difficult to achieve dissolution of nitrogen using molecular nitrogen under thermal activation. This is due to the high pressures (20 atm) and high temperatures (2000 K) required when one uses molecular nitrogen for the following reaction:

$$Ga\ (l) + \tfrac{1}{2}\,N_2\ \rightarrow\ GaN\,(s) \tag{10.1}$$

The atomic nitrogen allows the above reaction with Ga to be conducted at all pressures (subatmospheric to atmospheric) and temperatures above 750°C [1]. The atomic nitrogen can be obtained using various gas phase activation sources such as hot filaments and radiofrequency (RF) or microwave (MW) discharges.

$$Ga\ (l) + N\ \rightarrow\ GaN\,(s) \tag{10.2}$$

Ammonia decomposes at low temperatures (<650°C) and thus one can achieve nitridation at similar pressures and temperatures as that of atomic nitrogen.

$$Ga\ (l) + NH_3\ \rightarrow\ GaN\,(s) + 3/2\ H_2 \tag{10.3}$$

So, many synthesis studies for nanowires typically use either ammonia or nitrogen plasma for nitridation reactions irrespective of the use of foreign metal catalysts. Regardless of the type of nitrogen source, the synthesis techniques can be classified into either catalyst-assisted or catalyst-free techniques. In this section, the process details for synthesizing belt-like morphologies versus nanowires are also described.

10.2.1 Catalyst-Assisted Synthesis

GaN is the most widely studied material of all the III-nitrides. Hence, a generic synthesis procedure for III-nitrides is presented using GaN as a model material system. A typical experimental setup for catalyst-assisted synthesis for GaN would include a quartz tube reactor as illustrated in Figure 10.1. The reacting (NH_3) and carrier/dilution (N_2/H_2) gases enter the tube from the upstream side with the downstream end connected to a vacuum pump. The catalyst-coated substrates are typically placed downstream to the source for growth species. The vapor phase reacting species are transported onto the substrate by a carrier gas.

A set of experiments to produce p-type GaN nanowire arrays is described below [2]. In these experiments, the gallium metal was directly used as the source material for gallium. Mg_3N_2 powder was mixed with Ga source to provide a controlled supply of Mg as p-type dopant during growth of GaN nanowires. Mg_3N_2 has high thermal stability and decomposes into Mg and N_2 at high temperatures, providing low vapor pressures for Mg in the

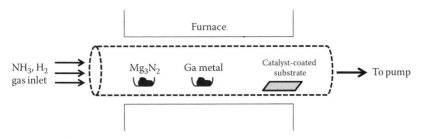

FIGURE 10.1

Schematic of a generic horizontal quartz tube furnace reactor used III-nitride nanowire growth. The relative positions of source and substrate are indicated along with the carrier/reacting gases.

vapor phase. Nickel-coated c-sapphire wafers are used as substrates for nanowire growth. The wafers were cleaned via ultrasonication in alcohol, rinsed in deionized (DI) water, and dried using Ar/N_2 gas. The catalyst layer (nickel) was deposited by evaporating a drop of 0.02 M solution of $Ni(NO_3)_2$ in ethanol on the wafer. The alcohol evaporates first and the $Ni(NO_3)_2$ decomposes to form nickel metal clusters. Nickel can also be coated on the substrate via sputtering or e-beam evaporation. The reactor tube is evacuated and purged with N_2/H_2 for a couple of cycles to remove any oxygen from the system. The experiments were conducted by flowing 30 sccm NH_3 and 60 sccm H_2. The temperature was ramped up and kept at 950°C for 20 min. Dual or three zone furnaces can also be used to maintain the source and substrate at different temperatures independently. In the initial stages of temperature ramping, the exposure of nickel layer to hydrogen-containing gas phase results in the creation of nickel clusters of sizes around few tens of nanometers. The one-dimensional growth of GaN occurs with molten alloy (GaN + Ni) cluster at its tip and due to enhanced dissolution kinetics of Ga and N into molten alloy as seen in Figure 10.2. Doping of magnesium into GaN nanowires occurs in situ. It is not clear whether doping occurs through the molten alloy droplet or through attachment kinetics at the molten alloy and the nanowire interface.

Nanowires synthesized via this technique can be epitaxially aligned with respect to the substrate. The resulting epitaxial nanowires have diameters in the 20–100 nm range and are 10–40 μm long. The wires were oriented in the <0001> direction, which indicates a direct epitaxial relation with the <0001> oriented c-sapphire substrate. The absence of tapering of the nanowires at the base can be attributed to the controlled supply of gallium, which prevents any addition of gallium adatoms to the nanowire from the base via surface diffusion.

Another very common approach for III-nitrides is MOCVD, which employs metal organic precursors as the material sources. The horizontal quartz tube reactor setup is described in Figure 10.3. Here, a precursor such as trimethyl gallium (TMGa) is used as gallium source. TMGa is stored in an external

FIGURE 10.2
SEM micrograph of epitaxially aligned magnesium doped GaN nanowire arrays synthesized via the reaction of gallium and nitrogen The presence of catalyst metal at the tip is evident. The scale bar is 10 μm. (Reproduced from Zhong, Z. et al., *Nano Lett.*, 3, 343, 2003. With permission.)

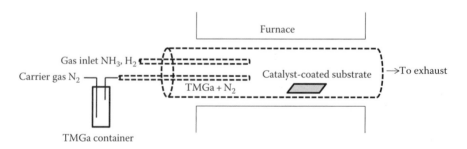

FIGURE 10.3
Schematic of the modified setup for quartz tube reactor for MOCVD growth. The source precursors are transported using a carrier gas bubbling through the container and the reactant gases are introduced into the reaction tube with the catalyst-coated substrate placed downstream. The tube is heated using a programmable furnace.

container at low temperatures (–10°C) to maintain low vapor pressures. A carrier gas, typically nitrogen at a high flow rate of 250 sccm, is bubbled through the TMGa cylinder to carry enough Ga precursor into the chamber. Also, ammonia diluted in hydrogen is typically used as a nitrogen source. A set of experiments using about 150 sccm of ammonia/hydrogen mixture, ~10 nm gold film coated *c*-plane oriented sapphire substrate at 800°C for 30 min results in GaN nanowire arrays with gold clusters at their tips as shown in Figure 10.4 [3]. These nanowires are straight, oriented in [210] direction, and exhibit a triangular cross section as indicated in the inset in Figure 10.4. The diameter of the resulting nanowires ranged from 15 to 100 nm and lengths ranged from 1 to 5 μm. The observed [210] growth direction of the GaN nanowires is unlike the previously observed <0001> growth direction that occurs by a direct epitaxy with the <0001> c-sapphire substrate, causing vertical growth. So, in these experiments, the nanowires are inclined at an angle to the substrate due to epitaxy of the [210] orientation. This is a good example to show that even under epitaxial growth conditions, the growth direction of the resulting nanowires can be different due to the prevailing process conditions such as material flux and the reaction temperature, etc.

10.2.1.1 Choice of Precursors

A variety of source materials ranging from pure Ga metal, GaN powder, and Ga_2O_3 mixed with carbon can be used to create Ga-containing vapors

FIGURE 10.4
SEM micrograph of aligned GaN nanowire arrays synthesized by the reaction of trimethylgallium with ammonia on gold-coated c-sapphire substrate. The nanowires 15–100 nm in diameter and 1–5 μm long appear to grow epitaxially on the substrate. The nanowires have a triangular cross-section, which is clearly indicated in the top view SEM micrograph in the inset along with the presence of a gold catalyst droplet at the tip. The scale bar is 10 μm in the figure and 100 nm in the inset. (Reproduced from Kuykendall, T. et al., *Nano Lett.*, 3, 1063, 2003. With permission.)

for reaction with ammonia downstream on the catalyst-coated substrates. In some cases, it is possible to create $GaCl_3$ vapors by passing HCl gas over Ga powder and then react with ammonia downstream. Each source choice has its own implications on both source and substrate temperatures. As mentioned above, metalorganic precursors can also be used through a showerhead onto the wafer coated with catalyst. The use of Ga metal poses some problems; for example, if the flux of Ga on the catalyst is high (achieved at higher temperatures), excessive formation of Ga droplets on the substrate can submerge the catalyst layer and eliminate its catalytic action. Growth then will occur via direct reaction schemes as discussed in Section 10.2.2. The use of high temperatures for source and use of chlorine place stringent requirements on the reactor parts. For instance, quartz starts to soften and decompose at 1200°C and silicon has a high solubility in Ga metal at 1000°C, which could lead to Si nanowire growth form Ga metal during the reaction. The use of metalorganic vapors may induce unintentional carbon doping into nanostructures.

MBE systems, on the other hand, run at very low pressures ~10^{-7} Torr and utilize pure Ga metal in effusion cells. The low species concentration complements the low operating pressures and allows for a very controlled, extremely slow growth of high-quality nanowires due to very selective dissolution of species into the catalyst cluster. But, the use of such high-vacuum environment makes the process costly both from capital equipment and operation points of view. A typical growth run uses using Ga supply from an effusion cell and activated nitrogen species using a RF plasma source in a MBE chamber. The system is maintained oxygen-free at 730°C and at very low pressures with a 0.3 nm Ni cluster–coated c-sapphire wafer as the substrate. At low Ga flux, <0001> oriented, epitaxial GaN nanowire arrays with 50–70 nm diameters and 2 μm long are obtained. The growth rates are low ranging from 0.2 to 0.4 nm/h. When the Ga flux is high, GaN gets deposited as a film on the substrate instead of nanowires with or without Ni catalyst, which confirms the need to control the Ga flux for nanowire synthesis.

10.2.1.2 Substrates for Epitaxial Array Growth

The choice of a substrate for the growth of GaN nanowires is dictated by the issues such as synthesis temperature, compatibility with Ga, and lattice mismatch for epitaxial growth. For example, the conducting FTO-coated (fluorinated tin oxide, which is stable up to 550°C) glass may not be suitable for GaN nanowire growth, but is suitable for InN due to lower synthesis temperatures. In the case of epitaxial array growth, the substrates with close lattice plane spacing (lattice match) can help. For example, the lattice plane spacing of crystalline MgO substrate (0.292 nm) closely matches the lattice spacing of c-plane-oriented GaN (0.313 nm). This makes MgO a good epitaxial substrate for GaN c-plane oriented nanowires. This is not the same for InN nanowires, which has a lattice parameter of 0.353 nm for the same orientation. In the case of nanowires with smaller diameter (less than few tens of nanometers), the lattice mismatch may not play an important role. Of course, the epitaxial growth requires clean substrates and a specific set of growth conditions that promote slow growth rates to initiate epitaxial growth in the initial stages. Other than MgO, single-crystal substrates of Si, sapphire, and oxides such as $LiAlO_2$ have been used for growing epitaxial GaN nanowire arrays. Unaligned GaN nanowire growth has been observed on crystalline silicon substrates irrespective of the synthesis technique due to the presence of a native oxide layer on silicon. Similar lack of alignment has also been observed in the case of InN nanowire growth on Au-coated silicon substrates.

Aligned nanowire growth has been achieved on sapphire substrates via epitaxial growth on Ni-coated GaN mesas using TMGa and 40 sccm NH_3 at 900°C [4]. Other studies involving epitaxial nanowire arrays include thermal evaporation of Ga onto Ni catalyst coated c-sapphire substrates at 950°C [2] and MOCVD growth using TMGa onto Au-coated sapphire substrate at 1000°C in NH_3 [3].

10.2.1.3 Choice of Catalysts and Process Variables

Most of the studies have primarily utilized metals such as Au, Ni, and Fe as catalysts. The ternary phase diagrams involving III-nitrides with many metals are not readily available. All of the experimental observations suggest that the catalyst clusters are molten during the synthesis of III-nitrides. So, it is possible to use any other foreign metal as the catalyst since the nitridation reaction can occur readily at temperatures above 650°C. In terms of process variables for III-nitrides, the substrate temperature plays the most important role. The choice of substrate temperature depends upon the type of Group III-nitride being synthesized. As a rule of thumb, the substrate temperature must be carefully chosen to be between the decomposition temperature of the respective III-nitride and the reaction temperature between the respective Group III metal and activated ammonia. In the case of GaN nanowire growth using any type of catalysts, a temperature between 800°C and 1000°C is reasonable. Similarly, InN nanowires using catalysts can be grown over a temperature range of 450°C–650°C. The substrate temperature in the initial stages is also important. In fact, the temperature ramp becomes more important with some substrates such as Si. For instance, using silicon substrate with Au catalyst for AlN nanowire synthesis at 1000°C might lead to dissolution of silicon into the catalyst as silicon has a eutectic with Au at lower temperatures, and interfere with AlN growth. In terms of the precursors, source used for one system might not be the right choice for other. Ga_2O_3 and In_2O_3 are used as Group III metal sources but it is very difficult to use Al_2O_3 due to its very high thermal stability and low vapor pressures. However, the use of oxides can lead to thin oxide sheath and oxygen doping. Also, substrates with epitaxial relation with GaN, for instance, might be unsuitable for InN/AlN due to the difference in lattice parameters and hence growth using the same substrate might not lead to aligned nanowire arrays.

10.2.1.4 Control of Nanowire Growth Direction

The control on growth direction for nanowires using catalyst-assisted growth techniques under nonepitaxial growth conditions is not clear for any materials systems. In the case of III-nitride systems, many experiments using catalysts resulted only in *a*-plane oriented nanowires. It has been suggested that epitaxial nanowire growth on different substrates with different orientations could be used to control the growth direction of the resulting nanowires. For example, MOCVD growth of GaN on lattice matching MgO and LiAlO$_2$ is used to grow epitaxial nanowire arrays [5]. GaN has lattice constants of *a* = 0.519 nm and *c* = 0.319 nm. (100) plane of LiAlO$_2$ has lattice constants of *a* = 0.517 nm, which matches well with GaN, and *c* = 0.628 nm, which is nearly twice the lattice parameter *c* for GaN. Similarly (111) plane of MgO has a lattice constant of 0.298 nm, which matches well with 0.319 nm and both have similar threefold symmetry. This close matching of lattice parameters makes lattice-matched epitaxial growth preferential, i.e., under the same synthesis conditions with only substrates being different, <001> GaN nanowires (with hexagonal cross-section) are formed on (111) MgO substrates while <1–10> oriented GaN nanowires (with triangular cross section) are formed on the lattice-matching LiAlO$_2$ substrate as shown in Figure 10.5. This method has some merit and can be extended to other substrates but using extreme control over epitaxial growth conditions. But, a major limitation of this technique is the limited availability of lattice-matched substrates for the various material systems of interest.

Combinations of substrates, precursors, and process conditions have been employed to yield unaligned as well as aligned III-nitride nanowires. Single crystalline <100> oriented GaN nanowires have been synthesized

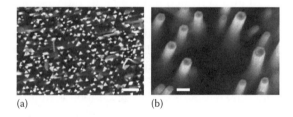

(a)   (b)

FIGURE 10.5
SEM micrograph of top view of epitaxial growth of GaN nanowires arrays via MOCVD process to demonstrate the epitaxial substrate based control on growth direction. (a) <1–10> oriented nanowires with triangular cross-section are grown on <1–10> lattice matching LiAlO$_2$ substrate and (b) <001> oriented nanowires with a hexagonal cross-section on MgO substrate with lattice spacing match in the <0001> direction of GaN. The scale bars are 100 nm. (Reproduced from Kuykendall, T. et al., *Nature*, 3, 524, 2004. With permission.)

using vaporization of a composite GaN and Fe target with a pulsed laser [6]. Nanowires formed on the walls of the quartz tube and did not have any preferred epitaxial relation with the substrate. GaN nanowires with <001> growth direction have been synthesized by PLD process using a gold-coated sapphire substrate [7]. Direct reaction of gallium metal with ammonia onto a gold catalyst–coated silicon substrate in a tube furnace reactor [8] resulted in high density of [100] GaN nanowires. GaN nanorods with <002> as the preferred orientation were synthesized via direct reaction of gallium with ammonia and nitrogen onto a gold-coated silicon substrate in a tube furnace reactor [9]. Direct reaction of gallium droplet with ammonia on a nickel-coated silicon substrate [10] resulted in GaN nanowires. <100> oriented GaN nanowires were also synthesized using direct reaction of gallium droplet on a nickel-coated silicon substrate with ammonia in a vertical furnace reactor [11]. Nanowires grown by the methods described so far do not have any epitaxial relationship to the substrate. The behavior seems to be substrate-independent as quartz, silicon, and sapphire show similar growth. Partially aligned GaN nanowires have been synthesized by the reaction of a mixture of gallium oxide and gallium nitride powders with ammonia onto gold-coated sapphire substrate in a quartz tube reactor [12].

Catalyst-assisted growth of InN nanowires has been realized via controlled evaporation of In metal onto a gold-coated substrate under ammonia flow [13]. Nanowires synthesized in this case did not have any alignment with the substrate. Similar lack of alignment with the substrate has also been observed for aluminum nitride nanowires synthesized via different routes such as Ni/Au catalyst–assisted MOCVD growth [14] and using carbon nanotubes as templates with aluminum and alumina powders [15].

10.2.2 Direct Reaction and Self-Catalysis Schemes

There has been tremendous interest to grow III-nitride nanowires without the use of any catalyst to facilitate the study of properties of pure III nitrides without the interference of foreign contaminants and develop synthesis techniques, which could be integrated with the current fabrication systems in the electronics industry. Any such techniques must avoid the use of gold, iron, and nickel, which can be present in the form of deep impurities in Si and other semiconductors. Direct reaction schemes, without the use of a catalyst, such as nitridation of Group III metal, metal oxide, and epitaxial growth via MOCVD and MBE techniques have been employed for the growth of III-nitride nanowires. In this section, various direct reaction schemes to synthesize GaN nanowires are described.

The schematic of a reactor setup for GaN nanowire synthesis via direct reaction is shown in Figure 10.6. Although the reactor setup for catalyst-assisted growth and direct reaction schemes are similar, major differences exist with the way the nanowires grow. In the case of direct nitridation of large Ga clusters, the amorphous quartz or boron nitride substrates are first

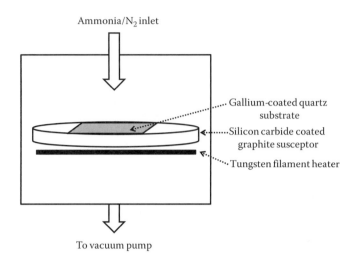

Ammonia/N$_2$ inlet

·····Gallium-coated quartz
substrate

·····Silicon carbide coated
graphite susceptor

·····Tungsten filament heater

To vacuum pump

FIGURE 10.6

Schematic of the reactor setup for direct nitridation of gallium in ammonia. The tungsten fila-ment heater radiatively heats up the susceptor and also decomposes NH$_3$ into radicals. Reaction of Ga and NH$_3$ occurs on the substrate leading to the growth of GaN nanowires.

cleaned in acetone/alcohol, rinsed with DI water, and dried in argon. A thin film of Ga is either smeared or evaporated onto the substrate and is cleaned with hydrochloric acid to remove any native oxide on the surface. A silicon carbide coated susceptor holding the substrate is radiatively heated from underneath using a tungsten filament heater. The chamber is evacuated to 10 mTorr and the susceptor is heated to 850°C using the tungsten heater at 1600°C with 50 sccm NH$_3$ flow at 20 Torr for 2 h. High densities of *c*-plane <0001> oriented GaN nanowires with sizes ranging from 15 to 40 nm and lengths around 100 μm are obtained from large Ga droplets as shown in Figure 10.7 [16]. As explained in Chapter 9, the high densities of nanowires from large Ga droplets result with bulk nucleation of III-nitride nuclei and their subsequent growth. The growth of individual nanowire occurs through basal attachment as illustrated in Figure 10.8. Here, the growth rates tend to be few tens of microns per hour compared to hundreds of microns per hour using catalyst-assisted growth.

An important feature in this case is the use of tungsten heater to radiatively heat the susceptor supporting the Ga-coated quartz substrate. In addition to radiative heating, the filament at high temperatures (1600°C or higher) can decompose ammonia completely into H, N, NH, and NH$_2$ radicals similar to an efficient plasma. Interestingly, the direct reaction between large Ga drop-lets and ammonia, leading to nanowire growth, happens only with the use of filament heater underneath or the use of plasma above. Similar synthesis conditions, with a boron nitride heater, where the substrate is placed directly on the heater, do not result in the growth of nanowires. Also, the use of pure

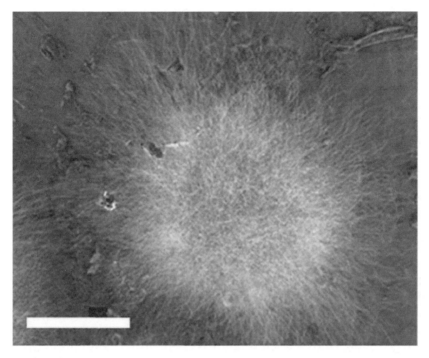

FIGURE 10.7
SEM micrograph of GaN nanowires synthesized via direct nitridation of gallium droplets in ammonia. Growth of high density of <0001> oriented GaN nanowires out of a gallium droplet takes place. The nanowires have 15–40 nm diameter range and are *ca.* 100 μm long. The scale bar is 100 μm.

Nitrogen vapor		
Gallium metal droplet	Spontaneous multiple nuclei of GaN	Multiple GaN nanowire growth from one gallium droplet

FIGURE 10.8
Schematic describing the growth mechanism of GaN nanowires via multiple nucleation and growth. Extremely low solubility of nitrogen in gallium leads to spontaneous multiple nucleation of GaN crystals on Ga droplet surface. Further growth of GaN nanowires occurs via basal attachment.

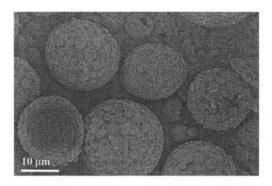

FIGURE 10.9
SEM micrograph of the polycrystalline GaN crust formed on gallium film treated with pure nitrogen plasma at 800°C. The formation of a GaN crust is clear indication of the role of hydrogen radicals in the gas phase for the growth of one-dimensional structures. (Courtesy of H. Chandrasekaran.)

nitrogen plasma and a moderate temperature of 850°C results in the formation of only GaN crust as shown in Figure 10.9. The use of hydrogen along with nitrogen in an electron cyclotron resistance (ECR) plasma experiment also results in nanowire growth from large Ga droplets. So, the use of either hydrogen or high substrate temperature is essential to achieve bulk nucleation and growth of GaN nanowires from Ga droplets with direct reaction schemes. Also, it must be noted that this technique does not produce epitaxial nanowires or nanowire arrays. But, these techniques of growing high densities of nanowires from large Ga droplets are good for producing large quantities of nanowires and nanowire thin films.

In a similar reactor setup as shown in Figure 10.6, GaN nanowires can also be synthesized in a slightly different fashion, i.e., using vapor transport of Ga in a reactive environment onto substrates kept at the synthesis temperature

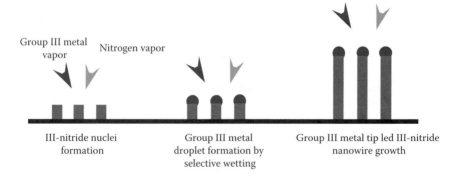

FIGURE 10.10
Schematic of III–V nanowire growth via the reactive vapor transport mechanism. The formation of III–V crystal nuclei, selective wetting of the nuclei by Group III metal forming a droplet and the tip-led growth of III–V nanowire aided by liquid phase epitaxy in illustrated.

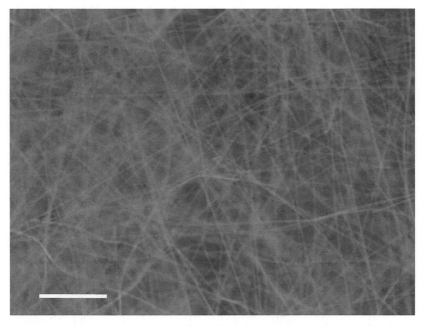

FIGURE 10.11
SEM micrograph of GaN *a*-plane <10–10> oriented nanowires synthesized via the reactive vapor transport mechanism. The nanowires are 20–50 nm in diameter and 100s of microns long. The scale bar is 1 µm. (Courtesy of H. Li.)

(see the experimental setup in Figure 10.10). GaN powder placed on a silicon carbide coated susceptor is used as the Ga source and the quartz substrate is kept 1–2 mm above the powder. After purging the chamber, the heater is ramped up slowly to 1000°C using a tungsten heater at 20 Torr and 150 sccm NH$_3$ for 3 h [16]. Such experiments typically yield unaligned *a*-plane-oriented nanowires as shown in Figure 10.11. The nucleation and growth mechanism of GaN nanowires is completely different from the bulk nucleation and growth from large, molten metal droplets. Here, in the initial stages, the Ga vapor and ammonia react on the substrate to create GaN crystal nuclei on the substrate first. The accumulation of Ga adatoms leads to Ga droplets on the GaN crystals. The Ga vapor and atomic nitrogen species dissolve into the Ga droplets, react on the molten metal surfaces, diffuse into the bulk, and start to precipitate selectively at the molten metal droplet and the crystal interface. This leads to one-dimensional growth with molten metal droplet at its tip. The growth mechanism is similar to that of catalyst cluster-assisted growth but the droplet at the tip is made of the same metal. Such schemes are termed as "self-catalysis" schemes. The experiments with GaN are typically performed at relatively high temperatures of 850°C or higher. So, no droplets are seen at the tips of the resulting thin nanowires. On the other hand, the InN nanowire growth can occur at much lower temperatures of 550°C. So, direct observations of self-catalysis schemes can be made.

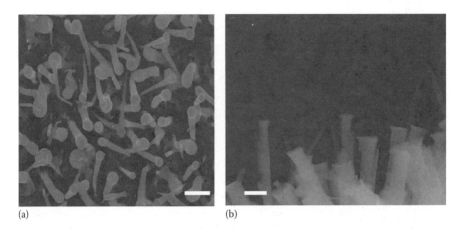

(a)   (b)

FIGURE 10.12

(a) InN nanowires synthesized on quartz substrate via the reactive vapor transport mechanism using thermal evaporation of indium metal onto the substrate in a 50 sccm flow of NH_3 at 500°C and 150 mTorr for 2 h. The scale bar on the image is 500 nm. (b) SEM micrograph of AlN nanowires synthesized via the reactive vapor transport mechanism using AlN powder onto a quartz substrate in 150 sccm NH_3 at 1100°C and 20 Torr for 3 h. Tapered wires as well as wires with hexagonal growth at the tip are synthesized. The scale bars are 1 μm. (Courtesy of H. Li and C. Pendyala.)

InN nanowire growth has also been observed to follow the reactive vapor transport mechanism as illustrated in Figure 10.10. Here, a boron nitride crucible filled with indium metal and covered with a quartz substrate is heated to 550°C and performed reactive vapor transport is carried out in 50 sccm NH_3 at 20 Torr pressure for 1 h [17]. This led to the formation of InN nanowires with 50–200 nm diameter and 1–5 μm long as shown in Figure 10.12a. The indium nitride nanowires have a large indium droplet at the tip as in the case of self-catalysis schemes. This technique can also be used to synthesize AlN nanowires at 1100°C, 30 Torr, and 150 sccm NH_3 as nitrogen source, AlN powder is heated onto a quartz piece supported on graphite blocks. Both tapered structures and nanowires with uniform diameter and hexagonal cross-section are formed as shown in Figure 10.12b.

One of the most important features of reactive vapor transport or "self-catalysis" schemes is that these techniques are quite similar to that using foreign catalyst metals except for the initial stages. So, these techniques, depending upon how the metal droplets are created, can actually lead to growth of epitaxial and nonepitaxial nanowire arrays on a variety of substrates. In the initial stages, the growth conditions must lead to epitaxial crystal nuclei formation on single-crystal substrates. An increase in Group III metal flux can lead to the nucleation of metal clusters, which further will lead the growth of epitaxial nanowire arrays.

There are several other evidences in the literature with self-catalysis schemes (or the foreign metal catalyst-free approaches) for epitaxial III-nitride nanowire arrays. Two examples using MBE and templated MOCVD are described in the following. Pure Group III metal such as Ga, In vapors using Knudsen effusion cells, and activated nitrogen species using 500 W RF nitrogen plasma were supplied to a silicon substrate heated to 750°C at 10^{-6} Torr pressure [18]. The very low flux of species at the low pressures creates conditions for controlled growth, leading to aligned GaN and InN nanowires. Aligned nanowire arrays were also synthesized using etching and growth on GaN films via masked etching and masked MOCVD growth, respectively. A mask of porous anodized aluminum is created on a GaN film via anodic oxidation of a thin aluminum layer coated on <0001> oriented GaN film. Inductively coupled plasma (ICP) comprising of N_2 and Cl_2 gases was used to etch off GaN exposed through the pores of anodized aluminum mask [19]. Silicon nitride layer was coated on GaN film and patterned openings with the required size and shape were etched on to the silicon nitride layer via lithography. MOCVD growth of GaN using TMGa and NH_3 at 1050°C at 100 Torr pressure in a commercial MOCVD reactor with a pulsed supply of 10 sccm TMGa for 20 s and 500 sccm NH_3 for 30 s on the patterned films led to the formation of <0001> oriented GaN with hexagonal cross section [20]. Pulsed supply of species resulted in uniform diameter nanowires while a continuous supply of the species resulted in a film-like morphology. Templated growth has a unique advantage in terms of precise control on the diameter and spacing of the nanowire arrays independently as per requirement.

10.2.2.1 Control of Growth Direction

It is possible to control growth direction in direct reaction schemes without the use of single-crystal templates. Bulk nucleation and growth of nanowires from large Group III droplets lead to III-nitride nanowires in <0001> or *c*-plane oriented nanowires with hexagonal cross-section. This is because, the hexagonal GaN nuclei on molten Ga surface tends to be stable with *c*-plane parallel to the molten Ga surface. Further growth of such a nuclei with *c*-plane parallel to the molten Ga surface leads to *c*-plane-oriented nanowire. In the case of self-catalytic growth of GaN nanowires using Ga droplets, the resulting nanowires are either in <10–10> direction (*a*-plane) or in <10–11> direction with rectangular cross-sections. The reasons for observed growth directions are the following: (a) the growth rates under self-catalysis schemes using Group III metal droplets can be one to two orders of magnitude higher than the bulk nucleation, and fast growth kinetics lead to growth in *a*-plane oriented nanowires compared to *c*-plane oriented nanowires; and (b) the wetting properties of Group III metals on different surfaces at different temperatures. Similar results were obtained with carbothermal reduction of Ga_2O_3 under continuous flow of ammonia. In these experiments, a mixture of Ga_2O_3 and carbon pressed into a solid bar was heated using a tungsten filament heater from underneath onto a graphite substrate (cleaned in alcohol and DI water) at 900°C at 200 Torr under 100 sccm NH_3. The region directly over the heating source had <10–11> oriented GaN nanowires and the periphery of the substrate developed <0002> oriented nanowires. The differences in the Ga flux and substrate temperature led to two different scenarios for the observed growth directions: bulk nucleation and growth from Ga droplets and self-catalytic growth using Ga droplets.

Catalyst-free synthesis techniques for III-nitrides cause both nonepitaxial and epitaxial growth. Laser ablation of a target mixture of GaN and Ga_2O_3 resulted in <0001> oriented wurtzite and cubic <111> oriented GaN nanowires [21]. Synthesis of GaN nanowires has been reported for in a hot filament CVD approach via the reaction of Ga_2O_3, C, and ammonia heated by a tungsten filament [22]. InN nanowires with <001> growth direction have been synthesized in a tube furnace reactor using nitridation of indium wires [23]. Nitridation of In_2O_3 powders has also been used to yield <001> oriented InN nanowires [24].

In catalyst-free synthesis MBE, lithography and templated synthesis routes have allowed for the synthesis of aligned GaN nanowire arrays. Aligned {0001} oriented GaN nanorods arrays are synthesized on silicon substrates with a SiN buffer layer via MBE [25]. Epitaxial growth of aligned GaN nanowire arrays via pulsed MOCVD with controlled position and diameter on presynthesized GaN films has been demonstrated with the use of patterned SiN masks [26]. In case of templated growth, ordered pores of anodized aluminum were used to transfer pattern on to a silicon layer deposited on a GaN film via reactive ion etching [27]. MOCVD growth of GaN into the porous silica template resulted in the formation of aligned GaN nanowire arrays.

10.2.3 Synthesis of Nanotubes

III-nitrides do not form layered structures similar to graphite. So, it is not natural to think that these materials will form tubes similar to carbon nanotubular structures. However, tubular morphologies with hollow core have been synthesized using two types of techniques: (a) using epitaxial growth onto sacrificial nanowire arrays and (b) direct reaction schemes.

In the sacrificial nanowire array template assisted technique, a ZnO nanowire array grown on a clean, (110) sapphire substrate was used as a template. First, the ZnO nanowire arrays were synthesized using gold catalyst layer and using Zn vapor from carbothermal reduction of ZnO and carbon mixture placed upstream in a quartz tube reactor at 850°C under 25 sccm flow of argon for 30 min. The resulting ZnO nanowires show a wide diameter distribution over 20–120 nm. Using the ZnO nanowire array as the template, MOCVD using TMGa as Ga precursor was performed with 250 sccm N_2 as carrier gas and 155 sccm of NH_3 and H_2 at 650°C for 30 min. The experiments resulted in heteroepitaxial growth of GaN over ZnO nanowire array. Subsequent removal of ZnO template by simple heating to high temperatures resulted in the formation of aligned GaN nanotubes arrays [28] as shown in Figure 10.13.

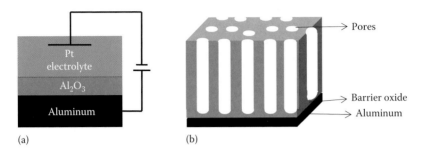

FIGURE 3.3
Simplified schematics illustrating (a) an anodization process and (b) a nanoporous alumina channel.

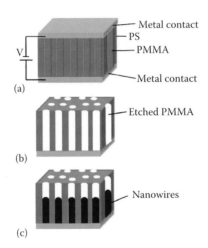

FIGURE 3.5
Schematic showing the steps in nanowire electrodeposition using a template made using a diblock copolymer: (a) formation of vertically oriented hexagonal array of cylinders of diblock copolymer under the applied electric field and heating at temperatures above their glass transition temperatures; (b) removal of PMMA by UV exposure followed by rinsing in acetic acid to form a nanoporous structure; and (c) the resulting PS matrix with nanopores is filled by electrodeposition to form nanowires in a polymer matrix. (Recreated from Thurn-Albrecht, T. et al., *Science*, 290, 2126, 2000. With permission.)

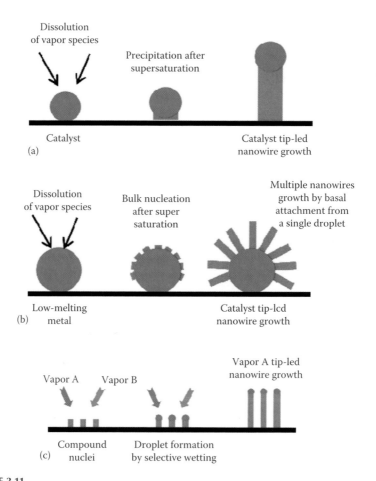

FIGURE 3.11
Schematics illustrating the growth mechanism of nanowires: (a) classical VLS growth mechanism; (b) low-melting metal mediated growth; and (c) reactive vapor transport mechanism.

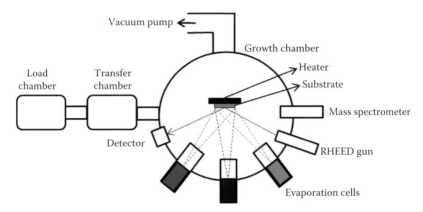

FIGURE 3.15
Schematic of a typical reactor setup used for MBE.

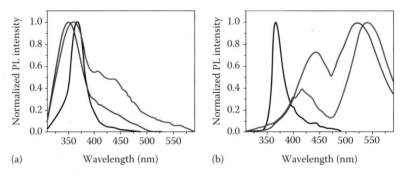

FIGURE 10.28

Photoluminescence spectra (a) *a*-plane-oriented GaN nanowires and (b) polycrystalline GaN at 0–100 ps (black), 500–600 ps (blue), and 1–5 ns (red) after laser excitation. The blue shift from 365 to 350 nm in case of *a*-plane-oriented nanowires is different from the emission observed in polycrystalline GaN (*c*-plane nanowires show the same behavior). (Reproduced from Chin, A. et al., *Nano Lett.*, 7, 626, 2007. With permission.)

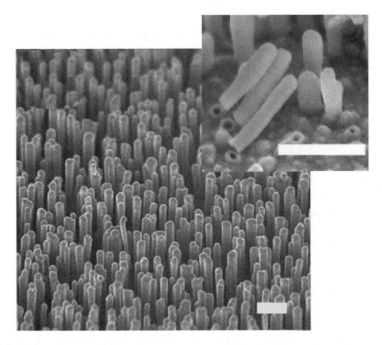

FIGURE 10.13
SEM micrograph of GaN nanotubes synthesized on ZnO nanowires by epitaxial casting. GaN film is grown on presynthesized zinc oxide nanowires to form a conformal layer. Thermal treatment removes the zinc oxide resulting in the formation of GaN tubes. Inset shows the close up of GaN nanotubes with the internal channels. The scale bars are 1 μm. (Reproduced from Goldberger, J. et al., *Nature*, 422, 599, 2003. With permission.)

Direct nitridation of molten gallium pools in nitrogen and hydrogen plasmas led to the formation of micron-sized GaN hexagonal rods and sometimes star-shaped or hexagonal tubes as seen in Figure 10.14a and b, respectively. Star-shaped tubes were synthesized when Ga metal smeared

(a)   (b)

FIGURE 10.14
(a) SEM micrograph of hexagonal star–shaped GaN tubes synthesized via the exposure of Ga metal pool to N_2 plasma at 1000°C for 5 h after N_2 plasma pretreatment. (b) SEM micrograph of hexagonal tubes synthesized via the N_2 plasma treatment of gallium film on quartz substrate at 700°C for 2 h after H_2 plasma pretreatment. The irregular and incomplete growth is an indication of self-assembly of GaN crystal nuclei. The scale bars are 1 μm. (Courtesy of H. Li.)

on quartz was subjected to H_2 plasma pretreatment and subsequent 100 W N_2 plasma treatment of the sample with 20 sccm N_2 flow at 70 mTorr and 1000°C for 5 h. Hexagonal hollow tubes were obtained when the gallium-smeared substrate was subjected to N_2 plasma for 30 min initially. This is followed by a N_2 plasma treatment under 30 sccm N_2 flow at 100 mTorr and 700°C for 2 h. The role of nitrogen plasma in the low-pressure synthesis of GaN crystals has been discussed before [1]. The observations for such tubular growth can be explained as follows: (a) self-assembly of GaN crystal nuclei on molten Ga surface can lead to hexagon-shaped platelets or hollow rings and (b) the basal growth onto such patterns leads to one-dimensional growth shapes such as hexagonal rods and hexagonal-shaped tubes. The assembly process can easily be confirmed through observations of various patterns: incomplete hexagonal patterns and irregular hexagonal patterns. However, the conditions to control such self-assembly process for nuclei in the initial stages are not completely understood. In the case of one-dimensional growth of hollow rods, the continuous flow of gallium through the hollow core due to capillary effect has been observed. Such Ga flow can provide large material flux for growth on the outer walls and can influence the faceting to become star-shaped.

In the case of InN, the tubular structures can easily be formed by decomposing larger size (>200 nm size) InN rods in a controlled atmosphere. In these experiments, the quartz substrate coated with InN rods was placed on a boron nitride heater in a cold wall stainless steel vacuum chamber. The heater was ramped up to 650°C under 50 sccm NH_3 flow at 150 mTorr for 30 min [32]. This resulted in the formation of InN nanotubes as shown in Figure 10.15. The transformation of thick nanowires to thin walled, hollow tubular structures is believed to take place as the following: simultaneous decomposition of InN into indium and nitridation to form InN shell formation. See the schematic shown in Figure 10.16.

GaN nanotubes have also been synthesized by a variety of other methods ranging from nitridation of Ga_2O nanotubes [33], Ga_2O_3 powders [33,34], and indium-assisted growth [35] to templated synthesis [36] and ICP etching of GaN films [37]. Direct nitridation of amorphous Ga_2O nanotubes in a vertical induction furnace [30] resulted in the formation of single-crystal hexagonal GaN nanotubes. Rectangular self-assembly of GaN nuclei on the substrate under the growth conditions and attachment of subsequent species on the pattern is believed to result in the formation of rectangular nanotubes [34].

Carbonitridation of multiwalled carbon nanotubes (MWCNTs) and graphite powders was employed to synthesize straight and conical <010> oriented InN nanotubes [38]. Crystalline <001> oriented InN nanotubes have been synthesized by the nitridation of indium wires in a horizontal tube furnace [23]. InN nanotubes have also been synthesized by the decomposition of presynthesized InN nanowires. InN nanowires have indium droplets at the tips, which enhance selective decomposition of the nanowire core, leading to the formation of nanotubes [17].

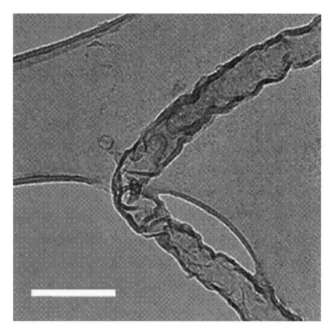

FIGURE 10.15
TEM micrograph of InN nanotubes synthesized by the decomposition of presynthesized InN nanowires. The internal hollow of the nanotube along with the thin wall is clearly visible. The scale bar is 200 nm. (Reproduced from Yoon, J.W., *Thin Solid Films*, 471, 273, 2005. With permission.)

Cubic AlN nanotubes are synthesized by nonequilibrium dc-arc plasma method [39] and hexagonal <0001> oriented AlN nanotubes by the nitridation of aluminum powder with cobalt sulfate in a tube furnace [40].

10.2.4 Micro/Nanomorphologies

Two types of morphologies that extend from micron scale to nanoscale for III-nitrides are described: belt-shaped morphologies and tapered morphologies.

10.2.4.1 III-Nitride Nanobelts

Belt-like morphologies are another one-dimensional morphological form commonly found in the case of many oxides and nitrides. However, unlike in the case of nanowires, the underlying growth mechanism has been elusive for belt-shaped morphology. In almost all the experiments, the belt-shaped morphologies occurred either at high temperatures or in experiments that used reactive vapor transport methods. In this section, a simple mechanism is described that explains the morphological evolution of belts especially in the case of Group III nitrides.

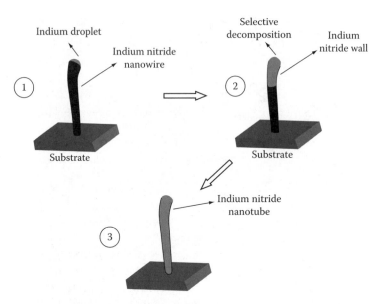

FIGURE 10.16
Schematic of the various steps in the formation of InN nanotubes from InN nanowires. The presence of indium at the nanowire tip (1) causes selective decomposition of the InN core (2) and leads to the formation of InN nanotubes (3). (Reproduced from Vaddiraju, S. et al., *Nano Lett.*, 5, 1625, 2005. With permission.)

Reactive vapor transport of Ga vapor using GaN powder as the source in NH_3 atmosphere yields 20 nm diameter and several microns long, *a*-plane oriented nanowires [16]. If the GaN powder in the above experiment is kept at 1000°C at 20 Torr to perform "homoepitaxy" onto the presynthesized *a*-plane oriented nanowires. Then belt-shaped morphologies emanate from the original *a*-plane oriented nanowires as seen in Figure 10.17a. The TEM micrograph of a nanobelt shown in Figure 10.17b also confirms that the growth direction of the original wire is along <10–10> and the growth direction of the nanobelt is <0001>. The *a*-plane oriented nanowires have rectangular facets; two of which are *c*-plane polar surfaces (Ga and N terminated surfaces). At low Ga flux, the exposed growth surface was determined to be the (0–110) plane with the growth direction of the belts being <0001> whereas at high Ga flux, kinetics favor <1–120> oriented growth with the polar (0001) planes exposed. In the case of GaN nanowire growth by the reactive vapor transport mechanism, saw tooth type growth of GaN nanowires with <01–11> orientation is also observed. As shown in Figure 10.18, the tooth edges have <0001> and <10–10> orientations.

Adatom diffusion lengths of the nonpolar surfaces are on the order of tens of microns and on the polar surfaces are in the range of 10–100 nm. Nonpolar surfaces provide preferential wetting sites for Ga. Also, it is important to note that the presence of mono layers of Ga enhances the adatom diffusion on the surface irrespective of polarity. Hence, it can be inferred that for *a*-plane oriented nanowires, the polar surfaces with lower diffusion lengths are

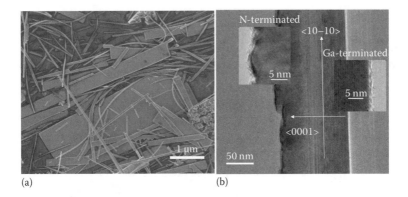

(a)   (b)

FIGURE 10.17

(a) SEM micrograph of the nanobelts grown by homoepitaxy on GaN *a*-plane oriented nanowires. The presence of large belts indicates a preferential growth on one surface. (b) TEM micrograph of a GaN nanobelt. The growth direction of the original nanowire is <10–10> and preferential epitaxial growth on the polar surfaces causes the belt to grow in <0001> direction. (Reproduced from Li, H. et al., *Adv. Mater.*, 18, 216, 2006. With permission.)

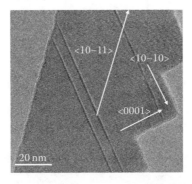

FIGURE 10.18

TEM micrograph of sawtooth-like <10–11> GaN nanowires. The tooth edges are <0001> and <10–10> oriented. Faster kinetics under the synthesis conditions for these planes could cause preferential growth in such fashion. (Courtesy of H. Li.)

preferential sites for epitaxial growth. It is also clear from Figure 10.18 that the edge facets are in fact the polar surfaces which substantiate the mechanism of growth on the polar surfaces. Since Ga-terminated surface is chemically more active than N terminated surfaces, growth occurs only on one polar face.

10.2.4.2 Tapered Morphologies

It is also possible to obtain tapered morphologies from micron-scale crystals to nanometer scale, one-dimensional, tapered morphologies. In the case of III nitrides, consider the following set of experiments using GaN [16]. The *c*-plane oriented nanowires were synthesized using direct nitridation of micron-scale Ga droplets using ammonia at 850°C. Homoepitaxy experiments were performed using controlled supply of Ga vapor using GaN powder source at 1000°C and ammonia onto presynthesized *c*-plane-oriented GaN nanowires. The homoepitaxial growth on *c*-plane-oriented GaN nanowires led to hexagonal prismatic island growth as shown in Figure 10.19. The

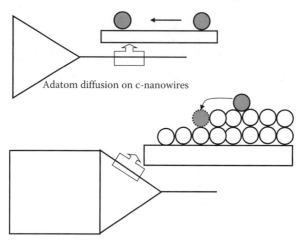

Adatom diffusion on c-nanowires

Steps generation and propagation on (1011) facets

FIGURE 10.19
Schematic of the growth mechanism of hexagonal islands growth on *c*-plane-oriented GaN nanowires. The adatoms diffuse on the nonpolar surfaces of the wire, form steps, which propagate along the pyramid toward the prism which leads to the observed island growth. (Reproduced from Li, H. et al., *Adv. Mater.*, 18, 216, 2006. With permission.)

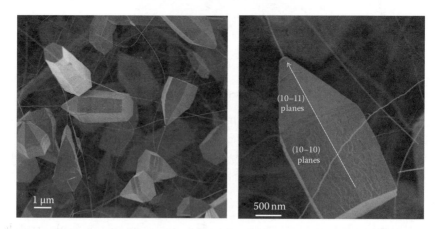

FIGURE 10.20
Hexagonal prism islands grown on *c*-plane-oriented GaN nanowires. The wires grow in <0001> direction with the walls being the six nonpolar facets. The side facets of the prism are (10–10) and the inclined facets are (10–11). The observed symmetry indicates a uniform growth on all the surfaces. (Courtesy of H. Li.)

islands grow with all the six parallel facets being nonpolar (10–10) and the inclined facets being (10–11) planes. The adatom ballistic diffusion length on smaller diameter nanowires can be several microns, leading to large spacing between islands. In fact, the islands tend to grow from ends of the nanowires and more islands develop on the wire with time. The island growth occurs with step nucleation and propagation toward the hexagonal prismatic end, leading to tapered morphologies as illustrated in Figure 10.20.

10.3 Branching of Nanowires

Branching of nanowires can be achieved with both catalyst-assisted or self-catalyzed schemes in the case of III-nitrides. In the case of catalyst-assisted schemes using gold, the catalyst particles need to be supplied onto growing nanowires to initiate a new branch. In this case, the diameter of the branch is determined primarily by the catalyst particle size. However, on the other hand, the branching can easily be achieved with III-nitrides as they exhibit self-catalytic growth using Group III metal droplets. In fact, it is easy to produce Group III metal clusters on the existing nanowires to initiate new branches. In this case, the diameter of the new branch depends upon the droplet diameter, which in turn depends heavily on the underlying nanowire diameter along with process conditions. Typically, the diameter of the Group III droplet is either same or less than the underlying nanowire. Several examples exist for both homo- and heterobranching of Group III nitrides.

10.3.1 Homobranching or "Tree-Like" Structures

Branching has been observed in the case of InN nanowires during synthesis via reactive vapor transport schemes as seen in Figure 10.21. Repetitive growth

 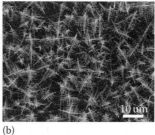

(a)   (b)

FIGURE 10.21
(a) SEM micrograph of branched InN nanowires grown via reactive vapor transport method on presynthesized InN nanowires. (b) SEM micrograph of multilevel branching in InN nanowires. (Courtesy of S. Vaddiraju.)

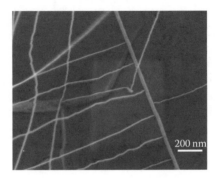

FIGURE 10.22

SEM micrograph of branched GaN *a*-plane oriented nanowires. GaN growth on presynthesized nanowires using GaN source at 900°C in NH$_3$ leads to the formation of GaN nanowire branches on GaN nanowires. (Courtesy of S. Vaddiraju.)

of InN on presynthesized nanowires leads to the formation of branches. Multiple branching steps can be performed to yield tree-like structures as shown in Figure 10.21b. In this case, indium droplets form on the nanowire surface due to preferential wetting, create liquid phase epitaxial conditions, and lead to nanowire growth with the vapor phase supply of group V species. This multilevel branching offers a unique advantage of increased surface area while retaining the properties of the material. In the case of InN, the diameters of subsequent branches are smaller than the underlying ones. Two-dimensional branching has also been observed in case of GaN *a*-plane-oriented nanowires as shown in Figure 10.22. In the case of *c*-plane-oriented nanowires, it is possible to observe branching from its six hexagonal facets (or in six directions around the wire).

10.3.2 Heterobranching

Heterobranching involves growth of nanowire branches of one material on nanowires of another material, which could potentially allow for the utilization of properties of both the materials in nanowire morphology. AlN branches have been synthesized on GaN nanowires via reactive vapor transport as shown in Figure 10.23. Synthesis procedure is the same as the growth of AlN nanowires with the substrate being GaN nanowire–coated quartz substrate rather than just a quartz piece. In this case, growth occurs via the nitridation of Al droplets that form on GaN nanowires, analogous to the mechanism for InN nanowire branching. Following a similar strategy, heteroepitaxial growth of InN on GaN nanowires following InN nanowire growth conditions and a GaN nanowire coated substrate leads to the formation of GaN nanowires decorated with InN beads as shown in Figure 10.24.

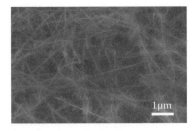

FIGURE 10.23
SEM micrograph of AlN nanowires heterobranching on GaN *a*-plane-oriented nanowires by reactive vapor transport method. Al droplets form on GaN wires at 1100°C and grow AlN nanowires in the presence of Al and NH₃. (Courtesy of S. Vaddiraju.)

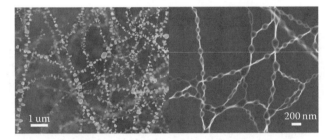

FIGURE 10.24
SEM micrograph of InN beads on GaN *a*-plane oriented nanowires. Indium droplets condense on the GaN nanowire used as substrates. The nitridation of the droplets leads to the formation of InN beads rather than InN nanowire branches on GaN nanowires. (Courtesy of S. Vaddiraju.)

10.4 Diameter Reduction of III-Nitride Nanowires

The true advantage of nanowires lies in the fact that the properties change drastically with size due to quantum confinement effects. Quantum confinement can be realized only when one of the dimensions of the nanowire is smaller than the Bohr radius, which is material-specific. So far, the synthesis of III-nitride nanowires via any technique has resulted in nanowires with diameters in the range of 20–100 nm. These sizes are not small enough to realize the true advantage that III-nitride nanowires can offer. The estimated Bohr radii values for III-nitrides range from 1 to 10 nanowires. For example, two different studies reported contrastingly different values for Bohr radius values for GaN as 2.8 nm [29] and 11 nm [30], respectively. In contrast, antimonides have a significantly larger Bohr radius; 65 nm has been reported for InSb [31]. Moreover, the synthesis of III-nitride nanowires with diameters in the range of 1–10 nm is very difficult as evident from the lack of reported studies. Catalyst metal droplets determine the size of nanowires and are the limiting factor in the synthesis of ultrathin nanowires. Until a reliable technique is developed, postsynthesis techniques can be used to reduce the diameter of nanowires. Controlled, postsynthesis decomposition of existing nanowires seems to be a promising route toward obtaining thinner nanowires.

The idea is to decompose III-Nitride nanowires under kinetic control which is demonstrated using ammonia as gaseous atmosphere compared to hydrogen. The decomposition of InN nanowires were performed at 650°C in 150 sccm NH_3 at 20 Torr for 15–60 min [32]. The decomposition of GaN nanowires was performed at a substrate temperature of 1100°C using 150 sccm of NH_3 at 20 Torr for 15–60 min. In the case of GaN nanowires, the wires with initial diameters in the range of 20–70 nm have been reduced uniformly to 5–20 nm after decomposition in NH_3 as shown in Figure 10.25d; in contrast, the wires

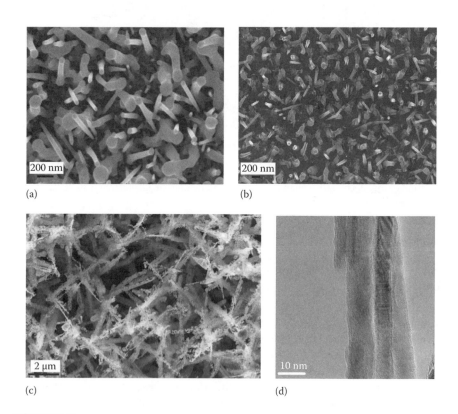

(a)   (b)

(c)   (d)

FIGURE 10.25
(a) SEM micrograph of as-synthesized InN nanowires with indium droplets at the tips. (b) SEM micrograph of InN nanowires after decomposition for 30 min clearly showing the reduction in diameter with the nanowires remaining intact. (c) SEM micrograph of the InN nanowires with large diameters (>200 nm) after 30 min decomposition showing local decomposition. (d) TEM micrograph of GaN nanowires after 30 min decomposition with intact wires. (Reproduced from Vaddiraju, S. et al., *Nano Lett.*, 5, 1625, 2005. With permission.)

decomposed completely in a H_2 ambient under identical experimental conditions. This is an indication of a uniform layered decomposition under ammonia atmosphere. The same mechanism of decomposition is observed in the case of InN nanowires of diameters <100 nm indicated by the Figure 10.25. The decomposition of GaN/InN produces Ga/In metal adatoms. In the case of thin nanowires, the adatoms undergo ballistic diffusion to the end of the nanowire, which in the case of thin GaN nanowires either evaporate or react and form GaN whereas in the case InN nanowires with indium tip, the indium adatoms merge with the existing In droplet. This mechanism of decomposition is illustrated in Figure 10.26b. The decomposition process is uniform, which is evidenced by the uniform reduction in nanowire diameter with the growth direction remaining the same as shown in Figure 10.25d.

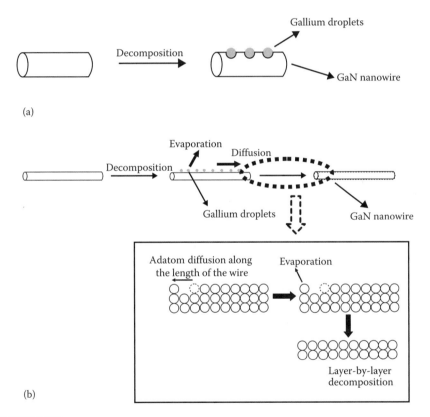

FIGURE 10.26
(a) A schematic illustrating the proposed decomposition mechanism in thick GaN nanowires. This mechanism can be extended to the decomposition of InN nanowires. In the case of thick InN nanowires, etch pits form on the nanowires, causing local decomposition of the structure. (b) A schematic illustrating the proposed decomposition mechanism in thin GaN nanowires. In this mechanism, the formation of very small droplets followed by their evaporation or ballistic diffusion along the length of the nanowire leads to the uniform thinning of the nanowires. (Reproduced from Vaddiraju, S. et al., *Nano Lett.*, 5, 1625, 2005. With permission.)

Decomposition of thicker InN nanowires creates a larger pool of indium metal on the nanowire surface, which does not diffuse very fast. Indium droplets on the nanowires cause selective etching forming etch pits, which are clearly indicated in the Figure 10.25c where the local etching of InN leads to the formation of porous structures. The mechanism for the decomposition of thicker nanowires is illustrated in Figure 10.26a. This technique should be applicable to other systems as well. A fine-tuned control on the decomposition process is a prerequisite for any such attempt.

10.5 Direction-Dependent Properties

Dependence of epitaxial growth on the nanowire orientation has already been established. The dependence of polarity and chemical nature with orientation gives rise to unique direction-dependent optoelectronic properties. In the case of GaN nanowires, a marked blue shift as seen in Figure 10.27 in UV photoluminescence for *a*-plane oriented compared to *c*-plane-oriented nanowires has been observed [5,16]. In the case of GaN *a*-plane-oriented nanowires, emission due to relaxation of excited electrons at 365 nm is transformed into a long-lived 350 nm emission, which is confirmed by time-resolved photoluminescence spectrum as seen in Figure 10.28a. This is unlike the emission in polycrystalline GaN as seen in Figure 10.28b, which is the same for *c*-plane-oriented GaN nanowires. Surface trapping processes due to oxidation are responsible for the observed change in emission. The Ga and N terminated polar surfaces in *a*-plane-oriented nanowires are susceptible to oxidation, which leads to the formation of a thin oxide layer on the nanowire surface. This can induce trap states into the material at the surface and cause

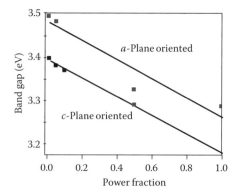

FIGURE 10.27
Consistent blue shifted (~0.4 eV) photoluminescence observed in *a*-plane nanowires compared to *c*-plane oriented GaN nanowires over a range of laser powers. (Reproduced from Li, H. et al., *Adv. Mater.*, 18, 216, 2006. With permission.)

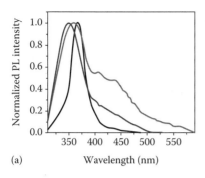
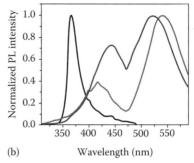

(a) Wavelength (nm) (b) Wavelength (nm)

FIGURE 10.28
(See color insert following page 240.) Photoluminescence spectra (a) *a*-plane-oriented GaN nanowires and (b) polycrystalline GaN at 0–100 ps (black), 500–600 ps (blue), and 1–5 ns (red) after laser excitation. The blue shift from 365 to 350 nm in case of *a*-plane-oriented nanowires is different from the emission observed in polycrystalline GaN (*c*-plane nanowires show the same behavior). (Reproduced from Chin, A. et al., *Nano Lett.*, 7, 626, 2007. With permission.)

the observations of blue shift of UV PL from GaN *a*-plane-oriented nanowires. *c*-Plane-oriented nanowires have nonpolar surfaces, thus eliminating any chances of oxidation or formation of trap states [41].

a-Plane- and *c*-plane-oriented GaN nanowires show entirely different gate-dependent conductivity behavior. While *a*-plane-oriented nanowires show variation in conductivity with varying gate voltages as shown in Figure 10.29, no variation in conductivity is observed in the case of *c*-plane-oriented nanowires as seen in Figure 10.29. The absence of variation of conductivity can be attributed to the presence of stacking faults in *c*-plane-oriented nanowires, which is not the case with the *a*-plane-oriented nanowires [42].

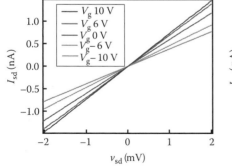
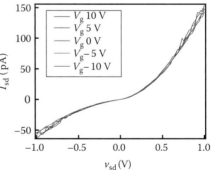

FIGURE 10.29
Gate-dependent I_{sd}–V_{sd} data recorded GaN nanowire FET (a) *a*-plane-oriented nanowire showing variation in conductivity with gate voltage and (b) *c*-plane oriented nanowire showing no variation in conductivity. (Reproduced from Sunkara, M.K. et al., *ECS. Trans.*, 3, 421, 2006. With permission.)

References

1. A. Argoitia, C. C. Hayman, J. C. Angus, L. Wang, J. S. Dyck, and K. Kash, *Appl. Phys. Lett.* 70, 179 (1997).
2. Z. Zhong, F. Qian, D. Wang, and C. M. Lieber, *Nano Lett.* 3, 343 (2003).
3. T. Kuykendall, P. Pauzauskie, S. Lee, Y. Zhang, J. Goldberger, and P. Yang, *Nano Lett.* 3, 1063 (2003).
4. K. Kim, T. Henry, G. Cui, J. Han, Y. Song, A. V. Nurmikko, and H. Tong, *Phys. Stat. Sol.* (b) 244, 1810 (2007).
5. T. Kuykendall, P. Pazauskie, Y. Zhang, J. Goldberger, D. Sirbuly, J. Denlinger, and P. Yang, *Nature* 3, 524 (2004).
6. X. Duan and C. M. Lieber, *J. Am. Chem. Soc.* 122, 188 (2000).
7. D. K. T. Ng, L. S. Tan, and M. H. Hong, *Curr. Appl. Phy.* 6, 403 (2006).
8. C. Cao, X. Xiang, and H. Zhu, *J. Crystal Growth* 273, 375 (2005).
9. Z. Yu, Z. Yang, S. Wang, Y. Jin, J. G. Liu, M. Gong, and X. Sun, *Chem. Vap. Dep.* 11, 433 (2005).
10. J. Xu, R. M. Wang, D. Yu, and X. Chen, *Adv. Mater.* 15, 419 (2003).
11. Y. H. Mo, K. S. Nahm, E. K. Suh, K. Y. Lim, and S.H. Lee, *phys. Stat. Sol.* (c) 1, 148 (2003).
12. B. Liu, Y. Bando, C. Tang, F. Xu, and D. Goldberg, *Appl. Phys. Lett.* 87, 073106 (2005).
13. C. H. Liang, L. C. Chen, J. S. Hwang, K.H. Chen, Y.T. Hung, and Y.F. Chen, *App. Phys. Lett.* 81, 22 (2002).
14. C. Foerster, D. Cengher, K. Tonisch, O. Ambacher, and V. Cimalla, *Phy. Stat. Sol.* (b) 243, 1476 (2006).
15. Y. Zhang, J. Liu, R. He, Q. Zhang, X. Zhang, and J. Zhu, *Chem. Mater.* 13, 3899 (2001).
16. H. Li, A. H Chin, and M. K Sunkara, *Adv. Mater.* 18, 216 (2006).
17. S. Vaddiraju, A. Mohite, A. Chin, M. Meyyappan, G. Sumanasekera, B. Alphenaar, and M. K. Sunkara, *Nano Lett.* 5, 1625 (2005).
18. R. Calarco and M. Marso, *Appl. Phys. A* 87, 499 (2007).
19. P. Deb, H. Kim, V. Rawat, M. Oliver, S. Kim, M. Marshall, E. Stach, and T. Sands, *Nano Lett.* 5, 1847 (2005).
20. S. Hersee, X. Sun, and X. Wang, *Nano Lett.* 6, 1808 (2006).
21. W. S. Shi, Y. F. Zheng, N. Wang, C.S. Lee, and S.T. Lee, *Chem. Phys. Lett.* 345, 377 (2001).
22. H. Y. Peng, N. Wang, X. T. Zhou, Y.F. Zheng, C.S. Lee, and S.T. Lee, *Chem. Phys. Lett.* 359, 241 (2002).
23. H. Y. Xu, Z. Liu, X. T. Zhang, and S.K. Hark, *Appl. Phys. Lett.* 90, 113105 (2007).
24. W. Zhou, Z. Zhang, L. Liu, X. Dou, J. Wang, X. Zhao, D. Liu, Y. Go, L. Song, Y. Xiang, J. Zhou, S. Xie, and S. Luo, *Small* 1, 1004 (2005).
25. Y. H. Kim, J. Y. Lee, S. H. Lee, J.E. Oh, and H.S. Lee, *Appl. Phys. A* 80, 1635 (2005).
26. S. D. Hersee, X. Sun, and X. Wang, *Nano Lett.* 6, 1808 (2006).
27. P. Deb, H. Kim, V. Rawat, M. Oliver, S. Kim, M. Marshall, E. Stach, and T. Stands, *Nano Lett.* 5, 1847 (2005).

28. J. Goldberger, R. He, Y. Zheng, S. Lee, H. Yan, H. Choi, and P. Yang, *Nature* 422, 599 (2003).
29. J. W. Yoon, T. Sasaki, C. H. Roh, S. H. Shim, K. B. Shim, and N. Koshizaki, *Thin Solid Films* 471, 273–276 (2005).
30. T. Paskova, *Nitrides with Nonpolar Surfaces: Growth, Properties, and Devices*, Wiley VCH, New York, 379 (2008).
31. Y. Yang, L. Li, X. Huang, G. Li, and L. Zhang, *J. Mater. Sci.* 42, 8, 2753–2757 (2007).
32. S. Vaddiraju, Chemical and Reactive Vapor Transport Methods for the Synthesis of Inorganic Nanowires and Nanowire Arrays, Ph.D. Dissertation, University of Louisville (2006).
33. Y. Bando, D. Golberg, Q. Liu, and J. Hu, *Angew. Chem. Int. Ed.* 42, 3493 (2003).
34. Y. Bando, J. H Zhan, F. F. Xu, T. Sekiguchi, D. Golberg, and J. Q. Hu, *Adv. Mater.* 16, 1465 (2004).
35. L. W. Yin, Y. Bando, Y. Zhu, D. Goldberg, and M. Li, *Appl. Phys. Lett.* 84, 3912 (2004).
36. J. Goldberger, R. He, Y. Zhang, S. Lee, H. Yan, H. J. Choi, and P. Yang, *Nature* 422, 599 (2003).
37. S. C. Hung, Y. K. Su, S. J. Chang, S.C. Chen, L.W. Ji, T.H. Fang, L.W. Tu, and M. Chen, *Appl. Phys. A* 80, 1607 (2005).
38. Y. Bando, D. Golberg, M. S. Li, and L. W. Yin, *Adv. Mater.* 16, 1833 (2004).
39. V. N. Tondare, C. Balasubramanian, S. V. Shende, D. S. Joag, V. P. Godbole, S. V. Bhoraskar, and M. Bhadbhade, *App. Phys. Lett.* 80, 4813 (2002).
40. Q. Wu, Z. Hu, and X. Wang, *J. Am. Chem. Soc.* 125, 10176 (2003).
41. A. Chin, T. Ahn, H. Li, S. Vaddiraju, C. Bardeen, C. Ning, and M. K. Sunkara, *Nano Lett.* 7, 626 (2007).
42. M. K. Sunkara, R. Makkena, H. Li, and B. Alphenaar, *ECS. Trans.* 3, 421 (2006).

11

Other Nanowires

Besides the elemental (Si, Ge) and compound (III–V) semiconductors, oxides, nitrides, and metals, a variety of other materials have been grown in the form of nanowires. The general motivation has come from the absence of lattice-matching requirement with the substrate, possibility of higher absorption and higher quantum efficiencies due to quantum confinement, band gap engineering allowing operation over wider wavelengths, and the overall desire to make smaller devices with higher integration density. The applications include optics and optoelectronics in the ultraviolet (UV), visible, and infrared (IR) regimes such as lasers, detectors, optical switches, and in other areas such as electronics, interconnects, sensors, thermoelectric devices, and field emitters. Reports of nanowire growth cover antimonides [1–6], selenides [7–24], tellurides [25–40], sulfides [41–65], silicides [66–83], boron [84–88], and its compounds such as boron carbide [89], silicon carbide [90–95], aluminum borate [96,97], magnesium boride [98], lithium fluoride [99], and others [100,101].

11.1 Antimonides

The band gap and optical properties can be tailored to meet application needs by either varying the diameter to realize quantum confinement or by alloying. The room temperature band gaps of GaSb and InSb are 0.72 and 0.17 eV, respectively. Ternary alloy nanowires, $Ga_xIn_{1-x}Sb$, can provide a continuous variation in band gap between 0.17 and 0.72 eV. The electron effective mass in both GaSb and InSb is low, $0.042m_o$ and $0.013,5m_o$ respectively. Therefore, varying the diameter, even in the 40 nm range, leads to a variation in band gap. For example, varying the diameter from 65 nm down to 10 nm can change the band gap of InSb from 0.17 to 0.4 eV. The importance of these materials has long been recognized for mid- and long-wave IR detection and lasing applications. In addition, InSb is useful for fabricating thermoelectric devices and magneto-resistive sensors.

Antimony dissolution into gallium or indium is negligible at room temperature and hence the supersaturation of gallium and indium with antimony should lead to the formation of respective nanowires. There are a couple of different approaches to synthesize these nanowires [5]. In direct antimonidization of gallium or indium, droplets of the metal can be exposed to

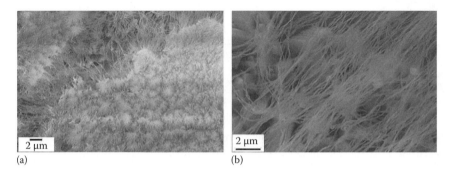

FIGURE 11.1
SEM images of GaSb synthesized using $SbCl_3$ as the antimony source. (Courtesy of S. Vaddiraju.)

antimony vapor. A thin layer of Ga or In on the substrate forms droplets at the growth temperature. Antimony can be supplied either as pure vapor or through $SbCl_3$. Alternatively, in the case of InSb growth, InSb powder can be sublimed to generate the source vapor and supplied to the indium droplets on the substrate. Since Ga and In tend to form oxides easily, growth is carried out at low pressures (150 mTorr to few Torr) under hydrogen flow. The growth temperature is 800°C–1050°C except when InSb is used to generate the source vapor which requires only 450°C.

Figure 11.1 shows GaSb nanowires grown using $SbCl_3$ using as the antimony source [5]. These nanowires are 20–30 nm in diameter and 5 μm long. The supersaturation of Ga with antimony leads to spontaneous nucleation of GaSb crystal nuclei on top of the droplets, and further growth of the nuclei occurs via basal attachment. This leads to the formation of GaSb nanowires growing outward from the fairly large size Ga droplet in all directions. The presence of chlorine (from $SbCl_3$) suppresses lateral growth while promoting growth of GaSb crystal nuclei in one dimension. High-resolution transmission electron microscopy (HRTEM) (not shown here) indicates a [110] growth direction. Figure 11.2 shows indium antimonide nanowires, which are about 100 nm in diameter and 2 μm long [5]. Dipping the indium-coated substrate in 30% HCl (by volume) quickly enhances the surface tension of the indium melt surface and helps to reduce lateral growth. XRD analysis indicates that the nanowires have a diamond cubic phase with a lattice parameter of 6.4959 Å. Raman spectrum of the nanowires shows two primary modes of InSb at 179 and 190 cm^{-1}. The transmission electron microscopy (TEM) analysis and diffraction pattern (not shown here) confirm the growth direction to be [110].

InSb can also be readily prepared using anodized alumina membrane (AAM) templates [1,4]. A typical electrolyte solution consists of 0.1 M $InCl_3$, 0.025 M $SbCl_3$, 0.15 M $C_6H_8O_7 \cdot H_2O$ and 0.06 M $K_3C_6H_5O_7 \cdot H_2O$ with the pH-value-adjusted to 2.2 by adding 7.0 M HCl [1]. A potential of −0.85 to −1.5 V is used at 16°C for 40 min to complete the deposition inside the AAM pores.

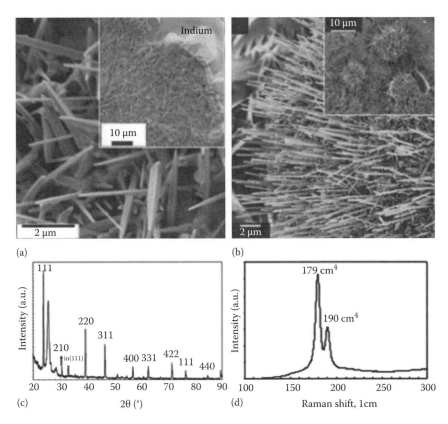

FIGURE 11.2
Indium antimonide nanowires synthesized (a) without the presence of chlorine and (b) with the presence of chlorine. (c) XRD of the InSb nanowires showing a diamond cubic crystal structure with a lattice parameter of 6.458 Å. (d) Raman spectrum showing modes corresponding to InSb. (Courtesy of S. Vaddiraju.)

The exciton Bohr radius of InSb is 65.5 nm and for diameters smaller than this, quantum confinement effects will be important. Figure 11.3 shows IR absorption spectra for 60 and 80 nm diameter InSb nanowires prepared using an AAM template approach [4]. When the diameter decreases, the optical absorption edge moves toward shorter wavelength region. Since InSb is a direct band gap material, the absorption coefficient α is related to the band gap E_g through:

$$(\alpha h\nu)^2 = A(h\nu - E_g) \tag{11.1}$$

where
   h is the Planck constant
   ν is the frequency
   A is a constant

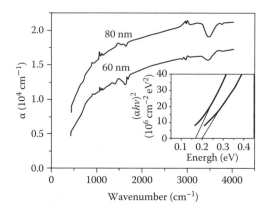

FIGURE 11.3
Infrared absorption spectra of InSb nanowires of different diameters. The inset is a plot of $(\alpha h\nu)^2$ versus $(h\nu)$. (Reproduced from Wang, Y. et al., *J. Mater. Sci.*, 42, 2753, 2007. With permission.)

The inset of Figure 11.3 is a plot of $(\alpha h\nu)^2$ versus $h\nu$ and the band gap is obtained by extrapolating the curves to the $\alpha = 0$, x-axis. The band gap is 0.17 eV for 80 nm nanowires, which is the same as that of bulk InSb. When the diameter is reduced to 60 nm, the band gap increases to 0.2 eV, indicating a blueshift due to quantum confinement.

11.2 Selenides

11.2.1 Zinc Selenide

ZnSe has a direct energy band gap of 2.7 eV at room temperature. In bulk and thin film form, this II–VI material has been extensively studied for applications in blue-green laser diode, mid-IR laser source for remote sensing, optical switching, modulated waveguides, photodetectors, and q-switches. The interest in the nanowire form has been strong [7–18] for further miniaturization of these devices with improved performance. One of the simplest routes to grow ZnSe involves the VLS technique using Zn and Se powders in two separate quartz boats (in a typical VLS reactor discussed throughout this book) heated to 520°C and 370°C, respectively [7]. A thin (~2 nm) gold film may serve as catalyst, and growth at reduced pressure (~100 Torr) and under hydrogen flow is necessary to prevent oxidation of the nanomaterials. The as-grown ZnSe nanowires appear as a yellowish layer on the silicon substrate. Compared to this moderate temperature process, the carbothermal reduction of ZnSe powder requires temperatures over 1250°C [8]. Figure 11.4 shows ZnSe nanowires grown on (001), (110), and (111) GaAs substrates by

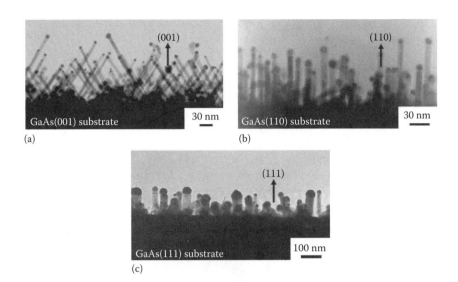

FIGURE 11.4

ZnSe nanowires grown by MBE on GaAs (a) (001), (b) (110), and (c) (111) substrates. (Reproduced from Cai, Y. et al., *Adv. Mater.*, 18, 109, 2006. With permission.)

molecular beam epitaxy (MBE) using a ZnSe compound-source effusion cell [15]. The substrates are coated with a very thin (0.3–0.6 nm) layer of gold and the growth temperature is maintained at 530°C. All the nanowires in Figure 11.4 are decorated with gold particle tips, indicating a VLS mechanism. The thin nanowires on the (001) substrate grow along two directions inclined approximately 35° to the normal of the substrate surface. Nanowires grown on GaAs (110) substrate are generally perpendicular to the substrate. The thick nanowires on (111) substrates prefer to grow along the <111> direction. The wires thinner than 20 nm contain very few defects whereas the thick nanowires exhibit stacking faults.

Because of the interest in optoelectronics applications, photoluminescence (PL) measurements are ideal to characterize the ZnSe nanowires. The nanowires from VLS growth using separate Zn and Se sources mentioned earlier show a very broad PL peak from 450 to 570 nm with the center at 505 nm [7]. The origin of this broad peak is from gold serving as a medium deep acceptor. The deep defect (DD) related emission peaks occur in the region of 1.8–2.4 eV (500–680 nm). The samples from carbothermal reduction of ZnSe mentioned above [8] also show a DD peak at 617 nm associated with the vacancies of Zn in ZnSe. What is ideal for applications is a strong band-edge (BE) emission at 2.68 eV (463 nm) and very weak or no DD emission peaks. A controlled study by Philipose et al. [17] indicates that the gas phase stoichiometry strongly influences optical properties. Figure 11.5 compares the room temperature PL spectra of stoichiometric, Zn-rich, and Se-rich ZnSe nanowires [17]. The nanowires are produced by VLS technique at 650°C using ZnSe powder and

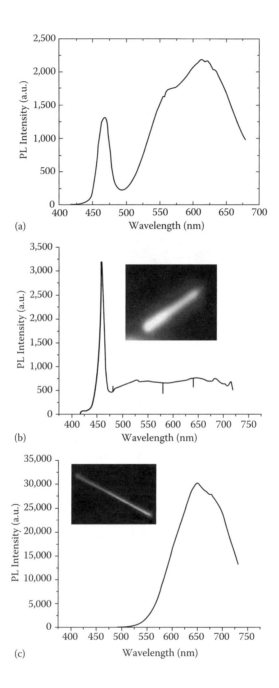

FIGURE 11.5
Room temperature of PL spectra of ZnSe nanowires. (a) Stoichiometric nanowires. (b) Zn-rich nanowires showing strong BE emission and a very weak DD emission. The inset shows blue luminescence from a single nanowire. (c) Se-rich nanowires showing strong DD emission and no BE emission. The inset shows red luminescence from a single nanowire. (Reproduced with permission from Philipose, U. et al., *J. Appl. Phys.*, 100, 084316, 2006. With permission.)

gold catalyst. An extra source is used for Zn- or Se-rich nanowire growth. Typically, the nanowires are 100 nm in diameter regardless of stoichiometry. The stoichiometric nanowire sample exhibits a BE emission peak at 463 nm and a broad DD peak from 500 to 680 nm, and the DD intensity is stronger than the BE intensity. In contrast, the Zn-rich sample shows a very strong BE peak and a very weak DD peak. Here, the low concentration of Zn vacancies accounts for the low intensity of DD peaks. Finally, with the Se-rich nanowires, DD emission dominates with BE peak being totally absent; it appears that these samples have a high density of point defects. As an alternative to obtaining high BE intensity from as-grown Zn-rich samples, posttreatment of stoichiometric ZnSe nanowires by annealing at 650°C in a Zn-rich atmosphere under argon flow achieves the same results. A 45 min annealing completely eliminates the DD peak leaving with a sample that shows only a very sharp BE peak [17].

11.2.2 Other Selenides

CdSe is another interesting II–VI material that has been grown in the nanowire form [19–22] and finds applications in light-emitting devices and photodetectors. An AAM template with a metal (such as silver) substrate as the cathode and a counterelectrode can be used to carry out electrodeposition to produce these nanowires. The electrolyte solution consists of 0.05 M/L of $CdCl_2$ and saturated elemental selenium in dimethylsulfoxide. A direct current (dc) density of 0.85 mA/cm² for 30–60 min. fills the pores, producing stoichiometric nanowires with diameters matching the pore size [19]. This approach produces CdSe nanowires with a stoichiometry very close to 1:1. The diffraction patterns confirm that the nanowire has a hexagonal CdSe crystal structure and additional evidence comes from an estimated interplanar spacing of 0.329 nm (from HTREM image), which corresponds to the {101} plane of the hexagonal system of CdSe [19].

Alternatively, MOCVD using dimethylcadmium and diisopropylselenide diluted in H_2 as sources (at 500 Torr) also yields CdSe nanowires with controllable orientations on a GaAs substrate [20–22]. The growth temperature is maintained in the range of 480°C–500°C with GaAs (100) and (110) substrates. PL spectra of the CdSe nanowires [22] show a strong peak centered at 710 nm (1.746 eV), which is assigned to the near-BE emission. The room temperature band gap of CdSe is 1.738 eV and the slight blueshift of the PL peak is thought to be the result of a strain due to the lattice mismatch between the nanowire and GaAs substrate instead of any dimension-led quantum confinement. Other selenides of interest include GaSe and PbSe, which can be readily grown by Au-catalyzed VLS using the respective powders at 800° and 700°C, respectively [23,24]. Low-pressure operation (several Torrs) under a reducing environment such as H_2 flow or mixture of N_2 and H_2 is critical to prevent the oxidation of nanowires.

11.3 Tellurides

11.3.1 Bismuth Telluride

Bi_2Te_3 and related compounds have been widely investigated for their potential in thermoelectric refrigeration and power generation. Bulk Bi_2Te_3 has a relatively high thermoelectric figure-of-merit, ZT (~1.0). The interest in the nanowire form [25–31] of bismuth telluride is to increase ZT due to higher density of states and increased phonon scattering (see Chapter 16 for further discussion on thermoelectric devices). AAM template based growth is one approach to produce Bi_2Te_3 nanowires [25] since an array of nanowires can be produced to generate a meaningful level of current transport in contrast to a single nanowire [27]. A typical electrolyte consists of 0.035 M $Bi(NO_3)_3 \cdot 5H_2O$ and 0.05 M $HTeO_2^+$; the latter can be obtained by reacting Te powder with 5 M HNO_3. The pH needs to be adjusted to 1 by adding 1 M HNO_3. A gold layer at the bottom of the anodized alumina membrane serving as the cathode and a graphite plate as counterelectrode constitute the electrochemical cell. A current density of 2.5 mA/cm² for 2 h produces Bi_2Te_3 nanowires filling the pores. Figure 11.6 shows these nanowires after etching the template for 15 min [25]. The nanowires are dense, vertical, and match the pore opening size, and the process yields a stoichiometry close to 2:3. The XRD patterns (not shown here) show the (110) and (220) peaks of the hexagonal Bi_2Te_3, indicating a preferred growth direction of [110].

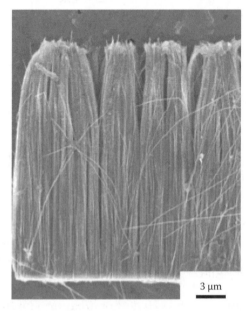

FIGURE 11.6

Bi_2Te_3 nanowires grown using an AAM. (Reproduced from Jin, C. et al., *J. Phys. Chem. B*, 108, 1844, 2004. With permission.)

11.3.2 Cadmium Telluride

CdTe is a key material in photovoltaics and CdTe/CdS junction cells have reached the commercial market to meet the current insatiable demand for solar cells. In addition to this, alloying CdTe with HgTe allows construction of IR detectors for various wavelengths. In spite of such potential, there is very little work on CdTe nanowires [32–34]. Growth of CdTe nanowires by the VLS technique appears to be rather difficult. A successful process starts with a sapphire substrate coated with polyvinyl alcohol (PVA), followed by annealing at 70°C to smooth out the alcohol layer [33]. Then a thin layer (2 nm) of bismuth is applied as catalyst by pulsed laser deposition. Growth of CdTe nanowires is carried out using a CdTe target at 365°C and 400 mTorr under a helium flow. Figure 11.7 shows CdTe nanowires from this approach, which are 100 nm in diameter and normal to the substrate, with a height of about 300 nm. The application of PVA appears to be critical, which allows creation of bismuth particles acting as seeds [33].

11.3.3 Other Tellurides

ZnTe is a II–VI semiconductor with a direct band gap of 2.26 eV at room temperature. It has applications in green LEDs, electro-optic detectors, thermo-electric devices, and solar cells and even the nanowire form has been of interest [35–38]. The exciton Bohr radius is 6.2 nm and even for nanowires less than 40 nm in diameter, a relatively strong quantum confinement effect is seen [37].

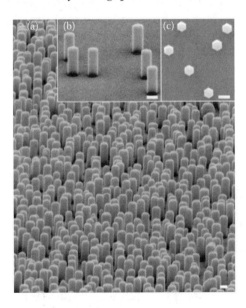

FIGURE 11.7
SEM images of CdTe nanowires. (a) 70° tilt side view. (b) High magnification 70° tilt side view. (c) Top view. Scale bar = 100 nm. (Reproduced from Neretina, S. et al., *Nanotechnology*, 18, 275301, 2007. With permission.)

These nanowires can be grown by VLS technique using ZnTe powder and gold catalyst film. The synthesis is carried out at 1000°C, which yields long (10 µm) nanowires 30–80 nm in diameter is about 30 min [37]. The XRD analysis of the sample indicates formation of zinc blende ZnTe nanowires. The nanowires are single crystals with the cubic structure in the direction of [111] as established by HRTEM. The PL spectra of these ZnTe nanowires at room temperature show BE emission at 2.26 eV (548 nm), which is indicative of high optical quality. Postgrowth copper doping can be achieved by immersing in $Cu(NO_3)_2$ solution for p-type doping ZnTe nanowires [38]. A ZnTe FET with p-type wires shows a channel mobility of 1 cm²/V s [38]. Other telluride nanowires include HgTe [39] for detectors and Sb_2Te_3 [40] for thermoelectric devices.

11.4 Sulfides

11.4.1 Zinc Sulfide

ZnS is a semiconductor with a direct band gap of 3.7 eV and a large exciton binding energy of 40 meV. It has been investigated previously for its potential in displays, electroluminescent devices, IR windows, and other optoelectronics applications. There has been extensive interest in growing ZnS nanowires [41–53] using VLS and AAM-template approaches. In the VLS approach, sublimation of ZnS powder at 900°C creates the source vapor and a gold thin film catalyst facilitates the growth of ZnS nanowires as shown in Figure 11.8 [41]. The product appears like a white sponge and consists of nanowires 30–60 nm in diameter. XRD analysis indicates wurtzite (hexagonal) structured ZnS nanowires with lattice constants of $a = 0.3825$ nm and $c = 0.627$ nm. The nanowires are stoichiometric as shown by EDS and the preferred growth direction is <110> [41]. The PL spectrum in Figure 11.8 shows

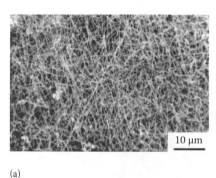

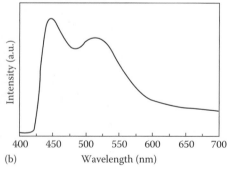

(a)   (b)

FIGURE 11.8

ZnS nanowire grown by VLS technique. (a) SEM image. (b) PL spectra with an excitation wavelength of 335 nm. (Reproduced from Wang, Y. et al., *Chem. Phys. Lett.*, 357, 314, 2002. With permission.)

two peaks at 450 and 520 nm. The blue emission is due to surface states and the broader visible peak is due to gold deep levels; this sample shows no BE emission. Another VLS study using ZnS powder and 2 nm Au colloids at a growth temperature 1000°C was able to get better quality ZnS nanowires confirmed by BE emission at 343 nm (3.61 eV) [44]. Instead of a single source, Zn and S powders can be loaded into two individual quartz boats for VLS growth; however, these need to be heated to different temperatures, 580°C for Zn and 90°C for sulfur. This approach produces ultrafine ZnS nanowires with diameters in the 5–10 nm range [52].

11.4.2 Other Sulfides

Other sulfide nanowires receiving attention include cadmium sulfide [54–60], lead sulfide [61–63], and bismuth sulfide [64–65]. CdS thin film is highly popular in constructing CdTe/CdS heterojunction solar cells. Its band gap is 2.54 eV at room temperature. To prepare CdS nanowires by AAM templating approach, the electrolyte may consists of equimolar (0.01 M) Na_2S and $CdCl_2$ [58], which yields nanowires of diameter 50–110 nm as seen in Figure 11.9. The nanowires are also fairly vertical and the optical properties of these nanowires show blueshifting of the band gap due to quantum confinement [58]. Lead sulfide is a narrow band gap material (0.41 eV) with an exciton Bohr radius of 9 nm. A carbothermal reduction process can be used to grow PbS nanowires by mixing sulfur, carbon, and $PbCl_2$ at a ratio of 4:1:1. This process yields black-sintered product of PbS nanowires at a temperature of 600°C [62]. Bismuth sulfide is a direct band gap material ($E_g = 1.3$ eV) and useful in photodiodes and thermoelectric devices. In an AAM-templated approach, the electrolyte to deposit Bi_2S_3 consists of 0.055 m/L of $BiCl_3$ and 0.19 m/L of elemental sulfur dissolved in dimethyl sulfoxide. This process yields nanowires of 40 nm in diameter with an adsorption band gap of 1.56 eV [64].

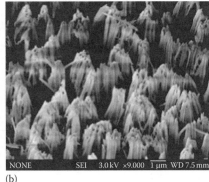

(a)　　　　　　　　　　　　　　　　(b)

FIGURE 11.9
CdS nanowires grown using an anodized alumina membrane. (a) Nanowires of diameters 50–60 nm. (b) Diameter 100–110 nm. (Reproduced from Mondal, S.P. et al., *Nanotechnology*, 18, 095606, 2007. With permission.)

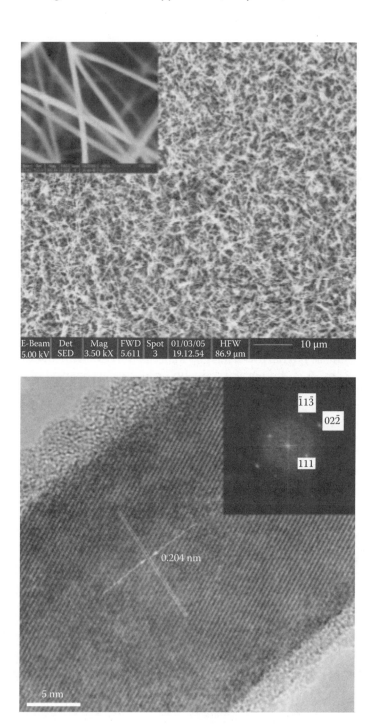

FIGURE 11.10
TiSi₂ nanowires prepared by physical vapor deposition. SEM and HRTEM images. (Reproduced from Xiang, B. et al., *Appl. Phys. Lett.*, 86, 243103, 2005. With permission.)

11.5 Silicides

Silicides of nickel [66–72], iron [73–75], cobalt [76,77], titanium [78,79], tantalum [80,81], manganese [82], and platinum [83] have been grown in the form of nanowires. Low resistance ohmic contacts in gate and interconnect formation are critical for high-performance devices [67] and nickel silicide has been widely investigated for this purpose due to its low resistivity and low formation temperature. Iron silicide exhibits several interesting phases which include α (metallic), β which is semiconducting at $E_g = 0.87$ eV with applications in 1.5 μm wavelength optical fiber communications, and ν and s phases, which are metastable [74]. The alloys of cobalt and iron silicides are ferromagnetic semiconductors useful in spintronics applications. A single-source organometallic precursor of $Fe(SiCl_3)_2(CO)_4$ appears to be effective to grow FeSi nanowires in a CVD setup without the aid of any catalyst metal. The key to nanowire growth is thought to be a thin (1–2 nm) SiO_2 layer without which FeSi nanowires do not grow [73].

Titanium silicide is also a low-resistivity material with the potential to meet interconnect needs in microelectronics. It can be prepared using an extremely simple physical vapor deposition method [78]. In a typical VLS-type setup discussed throughout this book, a silicon wafer is placed in an alumina boat loaded with pure Ti powder and the boat is placed inside a quartz tube heated by a tube furnace. The process yields a black layer of $TiSi_2$ nanowires on the silicon wafer (see Figure 11.10) at a temperature of 800°C and 150 Torr, and under a constant argon flow. The approach does not use any external catalysts and can be "self-catalytic" [78]. In the beginning of the process, the wafer is covered with Ti species, which dissolve into silicon and form tiny islands of Si–Ti alloy. Continuous incorporation of Ti into these islands leads to $TiSi_2$ nanowire growth. The nanowires in Figure 11.10 are about 40 nm in diameter and HRTEM image reveals the single crystalline nature of the nanowires with the growth direction along the <022> zone axis [78].

References

1. X. Zhang, Y. Hao, G. Meng, and L. Zhang, *J. Electrochem. Soc.* 152, C664 (2005).
2. A. Kuczkowski, S. Schultz, and W. Assenmacher, *J. Mater. Chem.* 11, 3241 (2001).
3. A.H. Chin, S. Vaddiraju, A.V. Maslov, C.Z. Ning, M.K. Sunkara, and M. Meyyappan, *Appl. Phys. Lett.* 88, 163115 (2006).
4. Y. Wang, L. Li, X. Huang, G. Li, and L. Zhang, *J. Mater. Sci.* 42, 2753 (2007).
5. S. Vaddiraju, M.K. Sunkara, A.H. Chin, C.Z. Ning, G.R. Dholakia, and M. Meyyappan, *J. Phys. Chem. C* 111, 7339 (2007).
6. Y.M. Lin, O. Rabin, S.B. Cronin, J.Y. Ying, and M.S. Dresselhaus, *Appl. Phys. Lett.* 81, 2403 (2002).

7. B. Xiang, H.Z. Zhang, G.H. Li, F.H. Yang, F.H. Su, R.M. Wang, J. Xu, G.W. Lu, X.C. Sun, Q. Zhao, and D.P. Yu, *Appl. Phys. Lett.* 82, 3330 (2003).

8. Y.C. Zhu and Y. Bando, *Chem. Phys. Lett.* 377, 367 (2003).

9. X.T. Zhang, Z. Liu, Y.P. Leung, Q. Li, and S.K. Hark, *Appl. Phys. Lett.* 83, 5533 (2003).

10. X.T. Zhang, K.M. Ip, Z. Liu, Y.P. Leung, Q. Li, and S.K. Hark, *Appl. Phys. Lett.* 84, 2641 (2004).

11. X.T. Zhang, Z. Liu, K.M. Ip, Y.P. Leung, Q. Li, and S.K. Hark, *J. Appl. Phys.* 95, 5752 (2004).

12. X. Zhang, Z. Liu, Q. Li, Y. Leung, K. Ip, and S. Hark, *Adv. Mater.* 17, 1405 (2005).

13. Y.F. Chan, X.F. Duan, S.K. Chan, I.K. Sou, X.X. Zhang, and N. Wang, *Appl. Phys. Lett.* 83, 2665 (2003).

14. S.K. Chan, Y. Cai, N. Wang, and I.K. Sou, *Appl. Phys. Lett.* 88, 013108 (2006).

15. Y. Cai, S.K. Chan, I.K. Sou, Y.F. Chan, D.S. Su, and N. Wang, *Adv. Mater.* 18, 109 (2006).

16. A. Colli, S. Hofmann, A.C. Ferrari, C. Ducati, F. Martelli, S. Rubini, S. Cabrini, A. Franciosi, and J. Robertson, *Appl. Phys. Lett.* 86, 153103 (2005).

17. U. Philipose, T. Xu, S. Yang, P. Sun, H.E. Ruda, Y.Q. Wang, and K.L. Kavanagh, *J. Appl. Phys.* 100, 084316 (2006).

18. A. Dong, F. Wang, T.L. Daulton, and W.E. Buhro, *Nano Lett.* 7, 1308 (2007).

19. D. Xu, X. Shi, G. Guo, L. Gui, and Y. Tang, *J. Phys. Chem. B* 104, 5061 (2000).

20. C.X. Shan, Z. Liu, and S.K. Hark, *Appl. Phys. Lett.* 87, 163108 (2005).

21. C.X. Shan, Z. Liu, and S.K. Hark, *Nanotechnology* 16, 3133 (2005).

22. C.X. Shan, Z. Liu, and S.K. Hark, *Appl. Phys. Lett.* 90, 193123 (2007).

23. H. Peng, S. Meister, C.K. Chan, X.F. Zhang, and Y. Cui, *Nano Lett.* 7, 199 (2007).

24. J. Zhu, H. Peng, C.K. Chan, K. Jarausch, X.F. Zhang, and Y. Cui, *Nano Lett.* 7, 1095 (2007).

25. C. Jin, X. Xiang, C. Jia, W. Liu, W. Cai, L. Yao, and X. Li, *J. Phys. Chem. B* 108, 1844 (2004).

26. W. Wang, X. Lu, T. Zhang, G. Zhang, W. Jiang, and X. Li, *J. Am. Chem. Soc.* 129, 6702 (2007).

27. A.L. Prieto, M.S. Sander, M.S. Martin-Gonzalez, R. Gronsky, T. Sands, and A.M. Stacy, *J. Am. Chem. Soc.* 123, 7160 (2001).

28. M.S. Sander, A.L. Prieto, R. Gronsky, T. Sands, and A.M. Stacy, *Adv. Mater.* 14, 665 (2002).

29. M. Martin-Gonzalez, A.L. Prieto, R. Gronsky, T. Sands, and A.M. Stacy, *Adv. Mater.* 15, 1003 (2003).

30. M. Martin-Gonzalez, G.J. Snyder, A.L. Prieto, R. Gronsky, T. Sands, and A.M. Stacy, *Nano Lett.* 3, 973 (2003).

31. L. Trahey, C.R. Becker, and A.M. Stacy, *Nano Lett.* 7, 2535 (2007).

32. Y. Wang, Z. Tang, X. Liang, L.M. Liz-Marzan, and N.A. Kotov, *Nano Lett.* 4, 225 (2004).

33. S. Neretina, R.A. Hughes, J.F. Britten, N.V. Sochinskii, J.S. Preston, and P. Mascher, *Nanotechnology* 18, 275301 (2007).

34. Y.P. Rakovich, Y. Volkov, S. Sapra, A.S. Susha, M. Doblinger, J.F. Donegan, and A.L. Rogach, *J. Phys. Chem. C* 111, 18927 (2007).

35. Y. Li, Y. Ding, and Z. Wang, *Adv. Mater.* 11, 847 (1999).

36. L. Li, Y. Yang, X. Huang, G. Li, and L. Zhang, *J. Phys. Chem. B* 109, 12394 (2005).

37. H.B. Huo, L. Dai, D.Y. Xia, G.Z. Ran, L.P. You, B.R. Zhang, and G.G. Qin, *J. Nanosci. Nanotechnol.* 6, 1182 (2006).
38. H.B. Huo, L. Dai, C. Liu, L.P. You, W.Q. Yang, R.M. Ma, G.Z. Ran, and G.G. Qin, *Nanotechnology* 17, 5912 (2006).
39. S. Rath, S.N. Sarangi, and S.N. Sahu, *J. Appl. Phys.* 101, 074306 (2007).
40. C. Jin, G. Zhang, T. Qian, X. Li, and Z. Yao, *J. Phys. Chem. B* 109, 1430 (2005).
41. Y. Wang, L. Zhang, C. Liang, G. Wang, and X. Peng, *Chem. Phys. Lett.* 357, 314 (2002).
42. M. Lin, T. Sudhiranjan, C. Boothroyd, and K.P. Loh, *Chem. Phys. Lett.* 400, 175 (2004).
43. X.M. Meng, J. Liu, Y. Jiang, W.W. Chen, C.S. Lee, I. Bello, and S.T. Lee, *Chem. Phys. Lett.* 382, 434 (2003).
44. X.T. Zhou, P.S.G. Kim, T.K. Sham, and S.T. Lee, *J. Appl. Phys.* 98, 024312 (2005).
45. R.A. Rosenberg, G.K. Shenoy, F. Heigl, S.T. Lee, P.S.G. Kim, X.T. Zhou, and T.K. Sham, *Appl. Phys. Lett.* 86, 263115 (2005).
46. R.A. Rosenberg, G.K. Shenoy, F. Heigl, S.T. Lee, P.S.G. Kim, X.T. Zhou, and T.K. Sham, *Appl. Phys. Lett.* 87, 253105 (2005).
47. J.S. Jie, W.J. Zhang, Y. Jiang, X.M. Meng, J.A. Zapien, M.W. Shao, and S.T. Lee, *Nanotechnology* 17, 2913 (2006).
48. D.D.D. Ma, S.T. Lee, P. Mueller, and S.F. Alvarado, *Nano Lett.* 6, 926 (2006).
49. Y.Q. Li, J.X. Tang, H. Wang, J.A. Zapien, Y.Y. Shan, and S.T. Lee, *Appl. Phys. Lett.* 90, 093127 (2007).
50. Y. Jung, D.K. Ko, and R. Agarwal, *Nano Lett.* 7, 264 (2007).
51. Z.G. Chen, J. Zou, G.Q. Lu, G. Liu, F. Li, and H.M. Cheng, *Appl. Phys. Lett.* 90, 103117 (2007).
52. Z. Zhang, H. Yuan, D. Liu, L. Liu, J. Shen, Y. Xiang, W. Ma, W. Zhou, and S. Xie, *Nanotechnology* 18, 145607 (2007).
53. H.Y. Sun, X.H. Li, W. Li, F. Li, B.T. Liu, and X.Y. Zhang, *Nanotechnology* 18, 115604 (2007).
54. D. Xu, Y. Xu, D. Chen, G. Guo, L. Gui, and Y. Tang, *Adv. Mater.* 12, 520 (2000).
55. K. Tang, Y. Qian, J. Zeng, and X. Yang, *Adv. Mater.* 15, 448 (2003).
56. Y. Liang, C. Zhen, D. Zou, and D. Xu, *J. Am. Chem. Soc.* 126, 16338 (2004).
57. Y. Long, Z. Chen, W. Wang, F. Bai, A. Jin, and C. Gu, *Appl. Phys. Lett.* 86, 153102 (2005).
58. S.P. Mondal, K. Das, A. Dhar, and S.K. Ray, *Nanotechnology* 18, 095606 (2007).
59. C.C. Kang, C.W. Lai, H.C. Peng, J.J. Shyue, and P.T. Chou, *Small* 3, 1882 (2007).
60. Y.F. Lin, Y.J. Hsu, S.Y. Lu, K.T. Chen, and T.Y. Tseng, *J. Phys. Chem. C* 111, 13418 (2007).
61. J.P. Ge, J. Wang, H.X. Zhang, X. Wang, Q. Peng, and Y.D. Li, *Chem. Eur. J.* 11, 1889 (2005).
62. H. Zhang, M. Zuo, S. Tan, G. Li, and S. Zhang, *Nanotechnology* 17, 2931 (2006).
63. D.V. Talapin, H. Yu, E.V. Shevchenko, A. Lobo, and C.B. Murray, *J. Phys. Chem. C* 111, 14049 (2007).
64. X.S. Peng, G.W. Meng, J. Zhang, L.X. Zhao, X.F. Wang, Y.W. Wang, and L.D. Zhang, *J. Phys. D: Appl. Phys.* 34, 3224 (2001).
65. H. Bao, X. Cui, C.M. Li, Y. Gan, J. Zhang, and J. Guo, *J. Phys. Chem. C* 111, 12279 (2007).
66. S.Y. Chen and L.J. Chen, *Appl. Phys. Lett.* 87, 253111 (2005).
67. J. Kim and W.A. Anderson, *Nano Lett.* 6, 1356 (2006).

68. Z. Zhang, J. Lu, P.E. Hellstrom, M. Ostling, and S.L. Zhang, *Appl. Phys. Lett.* 88, 213103 (2006).
69. Y. Song, A.L. Schmitt, and S. Jin, *Nano Lett.* 7, 965 (2007).
70. Y. Song and S. Jin, *Appl. Phys. Lett.* 90, 173122 (2007).
71. K.C. Lu, K.N. Tu, W.W. Wu, L.J. Chen, B.Y. Yoo, and N.V. Myung, *Appl. Phys. Lett.* 90, 253111 (2007).
72. J. Kim, D.H. Shin, E.S. Lee, C.S. Han, and Y.C. Park, *Appl. Phys. Lett.* 90, 253103 (2007).
73. A.L. Schmitt, M.J. Bierman, D. Schmeisser, F.J. Himpsel, and S. Jin, *Nano Lett.* 6, 1617 (2006).
74. S. Liang, R. Islam, D.J. Smith, P.A. Bennett, J.R. O'Brien, and B. Taylor, *Appl. Phys. Lett.* 88, 113111 (2006).
75. K. Yamamoto, H. Kohno, S. Takeda, and S. Ichikawa, *Appl. Phys. Lett.* 89, 083107 (2006).
76. H. Okino, I. Matsuda, R. Hobara, Y. Hosomura, S. Hasegawa, and P.A. Bennett, *Appl. Phys. Lett.* 86, 233108 (2005).
77. K. Seo, K.S.K. Varadwaj, P. Mohanty, S. Lee, Y. Jo, M.H. Jung, J. Kim, and B. Kim, *Nano Lett.* 7, 1240 (2007).
78. B. Xiang, Q.X. Wang, Z. Wang, X.Z. Zhang, L.Q. Liu, J. Xu, and D.P. Yu, *Appl. Phys. Lett.* 86, 243103 (2005).
79. H.C. Hsu, W.W. Wu, H.F. Hsu, and L.J. Chen, *Nano Lett.* 7, 885 (2007).
80. Y.L. Chueh, L.J. Chou, S.L. Cheng, L.J. Chen, S.J. Tsai, C.M. Hsu, and S.C. Kung, *Appl. Phys. Lett.* 87, 223113 (2005).
81. Y.L. Chueh, M.T. Ko, L.J. Chou, L.J. Chen, C.S. Wu, and C.D. Chen, *Nano Lett.* 6, 1637 (2006).
82. Z.Q. Zou, H. Wang, D. Wang, Q.K. Wang, J.J. Mao, and X.Y. Kong, *Appl. Phys. Lett.* 90, 133111 (2007).
83. B. Liu, Y. Wang, S. Dilts, T.S. Mayer, and S.E. Mohney, *Nano Lett.* 7, 818 (2007).
84. Y.Q. Wang, X.F. Duan, L.M. Cao, and W.K. Wang, *Chem. Phys. Lett.* 359, 273 (2002).
85. L.M. Cao, K. Hahn, C. Scheu, M. Ruhle, Y.Q. Wang, Z. Zhang, C.X. Gao, Y.C. Li, X.Y. Zhang, M. He, L.L. Sun, and W.K. Wang, *Appl. Phys. Lett.* 80, 4226 (2002).
86. L.M. Cao, H. Tian, Z. Zhang, X.Y. Zhang, C.X. Gao, and W.K. Wang, *Nanotechnology* 15, 139 (2004).
87. S.H. Yun, J.Z. Wu, A. Dibos, X. Gao, and U.O. Karlsson, *Appl. Phys. Lett.* 87, 113109 (2005).
88. S.H. Yun, J.Z. Wu, A. Dibos, X. Zou, and U.O. Karlsson, *Nano Lett.* 6, 385 (2006).
89. R. Ma and Y. Bando, *Chem. Phys. Lett.* 364, 314 (2002).
90. S.Z. Deng, Z.S. Wu, J. Zhou, N.S. Xu, J. Chen, and J. Chen, *Chem. Phys. Lett.* 356, 511 (2002).
91. W. Zhou, L. Yan, Y. Wang, and Y. Zhang, *Appl. Phys. Lett.* 89, 013105 (2006).
92. Z. Li, J. Zhang, A. Meng, and J. Guo, *J. Phys. Chem. B* 110, 22382 (2006).
93. S. Perisanu, P. Vincent, A. Ayari, M. Choueib, S.T. Purcell, M. Bechelany, and D. Cornu, *Appl. Phys. Lett.* 90, 043113 (2007).
94. H.W. Shim and H. Huang, *Appl. Phys. Lett.* 90, 083106 (2007).
95. K.F. Cai, Q. Lei, and A.X. Zhang, *J. Nanosci. Nanotechnol.* 7, 580 (2007).
96. C. Cheng, C. Tang, X.X. Ding, X.T. Huang, Z.X. Huang, S.R. Qi, L. Hu, and Y.X. Li, *Chem. Phys. Lett.* 373, 626 (2003).

97. Y. Liu, Q. Li, and S. Fan, *Chem. Phys. Lett.* 375, 632 (2003).
98. M. Nath and B.A. Parkinson, *Adv. Mater.* 18, 1865 (2006).
99. C.B. Jiang, B. Wu, Z.Q. Zhang, L. Lu, S.X. Li, and S.X. Mao, *Appl. Phys. Lett.* 88, 093103 (2006).
100. L. Wang, X. Zhang, X. Liao, and W. Yang, *Nanotechnology* 16, 2928 (2005).
101. D. Xue and X. Shi, *Nanotechnology* 15, 1752 (2004).

12

Applications in Electronics

12.1 Introduction

Progress in microelectronics industry has followed Moore's law in the last three decades with the computing power or the microprocessor speed doubling every 18 months. The International Technology Roadmap for Semiconductors (ITRS) predicts production of sub-10nm devices by 2015 [1]. Transistor scaling in silicon-CMOS-based computing is expected to reach its physical limit in about 15 years or less. The industry is facing enormous challenges to keep the momentum it has built up until now and these challenges span the entire spectrum: tools and infrastructure, scaling, performance, and power dissipation. While the tool capabilities in the areas of lithography, deposition, etching, metallization, etc. have kept up with and helped progress in each CMOS generation, significant difficulties are ahead with dimensions reaching into the nanoscale, and the tools may not work properly and reliably due to many physical limits. In addition, there has been an exponential increase in infrastructure costs of a new fab. Scaling faces serious difficulties in the nanoscale regime due to leakage issues arising from tunneling through the thin gate dielectric layer, band-to-band tunneling and short channel effects, all of which can cause logic functionality failure. Next, silicon itself is facing its intrinsic speed limit. Finally, as device size continues to shrink and performance continues to improve, power dissipation has become a serious bottleneck. Even today, the objective in chip design is to reduce power dissipation rather than seeking higher performance. This trade-off is forced upon the design engineer since power dissipation does not scale down with device size. This, combined with increased device density, makes heat dissipation a serious engineering challenge that can limit chip performance or cause failure.

All of the above have forced the industry to look into alternative, even nontraditional, technologies including new channel materials, devices, circuits, and architectures [2]. These include alternative silicon device configurations such as double-gated transistor, FinFET, etc., carbon nanotube (CNT) electronics, molecular electronics, single electron tunneling transistors, spintronics, and quantum computers. All emerging technologies must meet some basic criteria [3,4] to gain acceptance:

1. The alternative must be easier and cheaper to manufacture than the traditional silicon CMOS. Very low cost in the order of less than a μcent per transistor is a key economic metric.

2. Very high current drive with the ability to drive capacitances of interconnects of any length.

3. High level of integration, greater than 10^{10} transistors per circuit, must be possible.

4. High reproducibility, better than 95%.

5. High reliability, operating life time > 10 years.

6. Low power dissipation.

7. Timely availability of all the peripheral technologies, tools, and infrastructure.

It is not clear if any of the alternatives mentioned above can meet these criteria at present or in the near future. For example, single-wall CNTs have been widely investigated for electronics over the last decade; however, numerous challenges remain. A fundamental issue is that semiconducting nanotubes cannot be selectively grown for device fabrication; they are always mixed with metallic nanotubes. Even among the semiconducting nanotubes, the so-called chirality—which determines the diameter and band gap—is not a unique value in a batch of nanotubes; rather, there can be a range of chiralities with a corresponding variation in diameter and band gap. Postgrowth sorting, shown to have limited success, is not useful in large-scale integration. Transferring as-grown or purified nanotubes to the host wafer or any similar "pick and place" approach is also not viable. Progress to date involves rudimentary demonstration of device operation and basic circuit components, often with device feature size exceeding 1–2 μm. The real issue will be wafer-level demonstration if and when all the fundamentals warrant exploring at that level. This will be an enormous undertaking since a one-to-one replacement of silicon channel with a CNT-conducting channel in an otherwise fixed device/circuit/architecture configuration and fabrication scheme will not work. Several entirely new peripheral technologies, for example compatible materials for dielectrics, interconnects, metallization, etc., and processing tools, need to be developed. CNT band gap is strongly dependent on mechanical conditions such as strain, kinks, and twists along the length of the tube. Even if and when the chirality selection problem is solved, reliable and reproducible device performance will require control of the variation of the above parameters over millions of transistors across a wafer. Neither such issues critical in a CMOS-like operation of the CNT transistor nor any bold new architectures that may not be affected by such issues are under serious investigation. On the other hand, realizing the difficulties facing CNT electronics, there is a recent migration of interest toward graphene for electronics. At present, graphene technology is even more unproven. In this context, silicon and germanium nanowires provide a possible avenue to address many of the challenges discussed above. To begin with, the entire infrastructure built around the current silicon CMOS would still be useful and serve as the platform technology. Silicon and germanium nanowires are always semiconducting and pose no chirality selection issues like CNTs. Type of doping and concentration can be controlled during the nanowire growth process. Also, the $1/f$ noise in silicon nanowires is at least two order of magnitudes lower than that of CNTs [5]. The main source of noise is in the nanowire–bulk contact, which has room for optimization for even further reduction. Work to date includes basic device demonstration using long nanowires, back-gated configuration, thick dielectrics etc. to demonstrate the potential of nanowire technology. Early results from these attempts indeed are promising. Future work should involve development of nanowire synthesis approach that can be readily integrated into device fabrication sequence, alternative device configuration to suit the one-dimensional channel, and corresponding fabrication steps. Given the deep and broad knowledge of silicon technology in the industry, these tasks have a realistic chance of achievement.

Just as in logic technology, memory is also going through a stage of serious challenges [6]. The floating gate flash technology currently dominates the nonvolatile memory market. In the mean time, the rapid growth in applications such as digital camera, cell phones, MP3 player, and other personal devices has been pushing the demand for higher speed, lower power, higher endurance, and higher density memory devices. An ideal memory would combine the high speed of static random access memory (SRAM), nonvolatility of the flash and density of dynamic random access memory (DRAM) while being scalable and low cost. There are several candidates emerging at present including magnetic random access memory (MRAM), ferroelectric random access memory (FeRAM), and phase-change random access memory (PRAM or PCRAM). All these devices can also benefit from the nanowire form of the corresponding materials due to smaller size memory cell and improved properties at the nanoscale.

Finally, while pursuing scaling according to Moore's law, there is also a simultaneous push for what is dubbed as "more than Moore," which involves integration of several functional components [7]. For example, sensors, actuators, microelectromechanical systems, and passive components can be integrated with logic and memory devices on a single chip, leading to an entire system in a package (see Figure 1.1 and related discussion). Such functional integration may become the technology and economic driver beyond the current Moore's law based scaling era. Interestingly, silicon itself has been proven to be the potential building block to fabricate some of the above functional components. The nanoscale brings advantages in all those applications as well and wherever needed, other nanowires can be integrated with SiNWs to build the system.

12.2 Silicon Nanowire Transistors

Fabrication of both n-type and p-type FETs has been reported [8–19] using silicon nanowires grown by various techniques outlined in Section 6.2. In addition, silicon nanowire-like structures etched out of epitaxially grown silicon thin film have also been used in device fabrication but these are not considered in the discussion below. Cui et al. [10] reported p-FETs using boron-doped 10–20 nm diameter SiNWs grown by the VLS technique. The as-grown nanowires in a ethanol suspension are deposited onto an oxidized silicon substrate with 600 nm thermal oxide. Source and drain electrodes are defined 800–2000 nm apart by e-beam lithography followed by evaporation of Ti/Au (50/50 nm) contacts. Rapid thermal annealing at 300°C–600°C for 3 min is used to improve the contact quality. The back-gated structure has been shown to yield a transconductance 4–10 times larger than an identically sized planar silicon (on insulator) device.

Figure 12.1 describes the fabrication sequence from Ref. [16] for a SiNW FET with doped epitaxial source and drain contacts. Here also VLS-grown nanowires suspended in ethanol solution are transferred to a host wafer after removal of the gold catalyst. The SiNWs are not intentionally doped. The host wafer is a heavily doped p^{++} silicon substrate with a 4 nm SiO_2 and 15 nm Si_3N_4 layers. The entire wafer including the nanowire is first covered with SiO_2 and Si_3N_4 followed by patterning to expose source and contact holes. In the contact regions, a hydrofluoric acid cleaning removes any native oxide on the nanowire surface and then silicon epitaxy is conducted. For n-type device, doping involves a shallow P implantation followed by a deeper As implantation. For p-type device, boron doping is done in situ during epitaxy. Figure 12.1 also shows an SEM image of n-FETs built on a single silicon nanowire and measured I_d–V_{ds} characteristics as a function of gate voltage for a 25 nm SiNW and 400 nm source–drain separation. The estimated electron mobility is 346 cm²/V s and the hole mobility is 41 cm²/V s. The best subthreshold slope for these devices is 150 mV/decade.

The performance of these early devices is modest as evidenced from the two examples above but the potential is significant. There are several challenges currently facing SiNW device fabrication [12,16]. First, most SiNW FETs fabricated to date have used metal or silicide Schottky contacts. Instead it is desirable to use doped low-resistance contacts and eliminate undesirable ambipolar behavior [16]. However, varying doping from high to low and high again along the length of a nanowire is very difficult for reasonably short nanowires of interest in competitive MOSFET technology. Most reports to date have also used backgated configuration with oxide thickness well exceeding 100 nm. Source–drain separation and gate length have also been several times larger than current generation silicon technology. These issues need to be addressed to improve subthreshold slope, effective channel mobility, and other figures-of-merit. Chau et al. [12] also suggest benchmarking the

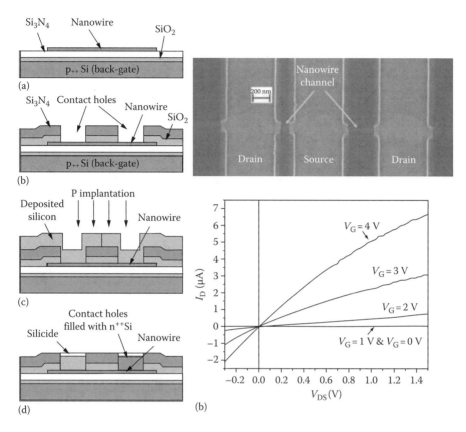

FIGURE 12.1
Silicon nanowire field effect transistor process flow to fabricate an n-SiNW FET (left), SEM image of two devices (top right) and current–voltage characteristics (bottom right). (Reproduced from Cohen, G.M. et al., *Appl. Phys. Lett.*, 90, 233110, 2007. With permission.)

emerging devices using four key metrics for high-performance, low-power logic applications: (1) intrinsic device speed versus gate length, (2) energy-delay product versus gate length, (3) subthreshold slope versus gate length, and (4) intrinsic device speed versus on/off current ratio.

In the last three decades, silicon device evolution has benefited greatly from modeling to understand the device physics, mechanisms for degradation and failure, and to develop new designs. Models have ranged from analytical to numerical simulation, drift-diffusion approximation for electron and hole currents, moments of the Boltzmann transport equation, Monte Carlo simulations, and finally quantum mechanical simulations. These types of efforts are necessary to understand SiNW FETs also [20–28]. Figure 12.2 shows the simulated effect of gate length variation from 5 to 15 nm in a SiNW FET (although such short gate lengths have not been demonstrated experimentally to date). The drain current versus gate voltage behavior is

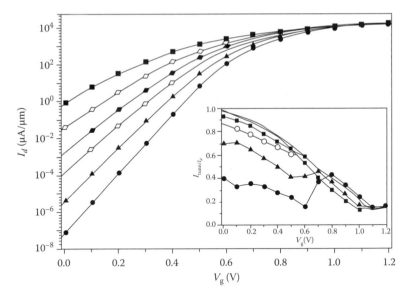

FIGURE 12.2
Simulation results showing the effect of gate length on drain current. The inset shows the ratio of tunneling current to drain current. (Reproduced from Shin, M., *IEEE Trans. Nanotechnol.* 6, 230, 2007. With permission.)

for a device that has a gate surrounding the nanowire. For all gate lengths, the drain current increases with gate voltage but the slope is smaller with decreasing gate length. This means the transconductance will be smaller with decreasing gate length. But more importantly, the tunneling current directly from source to drain increases with reduced gate lengths, which signifies the dominance of off-state by large tunneling currents.

12.3 Vertical Transistors

Silicon CMOS transistor is a planar, horizontal device today. The concept of a vertical transistor with the source at the bottom, drain on the top and gate surrounding or wrapping around the channel material has been around since the time of the first ITRS (see Figure 12.3). There are several advantages of this configuration. First, the critical dimension of source–drain separation is not defined by lithography, instead by the growth time. This eliminates any possible lithographic limit to device scaling. Second, the smaller footprint of the vertical device

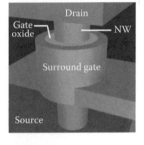

FIGURE 12.3
Schematic of a VSGT. (Courtesy of H.T. Ng.)

allows higher packing density than the conventional approach. It is also possible to eventually conceive of three-dimensional device structures. Finally, the surround gate is likely to have a better electrostatic control over the channel and can suppress the short channel effects. Indeed, there have been several attempts at fabricating vertical surround gate transistors (VSGTs) in the last two decades. However, all these efforts necessarily etched pillar-like structures out of epitaxially grown silicon thin film to create the conducting channel. Unfortunately, plasma etching leaves the surface quality too poor to make devices competitive or better than the planar transistors with the same critical dimension. For this reason, the vertical transistor idea has never reached commercial status. Now that the ability to grow SiNWs with high crystallinity and excellent surface quality has been demonstrated, there is interest in revisiting the old VSGT concept. Also from a growth point of view, it may be more advantageous to consider nanowire vertical transistors than planar transistors. In general, in situ growth is favored over "pick and place" approaches using bulk nanowire samples. It is much harder to grow horizontal nanowires of identical length (equivalent to source–drain separation) on specified locations than vertical nanowires.

A processing scheme to fabricate a VSGT is given below [29]. Though this work used ZnO nanowire as channel material grown on highly doped SiC substrate, the approach is applicable to SiNW VSGTs as well and for this reason, it is reproduced here and the intention is not to advocate ZnO in favor of silicon. SiC is chosen as substrate since a good interfacial lattice match between the major epitaxial crystal plane of the nanowires and the substrate is critical for the growth of vertically oriented single-crystal nanowires based on the VLS mechanism. The minimum lattice mismatch between ZnO (0001) and hexagonal SiC (0001) at 5.5% helps to promote vertically aligned growth of ZnO nanowires. SiC substrates can be heavily doped (highly p- or n-type) to obtain a conductive substrate for device architectures that require bottom electrical contact to the nanowires. A SiC substrate (5×5 mm) with a Au catalyst spot (~180 nm in diameter, 15 Å in thickness) patterned by e-beam lithography is employed for VSGT fabrication. This spot yields a single nanowire ~40 nm in diameter and 1 µm tall (not shown here, see Ref. [29]) using a VLS approach.

Figure 12.4 shows a generic process flow schematically outlining the major steps, leading to a functional p-VSG-FET. The ZnO nanowire functions as the active hole channel while the underlying p$^+$-SiC ($> 10^{18}$ cm^{-3}) epilayer serves as the source electrode of the p-VSG-FET. Next, a 20 nm thick SiO$_2$ layer is conformally deposited over the nanowire by CVD using tetraethoxysilane (TEOS) at 700°C (Figure 12.4b). This is a critical step since nonconformal coverage of the nanowire can lead to large gate leakage currents and eventual device failure. This is followed by a conformal ion-beam deposition of Cr metal (~40 nm) surrounding the SiO$_2$-encapsulated nanowire as shown in Figure 12.4c. The vertical directionality of the fabricated nanostructure is typically retained for nanowires with an aspect ratio ~20, beyond which the

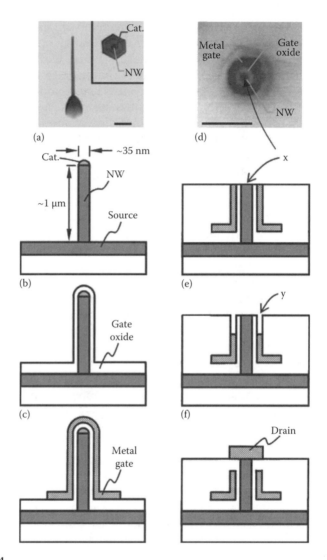

FIGURE 12.4

Process flow to fabricate a VSGT. (a) A vertical ZnO nanowire grown on p⁺-SiC/4H-SiC substrate. The underlying p⁺-SiC epilayer serves as the source contact while the vertical ZnO nanowire is the active hole channel. (b) Surround-gate oxide formation. A 20 nm SiO_2 conformally grown by TEOS CVD. (c) Surround-gate electrode formation. A 40 nm conformal Cr metal by ion beam deposition. (d) Active channel formation. SiO_2 deposition by TEOS CVD completely encapsulates the nanowire and excess SiO_2 is removed by CMP thus exposing the tip of the nanowire (denoted by x) and the Cr surround-gate electrode. Note that the Au catalyst head is removed during the CMP process. (e) Recess formation. A ~30 nm recess is introduced by selective Cr wet etching (denoted by y). (f) Formation of top Cr electrode (~100 nm thickness). The recess is filled with SiO_2 by TEOS CVD, followed by CMP to expose only the tip of ZnO nanowire and top Cr drain electrode. Top left SEM image is that of a single ZnO nanowire projecting from the dome-like ZnO buffer layer. The inset is the corresponding top view. Top right SEM image is taken prior to recess formation and deposition of Cr electrode. The surround gate oxide and chromium metal electrode are visible. (Courtesy of H.T. Ng.)

cladded nanostructures bend away from the substrate normal. A further SiO_2 CVD step followed by chemical mechanical polishing (CMP) is performed to planarize and fully expose the top of the nanowire (denoted x) and the Cr surround-gate electrode as shown in Figure 12.4d. The planarization step also removes the gold catalyst bead typically present on top of the nanowires and defines the vertical channel length of the p-VSG-FET. As mentioned earlier, difficult lithography steps and active channel processing schemes are eliminated in this process flow. Figure 12.4e shows the introduction of a ~30 nm recess (denoted y) in the Cr surround-gate by selective wet chemical etching of Cr. This is followed by SiO_2 tetra ethoxy saline (TEOS) CVD to fill in the recess and a CMP step to expose only the tip of the nanowire prior to depositing a top Cr electrode (100 nm thickness), which constitutes the drain electrode of the p-VSG-FET (Figure 12.4f). These steps avoid direct electrical contact between the Cr gate and the top drain electrode. Sufficient recess is necessary to minimize the overlap capacitance between the gate and drain electrodes. The performance of this ZnO nanowire vertical device will be discussed under Section 12.5. SiNW vertical surround gate devices can be fabricated with necessary modifications in the above processing scheme.

Fabrication of SiNW VSGTs has been reported in the literature although a single nanowire has not been contacted as discussed above; instead device measurements include an ensemble of nanowires. The approach by Schmidt et al. [18] uses a n-doped silicon nanowire grown on a p-type silicon substrate, which serves as the drain with metallization to create a backside electrical contact. The Au catalyst particle at the top serves as the Schottky contact with Al or Ti as source metal. The measurements reported are for 10^4–10^5 nanowires contacted in parallel over an area of 0.1 mm^2 with an on/off ratio of 6. Goldberger et al. [19] reported vertical transistors using 20–30 nm diameter SiNWs doped p-type and surrounded by 30–40 nm gate oxide and a Cr metal gate (length 500–600 nm). The channel length itself is about 1.0–1.5 μm. Nanowires are grown in place on a heavily doped p-type substrate by VLS growth using $SiCl_4$ precursor. Each drain contact in this case consists of anywhere from 8 to 269 nanowires. The on/off ratio for all devices is in the range of 10^4 to 10^6. The transconductance g_m given by

$$g_m = dI_{ds}/dV_{gs} \tag{12.1}$$

is 0.2–8.2 μS and when normalized to the total channel width (# of nanowires x nanowire diameter), this translates to 0.65–7.4 μS/μm. This compares favorably to silicon-on-insulator MOSFET performance of 5–12 μS/μm. The effective hole mobility can be extracted from the measured transconductance using

$$\mu = g_m L^2/(CNV_{ds}) \tag{12.2}$$

where
 L is the gate length
 C is the gate capacitance for an individual nanowire

N is the number of nanowires

V_{ds} is the source–drain voltage

The gate capacitance C is given by

$$C = 2\pi\varepsilon_0\varepsilon_{SiO_2}L/\ln\left(r_g/r_{nw}\right)$$ (12.3)

where

ε_0 is the vacuum permittivity

ε_{SiO_2} is the dielectric constant of the gate oxide

r_g is the inner radius of the gate electrode

r_{nw} is the nanowire radius

This yields a hole mobility of 7.5–102 cm^2/V s. The subthreshold slope of the vertical SiNW devices is 120 mV/decade for a 30 nm gate oxide thickness. This value can be lowered by reducing the gate oxide thickness and using higher-k dielectric materials.

12.4 Germanium Nanowire Transistors

As mentioned in Chapter 6, germanium offers higher electron and hole mobilities and larger Bohr radius compared to silicon. Germanium nanowires have been used in the fabrication of both n- and p-FETs [30–35].

Just as the industry standard silicon-on-insulator, GeNWs on insulator (GeNOI) can offer a nanomaterial based substrate platform for future device integration. The device fabrication discussed below [35] starts with VLS synthesis on a thin layer of SiO$_2$ in a silicon substrate using Ge powder as feedstock. Indium is used as catalyst at a growth temperature of 450°C. The preferred growth direction is [111] on SiO$_2$. The GeNOI FET is fabricated using photolithography patterning and metal lift-off process. Figure 12.5 shows an SEM image of a GeNOI with a backgate from a large array of such devices. Schottky contacts are used for both source and drain here. Though such contacts are known to be prone to junction leakage, the small cross-section of the nanowire junction minimizes this problem. Nickel monogermanide is the Schottky contact material used here, which has a barrier height of 0.5 eV for n-Ge and 0.16 eV for p-Ge. Its low anneal temperature prevents diffusion of Ni in Ge and maintains the integrity of the channel. The low-barrier Schottky contact eliminates the ambipolar conduction commonly reported for nanowire or CNT based FETs as seen in Figure 12.5. The p-type depletion mode device shows an on/off ratio of 10^4. The intrinsic gate delay given by CV/I, where C is the gate capacitance, V is the supply voltage, and I is the on-state drive current, is 50–200 ps for gate lengths of

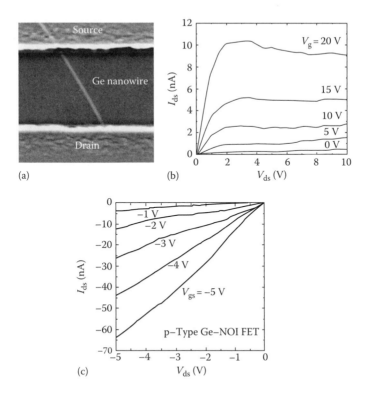

(a)

(b)

(c)

FIGURE 12.5

(a) A Ge back-gated FET. (b) Current–voltage characteristics of a N-channel GeNOI FET with a gate length of 2.8 μm and oxide thickness of 60 nm and (c) p-type Ge-NOI device behavior. (Courtesy of B. Yu.)

200–4000 nm. These values are well above the industry scaling trends due to thick gate oxides, nonohmic contacts, and nonpassivated nanowire surface but can be improved. While the above metric on switching speed is relevant for deeply scaled CMOS, a nanowire-FET may find applications in very low-power logic. The nanowire FETs operate at nA to μA on-state current with on/off ratio in the range of 10^4–10^6. The switching energy of a NW-FET is three to six orders of magnitude lower than a conventional planar FET and the standby power is negligible due to the pA level off-state leakage per nanowire.

Germanium nanowires grown with germane as source gas and doped with phosphine or diphorane have been used in a back-gated configuration [30] and shown to have field effect mobilities of 115 and 25 cm²/V s for holes and electrons respectively. As mentioned before, the use of backgating and unpassivated nanowire surface typically yields low mobilities and high sub-threshold slopes as reported by several other groups as well [31,32]. Wang et al. [31] obtained a hole mobility of ~600 cm²/V s compared to the best values for thin film Ge MOSFETs of $700\,cm^2/V\,s$. A surround gate (but not vertical) transistor using GeNWs cladded with a passivating nitride layer has been shown [32] to yield better results for on/off ratio (10^5) and subthreshold slope ($120\,mV/decade$).

12.5 Zinc Oxide and Other Nanowires in Electronics

Besides silicon and germanium, several other nanowires have been used in the fabrication of transistors, especially zinc oxide [29,36–44] and few other materials [45–47]. Though these materials are not advocated for logic or memory applications, they have potential in other areas. For example, thin film transistors (TFTs) using ZnO nanowires may be a replacement for conventional polysilicon TFTs since they offer flexibility and transparency for flexible electronics and displays. Another application in the transistor form for ZnO may be as gas sensors.

The device characteristics of a ZnO VSGT discussed in Section 12.3 [29] are shown in Figure 12.6. Figures 12.6a and b show the drain current (I_{ds}) versus drain voltage (V_{ds}) profiles for the n- and p-VSG-FETs respectively for different gate bias (V_{gs}) values. The drain current increases with increasing positive gate voltage for the n-VSG-FET, and this is characteristic of electron transport. The n-VSG-FET is normally on, and turns off at $V_{gs} = -3.5\,V$, which is the threshold voltage (V_{th}). This is attributed to unintentionally n-doped nature of the ZnO nanowire, possibly due to oxygen deficiency and/or interstitial Zn. In the case of p-VSG-FET, the drain current increases with increasing negative gate bias, and this is characteristic of hole transport. The p-VSG-FET has a V_{th} close to zero ($0.25\,V$) in contrast to the normally on n-VSG-FET. The absence of drain current saturation in both cases indicates that neither the traditional channel pinch-off nor the carrier velocity saturation is relevant in the VSG-FET operation. The transconductance per nanowire is 50 and $35\,nS$ at $V_{ds} = 1\,V$ for the n- and p-channels respectively. A high on-to-off current ratio of $>10^4$ is observed for the n-VSG-FET, while it is $>10^3$ for the p-VSG-FET.

Figure 12.6c shows I_{ds} versus $V_{gs} - V_{th}$ characteristics at $V_{ds} = 1.0\,V$ for the n-channel, and $V_{ds} = -1.0\,V$ for the p-channel devices where the drain current increases with $V_{gs} - V_{th}$. The former shows a linear dependency while the latter shows a strong nonlinearity. One possible reason is that not all charges induced by V_{gs} can contribute to charge transport. The surround cylindrical gate capacitance C_g is given by Equation 12.3. Figure 12.6d is a schematic showing the cross-section of the n-channel (left) and p-channel (right) of the VSG-FETs where the top drain and bottom source electrodes are not shown. The variation of gate-induced charge dQ due to dV_{gs} in the n-channel involves essentially electrons that are mobile in the channel.

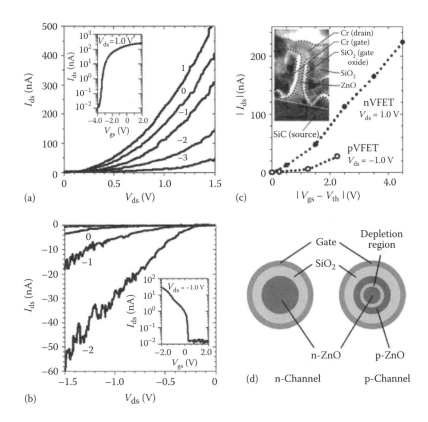

FIGURE 12.6

Device characteristics of ZnO nanowire-based n- and p-VSG-FETs. (a) I_{ds} versus V_{ds} characteristics for different V_{gs} of a n-VSG-FET. The inset shows its transfer characteristics at $V_{ds} = 1.0$ V. (b) I_{ds} versus V_{ds} characteristics for different V_{gs} of a p-VSG-FET. The inset shows its transfer characteristics at $V_{ds} = -1.0$ V. (c) I_{ds} versus $V_{gs} - V_{th}$ plots for both transistors. The inset shows a FE-SEM cross-sectional image of a VSG-FET with a channel length ~200 nm. Scale bar: 200 nm. (d) Cross-sectional images of n-type (left) and p-type (right) transistors (not to scale). Source and drain contacts are not shown. (Courtesy of H.T. Ng.)

The drain current is proportional to the electron density, leading to the observed linearity in Figure 12.6c. However, in the p-channel case, dQ involves both holes and ionized impurities in the depletion region. But only holes can contribute to charge transport and the ratio of holes to ionized impurities in the channel becomes higher as the magnitude of the gate bias becomes larger, which may explain the observed p-channel nonlinearity.

Using $|Q| = C_g |V_g - V_{th}|/l$ and with $r_g = 16$ nm, $r_{nw} = 40$ nm, and $\ell = 200$ nm in Equation 12.3, $Q/e = 1.9 \times 10^7$ cm^{-1} at $|V_{gs} - V_{th}| = 2$ V. Thus, the effective electron and hole mobilities are estimated as 0.53 and 0.23 cm^2/V s, respectively, using Equation 12.2. The above mobility values do not necessarily represent the crystal quality of the nanowires since the subthreshold gradient

S is 170 and 130 mV/decade for the n- and p-channel VSG-FETs, respectively. These values are comparable to those of CNT FETs published at that time but larger than that of the state-of-the-art silicon MOSFETs, and this indicates relevant contact effects on the mobility values. As mentioned before, higher values can be obtained by using a high-k dielectric material with reduced thickness, improving the channel transport and the source–drain contacts.

Conventional planar transistors using ZnO nanowires have been extensively reported [36–44]. Goldberger et al. [36] used a carbothermal reduction process to synthesize ZnO nanowires 50–200 nm in diameter and lengths up to 10 μm. A device configuration similar to the one in Figure 12.5 in Ref. [36] shows an electron mobility of 13 cm^2/V s and an on/off ratio of 10^5 to 10^7. Much higher field effect mobility values (928 cm^2/V s) along with an on/off ratio of 10^6 have been obtained [38] with a self-aligned gate configuration and a nanosize air-gap capacitor. The nanowire is suspended between the Nb source and drain electrodes. Nb contacts show ohmic transport at room temperature. Unlike many of the other ZnO-nanowire devices in the literature, this is an enhancement mode transistor with a zero current at zero voltage, which is turned on with positive drain voltages. It appears [40] that the device performance is seriously affected by the nanowire interface, both along the channel and at the contacts. Ozone treatment of the ZnO nanowires along with the use of a self-assembled organic gate insulator also provides high electron mobilities of 1175 cm^2/V s and an on/off ratio of ~10^7.

Planar transistors with a omega shaped gate are claimed [39] to be easy to fabricate and give enhanced mobility, transconductance, and on/off ratio by factors of 3.5, 32, and 10^6, respectively, when compared with the back-gated configuration for the same ZnO nanowire channel length. In$_2$O$_3$ nanowire has been used [46] as the channel material in a simple diode-like configuration with a conducting AFM tip serving as the movable gate in an investigation aimed at studying the effect of channel length easily. The mobility decreases from ~60 to 1 cm^2/V s when the channel length is decreased from 1 μm to 20 nm. In a short channel FET, most of the applied bias drops over the parasitic contact resistance, causing a mobility reduction. In contrast, the drain current and transconductance increase with a decreasing channel length. In$_2$O$_3$ nanowire channel has also been used to fabricate a vertical transistor with a top gate [45]. The nanowires are grown on a nonconducting optical sapphire substrate since it offers the best lattice match with indium oxide. First, a thin film of In$_2$O$_3$ is grown, which can serve as the source electrode and the nanowires are grown vertically on this film using VLS technique. Finally, CdS nanowires have also been used as channel material [47]. CdS nanowires are synthesized by VLS technique using pure CdS powder with indium as a shallow donor. The device with a Schottky Au gate and Ohmic In/Au source and drain electrodes functions as an enhancement mode FET. The on/off ratio is 5 × 10^3 for a device using a 200 nm diameter wire with a 3 μm gate length.

12.6 III–V Transistors

Most of the reported work on the fabrication of transistors using compound semiconductors are based on GaN nanowires [48–52]. As a wide band gap semiconductor, GaN nanowire devices may be valuable in high-temperature electronics. The group III nitrides are direct band gap materials covering the whole solar spectrum with InN at 0.7 eV and AlN at 6.2 eV. Back-gated GaN FETs using nanowires of various diameters (90–200 nm) show that the mobility is proportional to (diameter)$^{1.3}$ [49]. Typically, these devices are depletion mode transistors and because of the backgating configuration and thick oxide layers, very large negative voltages are needed to turn them off [51].

InAs and InP nanowires have also been investigated for electronics applications [53–55]. Figure 12.7 shows a VSGT using an InAs channel [55]. The nanowires are produced using molecular beam epitaxy as described in Section 6.5.2. Typical nanowire diameter is 80 nm and the length is ~3 μm. The nanowires are unintentionally doped n-type with a carrier concentration of 2×10^{17} cm^{-3}. As in the case of silicon VSGTs described in Section 12.3, an ensemble of 40 nanowires form the channel and a 60 nm silicon nitride layer is deposited as gate dielectric followed by Ti/Au gate metal sputtering. Selective wrapping of the gate location is guaranteed by spin coating an

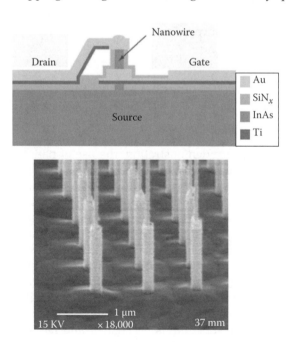

FIGURE 12.7
InAs VSGT. (a) Schematic and (b) SEM image of the nanowires other gate deposition. (Reprinted from Bryllert, T. et al., *IEEE Electron. Dev. Lett.*, 27, 323, 2006. With permission.)

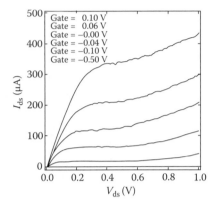

FIGURE 12.8

$I_{ds}-V_{ds}$ characteristics of the device from Figure 12.7. (Reprinted from Bryllert, T. et al., *IEEE Electron. Dev. Lett.*, 27, 323, 2006. With permission.)

organic thin film, which is etched back at the top of the wires. Gate finger definition is done using lithography and wet etching while the drain contact is fabricated by air bridge technology. InAs substrate provides the source contact. Device characteristics in Figure 12.8 for a channel cross-section of $0.2\,\mu m^2$ (40 nanowires) reveal a depletion mode transistor with current saturation. The transconductance is $2\,mS$ and the drive current is $100\,\mu A$ at zero gate voltage.

12.7 Memory Devices

Compared to logic device demonstration, the use of nanowires in memory devices has received less attention. Li et al. [56–59] used InO_2 nanowire transistors as a vehicle to demonstrate molecular memory using porphyrin molecules. This circumvents the difficulty of making direct contact to the molecules when a monolayer of redox-active molecules is applied onto the nanowire, which is contacted on either end by the electrodes [60]. However, the molecules themselves may be unstable at elevated operating temperatures.

The use of SiNW directly to construct a nonvolatile memory can provide a large memory window and reversible read, write, and erase operations with considerable retention and endurance [61]. The SiNW is grown on a stacked oxide/nitride/oxide layers (called ONO) as shown in the schematic in Figure 12.9. The p-type substrate is back-gated and has a 30 nm thermally grown oxide. A low stress Si_3N_4 (60 nm) is deposited by LPCVD at 650°C followed by a very thin (2 nm) oxide layer by PECVD (as tunneling oxide). The gold catalyst film (1 nm) is defined on top of the ONO stack using lithography

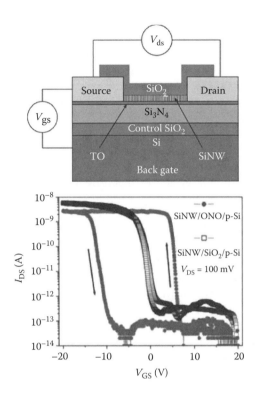

FIGURE 12.9
A schematic of silicon nanowire ONO (oxide/nitride/oxide) memory device (top) and the current versus gate voltage characteristics (bottom). (Reproduced from Li, Q. et al., *Nanotechnology*, 18, 235204, 2007. With permission.)

and lift-off. SiNW growth is carried out by VLS process using silane at 420°C and 500 mTorr. The nanowire diameter is in the range of 5–50 nm and the length is 10–60 μm. The final step involves patterning the contacts and encapsulating the device with an oxide layer. A device with a 20 nm diameter nanowire and 3 μm channel length shows a large hysteresis (memory) window. At a back gate voltage of 20 V, the electrons tunnel from the nanowire through the thin tunneling oxide to the dielectrics and get stored in the oxide/nitride interface. This induces a positive threshold voltage (V_{th}) shift and the ONO device is turned on. When the gate voltage is switched to –20 V, electrons tunnel out of the ONO stack, which induces a negative threshold voltage shift and the device is turned off. Continuous operation at ± 20 V with sampling (reading) at 100 mV leads to a permanent degradation of current and V_{th} after 10^5 cycles which is believed to be due to the low quality of the tunneling oxide. GaN nanowires can also be used to construct similar nonvolatile memory devices [62] and the radiation hardness of GaN is advantageous for harsh environment operation.

12.7.1 Phase-Change Random Access Memory

A detailed discussion on the preparation of phase change nanowires by VLS technique was provided in Chapter 7. Their potential is discussed in this section with the aid of a rudimentary memory device fabricated using a simple procedure [63,64]. First, the working concept of a PRAM is provided. Figure 12.10 shows a conventional thin film based PRAM device, which consists of a thin layer of a phase-change material (PCM) sandwiched between a pair of electrodes. The lower electrode shaped in the form of a resistive heater may have a dielectric surrounding it to isolate from its neighbor. The device can be driven by a bipolar or field effect transistor in a 1 transistor/1 resistor (1T/1R) configuration or by a diode in a 1 diode/1 resistor (1D/1R) configuration. The resistance of the PCM changes substantially between the amorphous and crystalline states, which is used to represent the two logic states. The change is induced by Joule heating of the PCM by passing electrical current pulses of different magnitudes. In the "reset" mode, a relatively high current (up to 1 mA) is passed through the cell, and during a brief period (5–10 ns), the temperature of the active region of the PCM rises above the melting point. When the pulse is suddenly switched off, the rapid heat dissipation leaves the PCM in an amorphous state since a much longer time would otherwise be needed to reach a long-range order. In the "set" mode, the crystalline state is recovered by passing relatively a lower current just to hold the active region of the PCM above the glass transition temperature but below the melting point. The reading is done by a fast current or voltage pulse at a much lower current level without triggering Joule heating during this stage. The current levels and pulse widths during operation are selected to enable cycling between the two states without irreversible breakdown.

The thin-film PRAM technology has made significant progress in the last decade and is on the verge of reaching the market. There are still concerns related to high programming current levels, intercell thermal interference, and the extent of scalability. Currently the footprint of the devices is large enough that the lithographic limit is not even a serious concern. In this regard, the nanowire-based PRAM (NW-PRAM) depicted in Figure 12.10 may offer solutions. First, as mentioned in Chapter 7, the memory cell volume

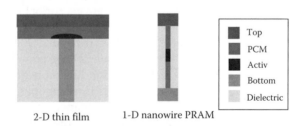

2-D thin film   1-D nanowire PRAM

■ Top
■ PCM
■ Activ
■ Bottom
□ Dielectric

FIGURE 12.10
Schematic of a conventional thin film based PRAM (left) and a NW-PRAM (right). (Courtesy of X.H. Sun and B. Yu.)

is smaller, which reduces the thermal budget. In addition, it has been shown in Chapter 7 that the melting point is reduced below the bulk value, and the extent of reduction is larger with smaller diameter wires.

The memory device shown in Figure 12.11 is fabricated by transferring the phase change nanowire onto a SiO_2-layered silicon substrate with a prepatterned molybdenum (Mo) pad array created by optical lithography. Focused ion beam (FIB) technique is used to directly write 150 nm-thick Pt interconnection lines between the nanowire and Mo probing pads. Ga ion beam (30 kV, 30 pA) is used to decompose the organometallic Pt precursor (trimethyl-methylcyclopentadienyl-platinum) to form metallic Pt. The figure also shows a SEM image of the fabricated GeTe nanowire device. The current–voltage (I–V) and resistance–voltage (R–V) characteristics of the nanowire

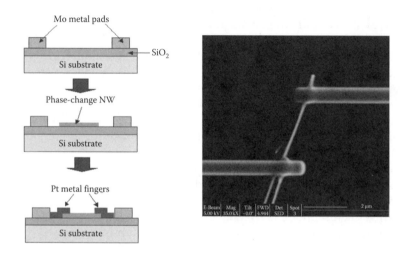

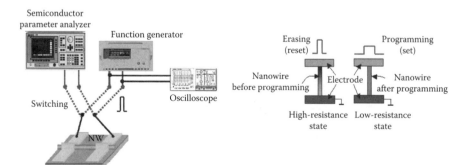

FIGURE 12.11
Process flow for the fabrication of the PRAM memory device (top left) and an SEM image of a fabricated device (top right). A schematic illustration of the test setup used for resistance measurement and the "set" and "reset" pulses are shown at the bottom. (Courtesy of X.H. Sun and B. Yu.)

memory device are measured using a probe station with HP 4156A semiconductor parameter analyzer. The voltage pulses for set and reset operations are generated by Agilent 33250A.

Figure 12.12 shows the measured device resistance for a GeTe nanowire device as a function of "reset/set" voltages. For this purpose, the GeTe nanowire device is subjected to a sequence of voltage pulses of constant width and varying magnitude, and the device resistance is measured at 0.1 V after each pulse. The initial low-resistance of the crystalline state is $5 \times 10^4 \, \Omega$. With a reset pulse width of 20 ns, the resistance is nearly constant till 2.0 V and then begins to increase sharply until reaching a high resistance state of $1.1 \times 10^8 \, \Omega$ at 2.5 V. The 20 ns voltage pulse (at voltages 2.5 V and above) raises the temperature of the nanowire (likely a portion of it) just above the melting point; the following rapid quenching action (with a 3 ns fall off) results in an amorphous state characterized by its very high resistance. For the device in the high resistance state initially as in Figure 12.12b, voltage pulses of 20 μs width up to 0.9 V maintain a constant resistance. Beyond 0.9 V, the temperature raises above the glass transition point and the transformation to crystalline state results in a sharp drop in resistance. At 1.1 V, the device recovers to its low resistance level of ~$5 \times 10^4 \, \Omega$. The two-resistance levels represent a high dynamic switching ratio of 2200. The power requirement at each stage is given by V^2/R, which is 125 μW for the reset and for a 20 ns pulse, the corresponding energy input is 25 pJ. The power requirement for the set operation is 11 nW with an energy input of 220 fJ for a 20 μs pulse.

The corresponding results for an In_2Se_3 nanowire memory device are more impressive than for the GeTe nanowire device due to its inherently larger resistivity (see Figure 12.13). The set operation is done at 5 V with 100 μs pulse width while the reset point is 7 V with a 20 ns pulse width. In both cases, the device resistance is measured at 0.2 V after each pulse. In the case of reset

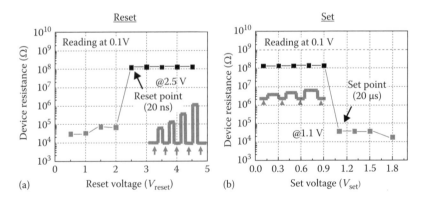

FIGURE 12.12
Germanium telluride phase change memory. (a) Reset operation performed with a 20 ns pulse width and (b) Set operation performed with a 20 μs pulse width. The device resistance is measured at 0.1 V. (Courtesy of X.H. Sun and B. Yu.)

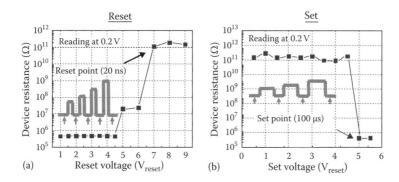

FIGURE 12.13

Switching behavior of In_2Se_3 phase change memory as a function of reset/set pulse voltage. Pulse width: (a) reset at 20 ns and (b) set at 100 μs. (Courtesy of X.H. Sun and B. Yu.)

operation, the low resistance state of $\sim 4 \times 10^5\,\Omega$ is maintained up to 4.5 V and then increases to the high resistance amorphous state ($\sim 10^{11}\,\Omega$) at 7 V. There is an intermediate state noticed in Figure 12.13a. In the set operation, the device at the high resistance state maintains a constant resistance up to 5 V when a 100 μs pulse is applied and beyond that, the resistance drops sharply. Here the dynamic switching ratio is 2×10^5. Similar performance improvement for GST (GeSbTe) based PRAM has also been reported previously [65,66].

Table 12.1 compares the performance of the GeTe and In_2Se_3 nanowire memory devices described above with reported results in the literature for In_2Se_3 thin film device [67], doped SbTe thin film device [68], and GST thin

TABLE 12.1

Performance Comparison of Various Nanowire and Thin Film Based PRAM Devices

Material/ Form	Reference	RSR	Reset Power, Energy, and Pulse Width	Set Power, Energy, and Pulse Width	Reset Current
GeTe nanowire	64	2200	125 μW, 2.5 pJ (20 ns)	11 nW, 220 fJ (20 μs)	50 μA
In_2Se_3 nanowire	63	2×10^5	80 μW, 1.6 pJ (20 ns)	0.25 nW, 25 fJ (100 μs)	11.7 μA
In_2Se_3 thin film	67	300	20 mW, 1.4 nJ (70 ns)	12 μW, 120 pJ (10 μs)	7.0 mA
Doped-SbTe thin film	68	200	300 μW, 15 pJ (50 ns)	1 μW, 100 fJ (100 ns)	0.25 mA
GST thin film	69	50	200 μW, 3 pJ (15 ns)	1 μW, 1.25 pJ (1.25 μs)	70 μA
GST thin film	70		6 mW, —	3.6 mW, —	—

RSR, resistance switching ratio.

film device [69,70]. Dynamic switching ratio, reset programming current, power for the set and reset operations, and the corresponding energy input for the chosen pulse widths are the metrics for comparison. Two interesting points are evident from this comparison. First, the In_2Se_3 nanowire device outperforms the GeTe nanowire device in terms of all the metrics. Reduction of programming current and power in conventional thin film GST devices is attempted by increasing the resistance by nitrogen doping. In contrast, In_2Se_3 is inherently a high resistivity material, enabling a lower programming current. In addition, as a single phase binary compound, phase segregation–related problems can be avoided with In_2Se_3 [67]. Next, it is clear in all cases that the nanowire devices require lower programming current and power for both "set" and "reset" operations than the thin film devices.

References

1. International Technology Roadmap for Semiconductors. Available online: http://www.public.itrs.net
2. J.A. Hutchby, G.I. Bourianoff, V. Zhirnol, and J.E. Brewer, *IEEE Cir. Dev. Mag.* 28 (2002).
3. R. Doering, NASA/SRC Workshop on Carbon Nanotubes Report (1999).
4. B. Yu and M. Meyyappan, *Solid State Electron.* 50, 536 (2006).
5. S. Reza, G. Bosman, M.S. Islam, T.I. Kamins, S. Sharma, and R.S. Williams, *IEEE Trans. Nanotechnol.* 5, 523 (2006).
6. J.E. Brewer, V.V. Zhirnov, and J. Hutchby, *IEEE Circ. Dev. Mag.* 13 (2005).
7. G.Q. Zhang, Strategic Research Agenda of "More than Moore," Proceedings of the 7th international conference on thermal, mechanical & multi-physics simulation and experiments in micro-electronics and micro-systems, ISBN: 1-4244-0275-1, Como, Italy, April 23–26, 2006, pp. 4–10.
8. Y. Cui and C.M. Lieber, *Science*, 291, 851 (2001).
9. Y. Huang, X. Duan, Y. Cui, L.J. Lauhon, K.H. Kim, and C.M. Lieber, *Science* 294, 1313 (2001).
10. Y. Cui, Z. Zhong, D. Wang, W.U. Wang, and C.M. Lieber, *Nano Lett.* 3, 149 (2003).
11. M.C. McAlpine, R.S. Friedman, S. Jin, K.H. Lin, W.U. Wang, and C.M. Lieber, *Nano Lett.* 3, 1531 (2003).
12. R. Chau, S. Datta, M. Doczy, B. Doyle, B. Jin, J. Kavalieros, A. Majumdar, M. Metz, and M. Radosavljevic, *IEEE Trans. Nanotechnol.* 4, 153 (2005).
13. J. Appenzeller, J. Knoch, E. Tutuc, M. Reuter, and S. Guha, IEDM 2006.
14. W.M. Weber, L. Geelhaar, A.P. Graham, E. Unger, G.S. Duesberg, M. Liebau, W. Pamler, C. Cheze, H. Riechert, P. Lugli, and F. Kreupl, *Nano Lett.* 6, 2660 (2006).
15. J.F. Dayen, A. Rumyantseva, C. Ciornei, T.L. Wade, J.E. Wegrowe, D. Pribat, and C.S. Cojocaru, *Appl. Phys. Lett.* 90, 173110 (2007).
16. G.M. Cohen, M.J. Rooks, J.O. Chu, S.E. Laux, P.M. Solomon, J.A. Ott, R.J. Miller, and W. Haensch, *Appl. Phys. Lett.* 90, 233110 (2007).
17. C.J. Kim, J.E. Yang, H.S. Lee, H.M. Jang, M.H. Jo, W.H. Park, Z.H. Kim, and S. Maeng, *Appl. Phys. Lett.* 91, 033104 (2007).

18. V. Schmidt, H. Riel, S. Senz, S. Karg, W. Riess, and U. Gosele, *Small* 2, 85 (2006).
19. J. Goldberger, A.I. Hochbaum, R. Fan, and P. Yang, *Nano Lett.* 6, 973 (2006).
20. J. Wang, E. Polizzi, and M. Lundstrom, *J. Appl. Phys.* 96, 2192 (2004).
21. J. Wang, A. Rahman, A. Gosh, G. Klimech, and M. Lundstrom, *Appl. Phys. Lett.* 86, 093113 (2005).
22. G. Liang, J. Xiang, N. Kharche, G. Klimeck, C.M. Lieber, and M. Lundstrom, *Nano Lett.* 7, 642 (2007).
23. O. Wunnicke, *Appl. Phys. Lett.* 89, 083102 (2006).
24. E.B. Ramayya, D. Vasileska, S.M. Goodnick, and I. Knezevic, *IEEE Trans. Nanotechnol.* 6, 113 (2007).
25. M. Shin, *J. Appl. Phys.* 101, 024510 (2007).
26. M. Shin, *IEEE Trans. Nanotechnol.* 6, 230 (2007).
27. A. Ghetti, G. Carnevale, and D. Rideau, *IEEE Trans. Nanotechnol.* 6, 659 (2007).
28. A. Bindal and S. Hamadi-Hagh, *Nanotechnology* 17, 4346 (2006).
29. H.T. Ng, J. Han, T. Yamada, P. Nguyen, Y.P. Chen, and M. Meyyappan, *Nano Lett.* 4, 1247 (2004).
30. A.B. Greytalk, L.J. Lauhon, M.S. Gudiksen, and C.M. Lieber, *Appl. Phys. Lett.* 84, 4176 (2004).
31. D. Wang, Q. Wang, A. Javey, R. Tu, H. Dai, H. Kim, P.C. McIntyre, T. Krishnamohan, and K.C. Saraswat, *Appl. Phys. Lett.* 83, 2432 (2003).
32. L. Zhang, R. Tu, and H. Dai, *Nano Lett.* 6, 2785 (2006).
33. E. Tutuc, J. Appenzeller, M.C. Reuter, and S. Guha, *Nano Lett.* 6, 2070 (2006).
34. E.K. Lee, B.V. Kamanev, L. Tsybeskov, S. Sharma, and T.I. Kamins, *J. Appl. Phys.* 101, 104303 (2007).
35. B. Yu, X.H. Sun, G.A. Calebotta, G.R. Dholakia, and M. Meyyappan, *J. Cluster Sci.* 17, 579 (2006).
36. J. Goldberger, D.J. Sirbuly, M. Law, and P. Yang, *J. Phys. Chem. B* 109, 9 (2005).
37. B. Pradhan, S.K. Batabyal, and A.J. Pal, *Appl. Phys. Lett.* 89, 233109 (2006).
38. S.N. Cha, J.E. Jang, Y. Choi, G.A.J. Amaratunga, G.W. Ho, M.W. Welland, D.G. Hasko, D.J. Kang, and J.M. Kim, *Appl. Phys. Lett.* 89, 263102 (2006).
39. K. Keem, D.Y. Jeong, S. Kim, M.S. Lee, I.S. Yeo, U.I. Chung, and J.T. Moon, *Nano Lett.* 6, 1454 (2006).
40. S. Ju, K. Lee, M.H. Yoon, A. Facchetti, T.J. Marks, and D.B. Janes, *Nanotechnology* 18, 155201 (2007).
41. T.L. Wade, X. Hoffer, A.D. Mohammed, J.F. Dayen, D. Pribat, and J.E. Wegrowe, *Nanotechnology* 18, 125201 (2007).
42. W.K. Hong, D.K. Hwang, I.K. Park, G. Jo, S. Song, S.J. Park, and T. Lee, B.J. Kim, and E. Stach, *Appl. Phys. Lett.* 90, 243103 (2007).
43. W. Wang, H.D. Xiong, M.D. Edelstein, D. Gundlach, J.S. Suehle, C.A. Richter, W.K. Hong, and T. Lee, *J. Appl. Phys.* 101, 044313 (2007).
44. H.D. Xiong, W. Wang, Q. Li, C.A. Richter, J.S. Suehle, W.K. Hong, T. Lee, and D.M. Fleetwood, *Appl. Phys. Lett.* 91, 053107 (2007).
45. P. Nguyen, H.T. Ng, T. Yamada, M.K. Smith, J. Li, J. Han, and M. Meyyappan, *Nano Lett.* 4, 651 (2004).
46. G. Jo, J. Maeng, T.W. Kim, W.K. Hong, M. Jo, H. Hwang, and T. Lee, *Appl. Phys. Lett.* 90, 173106 (2007).
47. R.M. Ma, L. Dai, and G.G. Qin, *Appl. Phys. Lett.* 90, 093109 (2007).
48. A. Motayed, M. He, A.V. Davydov, J. Melngailis, and S.N. Mohammad, *J. Appl. Phys.* 100, 114310 (2006).

49. A. Motayed, M. Vaudin, A.V. Davydor, J. Melngailis, M. He, and S.N. Mohammad, *Appl. Phys. Lett.* 90, 043104 (2007).

50. Y. Li, J. Xiang, F. Qian, S. Gradecak, Y. Wu, H. Yan, D.A. Blom, and C.M. Lieber, *Nano Lett.* 6, 1468 (2006).

51. M.H. Ham, J.H. Choi, W. Hwang, C. Park, W.Y. Lee, and J.M. Myoung, *Nanotechnology* 17, 2203 (2006).

52. M.H. Ham, D.K. Oh, and J.M. Myonng, *J. Phys. Chem. C* 111, 11480 (2007).

53. C. Thelander, T. Martensson, M.T. Bjork, B.J. Ohlsson, M.W. Larsson, L.R. Wallenberg, and L. Samuelson, *Appl. Phys. Lett.* 83, 2052 (2003).

54. E. Lind, A.I. Persson, L. Samuelson, and L.E. Wernersson, *Nano Lett.* 6, 1842 (2006).

55. T. Bryllert, L.E. Wernersson, L.E. Froberg, and L. Samuelson, *IEEE Elec. Dev. Lett.* 27, 323 (2006).

56. C. Li, J. Ly, B. Lei, W. Fan, D. Zhang, J. Han, M. Meyyappan, M. Thompson, and C. Zhou, *J. Phys. Chem. B* 108, 9646 (2004).

57. C. Li, W. Fan, D.A. Straus, B. Lei, S. Asano, D. Zhang, J. Han, M. Meyyappan, and C. Zhou, *J. Am. Chem. Soc.* 126, 7750 (2004).

58. C. Li, W. Fan, B. Lei, D. Zhang, S. Han, T. Tang, X. Liu, Z. Liu, S. Asano, M. Meyyappan, J. Han, and C. Zhou, *Appl. Phys. Lett.* 84, 1949 (2004).

59. C. Li, B. Lei, W. Fan, D. Zhang, M. Meyyappan, and C. Zhou, *J. Nanosci. Nanotech.* 7, 138 (2007).

60. X. Duan, Y. Huang, and C. Lieber, *Nano Lett.* 2, 487 (2002).

61. Q. Li, X. Zhu, H.D. Xiong, S.M. Koo, D.E. Ioannou, J.J. Kopanski, J.S. Suehle, and C.A. Richter, *Nanotechnology* 18, 235204 (2007).

62. H. Y. Cha, H. Wu, S. Chae, and M. G. Spencer, *J. Appl. Phys.* 100, 024307 (2006).

63. B. Yu, S. Ju, X Sun, G. Ng, T.D. Nguyen, M. Meyyappan, and D.B. Janes, *Appl. Phys. Lett.* 91, 133119 (2007).

64. B. Yu, X. Sun, S. Ju, D.B. Janes, and M. Meyyappan, *IEEE Trans. Nanotechnol.* 7, 496 (2008).

65. Y. Jung, S.H. Lee, D.K. Ko, and R. Agarwal, *J. Am. Chem. Soc.* 128, 14026 (2006).

66. S.H. Lee, Y. Jung, and R. Agarwal, *Nature Nanotechnol.* 2, 626 (2007).

67. H. Lee, Y.K. Kim, D. Kim, and D.H. Kang, *IEEE Trans. Mag.* 41, 1034 (2005).

68. M.H.R. Lankhorst, B.W.S.M.M. Ketelaars, and R.A.M. Wolters, *Nat. Mat.* 4, 347 (2005).

69. N. Takaura, M. Terao, K.Kurotsuchi, T. Yamauchi, O. Tonomura, Y. Hanoka, R. Takamura, K. Osada, T. Kawahara, and H. Matsuoka, *IEDM Digest* 897 (2003).

70. S.H. Lee, Y.N. Hwang, S.Y. Lee, K.C. Ryoo, S.J. Ahn, K.C. Koa, C.W. Jeong, Y.T. Kim, G.H. Koh, G.T. Jeong, H.S. Jeong, and K. Kim, *VLSI Technical Digest*, 2004.

13

Applications in Optoelectronics

13.1 Introduction

As we have seen in previous chapters, various elemental, oxide, nitride, and other nanowires have been grown in the band gap range of 0.2–5.0 eV, which covers from ultraviolet (UV) to near- and mid-infrared in the wavelength spectrum. These nanowires exhibit interesting optical properties and find applications in photodetectors [1–16], light emitting diodes (LEDs) [17–32], nanoscale lasers [33–51], photovoltaics [52–54], and others [55–57]. The basic optoelectronic demonstrations are also expected to have an impact on the development of optical switches, optical interconnects, optical waveguides, optoelectronic integrated circuits, electro-optic modulators, optical bio-sensors for lab-on-a-chip in security and biomedical fields, and analytical instruments. As in the case of electronics devices, modeling has been used to understand mechanisms and guide development [58–64]. The early results emphasize the potential of nanowires in various optoelectronics applications and as in all cases discussed in this book, much further work is needed in terms of material quality, reproducible processes, understanding device mechanisms, device fabrication, large wafer processing, and approaches amenable to low-cost production.

13.2 Photodetectors

Photodetectors typically take the form of photoconductors and photodiodes, which produce an electrical output in response to an optical input signal. The photoconductor setup in its simplest form uses a thin film or nanowire of the semiconductor material contacted by a pair of current-collecting electrodes. The photodiode is made of a p–n junction where the optical input produces carriers (electrons and holes) in the high field region surrounding the junction which get collected by the appropriate polarity contacts. In general, photo-diodes provide fast response time, high sensitivity, and high signal-to-noise ratio when compared to photoconductors. Most of the nanowire device demonstrations in the literature are of the photoconductor type [1,2,5–13,15,16] with a few reports on photodiodes [3,4,14]. The demonstrations cover visible [1–4], UV [6–15], and NIR [5,16] ranges using appropriate materials.

Ahn et al. [1] reported photoconductance (PC) response of silicon nano-wire field-effect transistors (SiNW FETs) showing that these can function as polarization-sensitive photodetector in the visible range. The SiNW FET is similar to the back-gated device discussed in Chapter 12 with both intrinsic and p-type devices (doped using B_2H_4). A scanning photocurrent measurement setup is used to record PC as a function of position along the nanowire axis when the device is illuminated with a 532 nm laser focused on a 500 nm spot. The device shows a large PC when the light is linearly polarized parallel to the nanowire axis and focused on the nanowire. The observed PC is much smaller when the polarization is perpendicular to the nanowire axis.

Another demonstration in the visible regime [2] involves the use of n-type CdS nanowires in a two-terminal configuration. The nanowires are deposited on highly doped silicon wafer capped with a 400 nm SiO_2. The Ti/Au electrodes are fabricated using e-beam lithography and evaporation followed by liftoff. A frequency-doubled Ti: sapphire laser provides the optical excitation at 1 KHz. Figure 13.1 shows the PC characteristics at 400 nm excitation and 0.7 W/cm² power. The photocurrent is about 10^5 times larger than the dark current. The shape of the curve is characteristic of a metal–semiconductor–metal structure with a back-to-back Schottky contacts between Ti and CdS. The electrodes themselves are opaque and do not yield any photoresponse under 800 nm, indicating that the *I–V* results in Figure 13.1 are due to the generation of free carriers upon illumination within the nanowire. The reported response time for the CdS photodetector circuit is about 15 µs [2].

Hayden et al. [3] also used n-CdS nanowires but crossed with a p-SiNW to create an avalanche photodiode (APD). Figure 13.2 shows the APD and its photocurrent response. The dark current shows a sharp increase at about −9 V

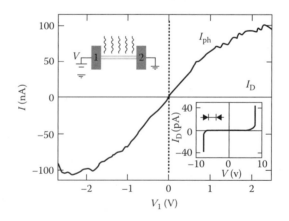

FIGURE 13.1

Photoconductive response of CdS nanowire under 400 nm illumination. The bottom right inset shows the *I–V* characteristics under dark conditions. The top left inset shows the device configuration for making measurements. (Reproduced from Gu, Y. et al., *Appl. Phys. Lett.*, 87, 043111, 2005. With permission.)

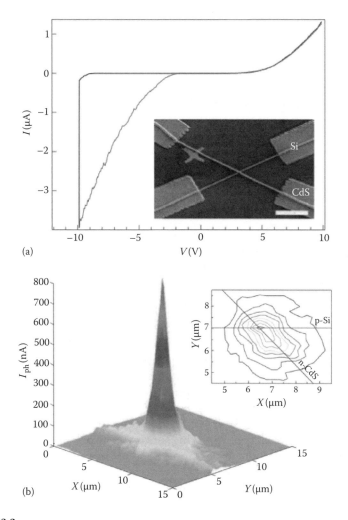

FIGURE 13.2

n-CdS and p-Si crossed nanowire p–n junction. (a) Photoresponse under dark and illuminated (488 nm source with 500 nW power) conditions. The inset is an SEM image of the crossed
nanowire device. (b) Spatially resolved photocurrent shown in the proportional mode using
a diffraction-limited laser. (Reprinted from Hayden, O. et al., *Nat. Mater.*, 5, 352, 2006. With
permission.)

and illumination leads to a bias-dependent photocurrent that increases with
increasing negative bias. The device exhibits a high gain of 7×10^4, which is
due to the amplification process at the p–n junction. Hayden et al. [3] report
4 pW as the lowest power measured above the noise level, which corresponds
to a detection limit of 75 photons. Instead of forming a p–n junction using
two crossed nanowires, it may be desirable to have distinctly doped p and n
regions within the same nanowire. Yang et al. [4] constructed such a p–i–n
complementarily doped single wire photodetector. Their SiNW fabrication is

based on a VLS process using silane as precursor and 20 nm gold colloids as discussed in Chapter 6. Diborane and phosphine are used for p- and n-type doping, respectively. The initial growth starts with diborane coflow with silane/H_2 for a nanowire length of 2.88 µm. Then the diborane is switched off for growing 1.08 µm long intrinsic region, followed by a coflow of phosphine with the source gases for growing 2.88 µm of n-type nanowire, segment. The photocurrent measurements show a large current in the middle region corresponding to the intrinsic segment of the nanowire, which is 500 times larger than the currents in the doped regions. Observed photocurrent is 100 times larger than the dark current and increases linearly with laser power.

Oxide nanowires with a wide band gap have been investigated for their photocurrent behavior in the UV regime. For example, Ga_2O_3 exhibits a band gap of 4.9 eV. The nanowires are readily produced by self-catalyzed VLS approach. Figure 13.3 shows a single nanowire connected with two gold electrodes [5]. The electrodes are deposited by e-beam deposition on a silicon substrate with a 500 nm thermally grown oxide. Figure 13.3 shows the dark current and photocurrent for a 40 nm diameter Ga_2O_3 nanowire, which is 400 nm long between the two gold electrodes. The dark current is a few pA, which increases with applied bias. When illuminated with a 254 nm light source, the current jumps to nA level. The *I–V* curve in Figure 13.3b is asymmetric, which is attributed to the poor ohmic contact [5]. Gallium oxide nanowires have extremely low free carrier density and in addition, the mismatch of the work function between the gold electrode and the nanowire is large. The real-time response of the Ga_2O_3 photodetector to 254 nm illumination on/off switching shows response and recovery times of 0.22 and 0.09 s, respectively. These values are better than for detectors using other oxides such as ZnO and In_2O_3, which have a much higher free electron density. In the case of other oxide nanowires, when the illumination is turned off, the free electrons recombine with excited holes and the PC decay is slow and even exponential. Slow adsorption of oxygen on the surface of the nanowire such as ZnO may

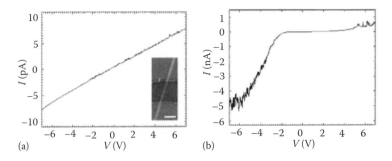

(a)   (b)

FIGURE 13.3
Photoconductive response of β-Ga_2O_3 nanowire. (a) The response under dark conditions. The inset is an SEM image of the nanowire bridging a pair of Au electrodes. (b) The current–voltage behavior under 254 nm light illumination. (Reproduced from Feng, P. et al., *Appl. Phys. Lett.*, 88, 153107, 2006. With permission.)

also be the cause of slow decay. In contrast, oxygen adsorption does not affect the conductance in the case of Ga_2O_3.

For UV photodetection, ZnO is the most widely investigated material in the literature [7–14]. It is a wide band gap semiconductor (3.37 eV at room temperature) with a large exciton binding energy of 60 mV. The large binding energy enables the stable formation of excitons at room temperature, thus making it attractive for optoelectronics applications. Kind et al. [7] first reported the potential of ZnO for photodetectors. Figure 13.4 shows the four terminal *I–V* measurements for a single ZnO nanowire under dark and illuminated conditions. The 60 nm wire is transferred from a VLS-grown sample directly onto prefabricated gold electrodes. The current transport is negligible under dark conditions, characterized by a resistivity of 3.5 MΩ cm. The current increases significantly with voltage upon illumination with a 380 nm UV lamp source. The resistivity now decreases by four to six orders of magnitude. The photodetector is highly sensitive, showing an increase in photocurrent with incident power. The behavior in Figure 13.4b can be fitted with a relation $I_{pc} \propto P^{0.8}$ where I_{pc} is the photocurrent and P is the incident laser power. Kind et al. [7] also showed that ZnO photodetectors exhibit excellent wavelength selectivity. For example, there is no observable photocurrent when illumination is done with a 532 nm source for 200 s; when illumination is immediately switched onto a 365 nm source, the photocurrent goes up by four orders of magnitude. An extensive investigation reveals that the response wavelength cutoff appears to be about 370 nm.

In ZnO nanowires, the photoconduction is governed by the surface effects characterized by the adsorption and desorption of oxygen. Normally, oxygen atoms get adsorbed onto the nanowire surface and combine with free electrons, which leaves behind a depletion layer. Upon exposure to the UV source, the holes produced by the light adsorption recombine with the O_2^- ions and the electrons released in the process give rise to the photocurrent.

13.3 Light-Emitting Diodes

Light-emitting diodes (LEDs) have a large demand in lighting, displays, and other applications. Conventional planar LEDs have some drawbacks such as poor light extraction, wide spectral width, and large output divergence [20]. Micron-sized array LEDs have been investigated as an alternative and found to provide higher efficiency. Fabrication of these arrays involves dry etching, which has its own limitations. In addition, the success of epitaxial growth of active layers is subject to obtaining good lattice match with the substrate, which is not always easy. Any significant mismatch leads to threading dislocations, which act as nonradiative trap centers and affect the LED performance [20]. In contrast, lattice mismatch is not a serious issue with the use of nanowires in the fabrication of LEDs and therefore, various materials such as GaN [18–25], ZnO [26–29], and InP [30–32] have been investigated.

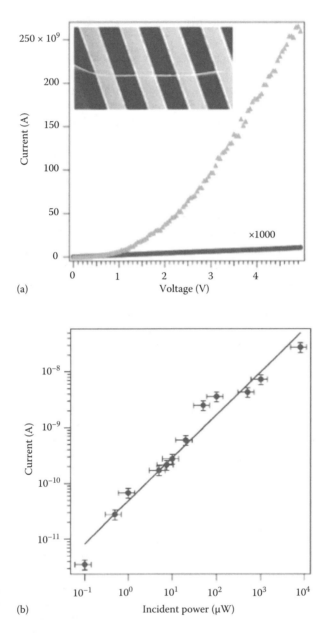

(a)

(b)

FIGURE 13.4
Photoconductive response of ZnO nanowire. (a) Dark current and photocurrent upon illumination with a 365 nm light source. The inset shows a 60 nm ZnO nanowire in a four-terminal measurement configuration. (b) Photocurrent variation with incident power. (Reproduced from Kind, H. et al., *Adv. Mater.*, 14, 158, 2002. With permission.)

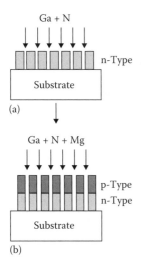

Ga + N

n-Type

Substrate

(a)

Ga + N + Mg

p-Type
n-Type

Substrate

(b)

FIGURE 13.5
Formation of a p–n junction within a nanowire. (a) n-Type GaN growth and (b) p-type GaN growth using Mg doping. (Reproduced from Kim, H.M. et al., *Adv. Mater.*, 15, 567, 2003. With permission.)

Kim et al. [19] reported the fabrication of GaN nanowire LEDs by forming a p–n junction within the same structure. Figure 13.5 shows schematically the formation of p–n junction. First, GaN nanowires (unintentionally n-doped) are grown by hydride vapor epitaxy method, which involves reacting GaCl and NH_3 at a substrate temperature of 478°C. GaCl is generated in situ by passing HCl (diluted in N_2) over Ga metal at 750°C. After some time, magnesium doping is carried out to change to p-type GaN. The carrier densities are estimated to be 10^{17} cm^{-3} for both types of nanowires. Figure 13.6 shows an SEM image of such a p–n junction within a GaN nanowire. The LED has contacts fabricated by e-beam evaporation followed by an annealing step. The n-type contact is Ti/Al (100/200 nm) and the p-type contact is Ni/Au (100/200 nm). At forward bias, the p–n junction produces light emission as seen in Figure 13.6b. The emission peak occurs at 3.179 eV (390 nm) independent of the injected current level [19].

Instead of using a single nanowire contacted by planar electrodes, Kim et al. [20] also fabricated an array of vertical nanorods incorporating six alternating layers of GaN/InGaN multiple quantum wells (MQWs). Figure 13.7 shows a schematic of the device fabricated on a sapphire substrate with a n-GaN buffer layer. Ga and trimethyl indium are the group III precursors while NH_3 is the nitrogen source. Co-flows of silane and Cp_2Mg are used as needed to obtain n- and p-dopings, respectively. The device consists of 0.5 μm tall nanorod arrays (NRAs) of n-doped GaN, followed by six periods of $In_{0.25}Ga_{0.75}N$/GaN and finishing with 0.4 μm tall p-GaN nanorods. The NRAs are then buried in spin-on-glass (SOG) to isolate individual rods. The n-contact consists of e-beam evaporated Ti/Al (20/100 nm) on the GaN buffer layer. Before the p-contact deposition, a thin Ni/Au (20/40 nm) is deposited on the entire surface of the SOG-NRA to facilitate current spreading and light emission. This is followed by e-beam evaporation of Ni/Au (20/200 nm) for the p-type probe pad. Figure 13.7b shows the light output of this array LED as a function of injected forward current and for comparison purposes, results are shown for a conventional broad area LED with the same materials and identical area. In the entire current range, the nanoarray LED has more than four times light output than the broad area LED. In the case of the arrays, the sidewalls also allow light extraction [20] and the nanowires provide a large surface area as expected.

Zhong et al. [22] crossed individual n-GaN and p-GaN wires and created a p–n junction. This device also exhibits UV-blue light emission at forward

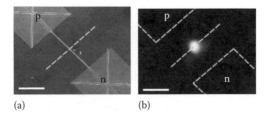

(a)   (b)

FIGURE 13.6

GaN LED fabricated using the process in Figure 13.5. (a) SEM image of the p-n GaN with respective contact pads. (b) Image of the emitted light at a forward bias of 3V. (Reproduced from Kim, H.M. et al., *Adv. Mater.*, 15, 567, 2003. With permission.)

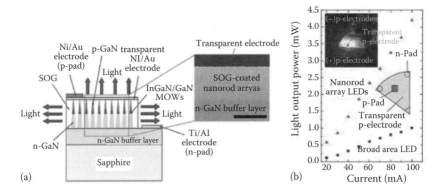

(a)   (b)

FIGURE 13.7

InGaN/GaN multiple quantum well nanorod array. (a) Schematic of the LED. (b) Light output as a function of forward bias current. The inset shows image of light emission at 20 mA dc current. (Reproduced from Kim, H.M. et al., *Nano Lett.*, 4, 1059, 2004. With permission.)

bias with a dominant peak at 415 nm. Instead of varying alloy composition axially as in Ref. [20], radial variation has been shown to be possible [23–25]. Here, GaN is first grown using VLS technique mediated by nickel catalyst clusters. The GaN core has a triangular cross-section and then an InGaN layer is grown on this core epitaxially by MOCVD. Multiple layers of alternating materials may be grown as desired to create a core/multiple shell structures. Devices constructed using such structures show higher quantum efficiencies than other nanoscale LEDs in the literature [25].

13.4 Nanoscale Lasers

Nanoscale laser is another key application for nanowires in optoelectronics for several attractive reasons. The nanowire diameter can be smaller than the emission wavelength in vacuum and the length as long as several tens of microns, thus offering the possibility of smallest lasers. They can form an optical cavity due to the difference in refractive index between the nanowire and the surroundings [44]. The large refractive index contrast between the nanowire and the surrounding allows strong mode confinement. Also, since nanowires can be grown in vertical arrays, two-dimensional laser arrays are possible in the future [58]. These, in addition to the cylindrical geometry, high-quality crystal structure, superior surface (unlike etched structures), and confinement of electrons, holes and photons, have prompted extensive investigations of nanoscale lasers in ZnO [33,44], GaN [45,46], ZnS [47], CdS [48,49], InN [50], and GaSb [51].

ZnO is by far the widely investigated nanowire for UV lasers. As mentioned before, ZnO has an exciton binding energy of 60 mV, which is larger than the thermal energy of 26 mV at 300 K. This enables excitonic recombination even at room temperature, which is inherently more efficient compared to the electron–hole plasma process commonly behind the operation of conventional lasers. The excitonic recombination makes it possible to obtain stimulated emissions at lower threshold power than what has been possible to date. Huang et al. [33] first reported the fabrication of room temperature ZnO nanowire UV lasers. An array of vertical ZnO nanowires (see Figure 13.8)

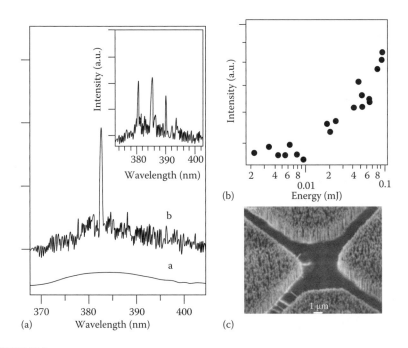

FIGURE 13.8
Room temperature ZnO nanowire–based UV laser. (a) Emission spectra collected at 20 (curve a), 100 (curve b), and 150 (inset) kW/cm². (b) Integrated emission intensity as a function of laser pumping energy. (c) SEM image of ZnO nanowire array prepared by VLS technique. (Reproduced from Huang, M.H. et al., *Science*, 292, 1897, 2001. With permission.)

grown by VLS technique through carbothermal reduction of ZnO power form the laser cavities ranging in diameter from 20 to 150 nm. Optical pumping of the nanowire samples is done at room temperature with the aid of Nd:yttrium-aluminum-garnet laser (266 nm and 3 ns pulse width) and the emitted light is collected normal to the nanowire end plane, i.e., along the nanowire axis. At low pump powers, the emission spectrum has a single broad peak at 3.23 eV as seen in Figure 13.8. The emission spectra display sharp peaks at pump fluences above the threshold of 40 kW/cm² (120 μJ/cm²). The integrated emission intensity rapidly increases with power, which is characteristic of stimulated emission.

While the above laser fabrication in Figure 13.8 uses an array of vertical nanowires, emission from a single ZnO wire dispersed on a quartz substrate has also been investigated [34]. But this laser has significantly higher threshold power (120 kW/cm² versus 40 kW/cm² in Figure 13.8), which may be due to the weak nanowire coupling and possible damage to the end faces of the nanowire during sample preparation. It is well known that VLS technique involves high-temperature growth and therefore low-temperature (220°C) hydrothermal techniques has been used to prepare ZnO nanowires for constructing UV lasers [38,39]. These hydrothermal samples exhibit lower emission thresholds of 8 kW/cm² [38] and 24 kW/cm². However, it is not clear if this reduction is due to any reduction in defects from the low-temperature preparation or differences in cavity length between various reports. The cavity length in Ref. [38] reporting 8 kW/cm² is ultralong, submillimeter versus the tens of microns in other works. The hydrothermal samples used by Hirano et al. [42] in contrast exhibit a higher lasing threshold of 70 kW/cm² compared to the 40 kW/cm² in Figure 13.8.

GaN nanowires have also been successfully used to construct UV lasers [45,46]. Gradecak et al. [46] reported the lowest room temperature threshold to date for GaN lasers at 22 kW/cm². The nanowires are grown by MOCVD on a sapphire substrate using trimethyl gallium and ammonia with silane coflow as n-type dopant. Figure 13.9 shows the results for an optically excited

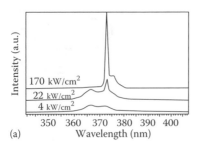
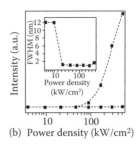

FIGURE 13.9
GaN nanowire laser (a) photoluminescence spectra recorded at various excitation power densities. (b) Intensity of the 373 nm lasing peak as a function of pumping fluence. The inset shows the variation in full width half maximum (FWHM) with power density. (Reproduced from Gradecak, S. et al., *Appl. Phys. Lett.*, 87, 173111, 2005. With permission.)

lasing in a 25 μm long n-type GaN nanowire [46]. The spontaneous emission centers around 365 nm at low pumping powers. A narrow peak begins to appear at 373 nm (3.33 eV) above the threshold power, and its intensity increases rapidly with power. The nanowires used in Figure 13.9 have a MOCVD growth temperature of 950°C. Interestingly, when a lower growth temperature (775°C) is used, these nanowires consistently show an order of magnitude higher laser threshold, emphasizing the importance of material quality, defect states, etc. on laser performance. The high-temperature nanowires are generally free from dislocations, which act as nonradiative recombination centers. The n-type doping also helps to lower the threshold compared to intrinsic nanowires by reducing the deep-level emission. Finally, the high-temperature growth yields more uniform diameter nanowires, thus improving the cavity properties [46].

Chin et al. [51] reported the first near-infrared nanowire based lasers at λ ~ 1550 nm using GaSb. At this wavelength, the wires are necessarily thicker (about 1 μm) than the accepted terminology of nanoscale and therefore, classified correctly as subwavelength-wire lasers. GaSb wires are produced by the spontaneous nucleation technique using a pool of Ga supported on a quartz substrate and a supply of Sb on the vapor phase from solid antimony with a 10% hydrogen in argon as carrier gas. Growth temperature is varied between 800°C and 1050°C. Figure 13.10 shows an SEM image of a well-faceted GaSb

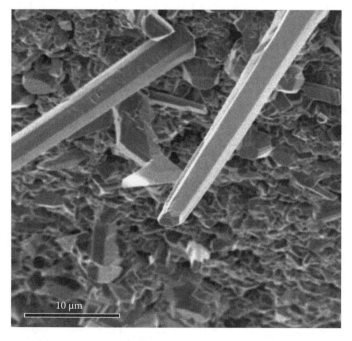

10 μm

FIGURE 13.10
GaSb wire grown by spontaneous nucleation showing well-faceted rectangular cross section. (Courtesy of S. Vaddiraju.)

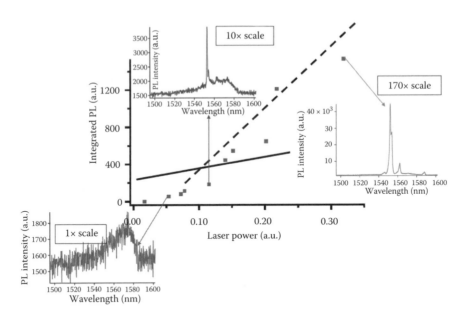

FIGURE 13.11

Integrated emission density as a function of laser excitation power for a GaSb sub wavelength wire laser. (Courtesy of C.Z. Ning.)

wire used for optical characterization. The wires grown by spontaneous nucleation approach show a rectangular cross-section consistent with the [110] growth directions. Figure 13.11 shows the optically pumped lasing results for a single GaSb wire. The sample is excited by a passively mode-locked Ti: sapphire laser (810 nm) with 150 fs pulse duration, 80 MHz repetition rate, and at 1.5 W average power. The spectra shows characteristic behavior of the lasers with only a broad spontaneous emission peak at low powers, appearance of a sharp peak at threshold and increase of intensity with increasing power. The laser threshold for this NIR laser is $50\,\mu J/cm^2$.

References

1. Y. Ahn, J. Dunning, and J. Park, *Nano Lett.* 5, 1367 (2005).
2. Y. Gu, E.S. Kwak, J.L. Lensch, J.E. Allen, T.W. Odom, and L.J. Lauhon, *Appl. Phys. Lett.* 87, 043111 (2005).
3. O. Hayden, R. Agarwal, and C.M. Lieber, *Nat. Mater.* 5, 352 (2006).
4. C. Yang, C.J. Barrelet, F. Capasso, and C.M. Lieber, *Nano Lett.* 6, 2929 (2006).
5. B. Polyakov, B. Daly, J. Prikulis, V. Lisauskas, B. Vengalis, M.A. Morris, J.D. Holmes, and D. Erts, *Adv. Mater.* 18, 1812 (2006).
6. P. Feng, J.Y. Zhang, Q.H. Li, and T.H. Wang, *Appl. Phys. Lett.* 88, 153107 (2006).

7. H. Kind, H. Yan, B. Messer, M. Law, and P. Yang, *Adv. Mater.* 14, 158 (2002).
8. Q.H. Li, Q. Wan, Y.X. Liang, and T.H. Wang, *Appl. Phys. Lett.* 84, 4556 (2004).
9. S.E. Ahn, J.S. Lee, H. Kim, S. Kim, B.H. Kang, K.H. Kim, and G.T. Kim, *Appl. Phys. Lett.* 84, 5022 (2004).
10. K. Keem, H. Kim, G.T. Kim, J.S. Lee, B. Min, K. Cho, M.Y. Sung, and S. Kim, *Appl. Phys. Lett.* 84, 4376 (2004).
11. Z.Y. Fan, P.C. Chang, J.G. Lu, E.C. Walter, R.M. Penner, C.H. Lin, and H.P. Lee, *Appl. Phys. Lett.* 85, 6128 (2004).
12. C.L. Hsu, S.J. Chang, Y.R. Lin, P.C. Li, T.S. Lin, S.Y Tsai, T.H. Lu, and I.C. Chen, *Chem. Phys. Lett.* 416, 75 (2005).
13. C.Y. Lu, S.J. Chang, S.P. Chang, C.T. Lee, C.F. Kuo, H.M. Chang, Y.Z. Chiou, C.L. ·Hsu, and I.C. Chen, *Appl. Phys. Lett.* 89, 153101 (2006).
14. J.H. He, S.T. Ho, T.B. Wu, L.J. Chen, and Z.L. Wang, *Chem. Phys. Lett.* 435, 119 (2007).
15. D. Zhang, C. Li, S. Han, X. Liu, T. Tang, W. Jin, and C. Zhou, *Appl. Phys. A* 77, 163 (2003).
16. J. Wang, M.S. Gudikson, X. Duan, Y. Cui, and C. Lieber, *Science* 293, 1455 (2001).
17. X.C. Wu, W.H. Song, W.D. Huang, M.H. Pu, B. Zhao, Y.P. Sun, and J.J. Du, *Chem. Phys. Lett.* 328, 5 (2000).
18. H.M. Kim, Y.H. Cho, and T.W. Kang, *Adv. Mater.* 15, 232 (2003).
19. H.M. Kim, T.W. Kang, and K.S. Chung, *Adv. Mater.* 15, 567 (2003).
20. H.M. Kim, Y.H. Cho, H. Lee, S.L. Kim, S.R. Ryu, D.Y. Kim, T.W. Kang, and K.S. Chung, *Nano Lett.* 4, 1059 (2004).
21. Y. Huang, X. Duan, Y. Cui, and C.M. Lieber, *Nano Lett.* 2, 101 (2002).
22. Z. Zhong, F. Qian, D. Wang, and C.M. Lieber, *Nano Lett.* 3, 343 (2003).
23. F. Qian, Y. Li, S. Gradecak, D. Wang, C.J. Barrelet, and C.M. Lieber, *Nano Lett.* 4, 1975 (2004).
24. Y. Huang, X. Duan, and C.M. Lieber, *Small* 1, 142 (2005).
25. F. Qian, S. Gradecak, Y. Li, C.Y. Wen, and C.M. Lieber, *Nano Lett.* 5, 2287 (2005).
26. R. Konenkamp, R.C. Word, and C. Schlegel, *Appl. Phys. Lett.* 85, 6004 (2004).
27. R. Konenkamp, R.C. Word, and M. Godinez, *Nano Lett.* 5, 2005 (2005).
28. J. Bao, M.A. Zimmler, F. Capasso, X. Wang, and Z.F. Ren, *Nano Lett.* 6, 1719 (2006).
29. M.C. Jeong, B.Y. Oh, M.H. Ham, S.W. Lee, and J.M. Myoung, *Small* 3, 568 (2007).
30. X. Duan, Y. Huang, Y. Cui, J. Wang, and C.M. Lieber, *Nature* 409, 66 (2001).
31. M.S. Gudiksen, L.J. Lauhon, J. Wang, D.C. Smith, and C.M. Lieber, *Nature* 415, 617 (2002).
32. E.D. Minot, F. Kelkensberg, M. van Kouwen, J.A. van Dam, L.P. Kouwenhoven, V. Zwiller, M.T. Borgstrom, O. Wunnicke, M.A. Verheijen, and E.P.A.M. Bakkers, *Nano Lett.* 7, 367 (2007).
33. M.H. Huang, S. Mao, H. Feick, H. Yan, Y. Wu, H. Kind, E. Weber, R. Russo, and P. Yang, *Science* 292, 1897 (2001).
34. J.C. Johnson, H. Yan, R.D. Schaller, L.H. Haber, R.J. Saykally, and P. Yang, *J. Phys. Chem. B* 105, 11387 (2001).
35. J.C. Johnson, H.Q. Yan, P.D. Yang, and R.J. Saykally, *J. Phys. Chem. B* 107, 8816 (2003).
36. H. Yan, R. He, J. Johnson, M. Law, R.J. Saykally, and P. Yang, *J. Am. Chem. Soc.* 125, 4728 (2003).

37. J.C. Johnson, K.P. Knutsen, H. Yan, M. Law, Y. Zhang, P. Yang, and R.J. Saykally, *Nano Lett.* 4, 197 (2004).

38. J.H. Choy, E.S. Jang, J.H. Won, J.H. Chung, D.J. Jang, and Y.W. Kim, *Appl. Phys. Lett.* 84, 287 (2004).

39. Z. Qiu, K.S. Wong, M. Wu, W. Lin, and H. Xu, *Appl. Phys. Lett.* 84, 2739 (2004).

40. H.C. Hsu, C.Y. Wu, and W.F. Hsieh, *J. Appl. Phys.* 97, 064315 (2005).

41. H.C. Hsu, C.Y. Wu, H.M. Cheng, and W.F. Hsieh, *Appl. Phys. Lett.* 89, 013101 (2006).

42. S. Hirano, N. Takeuchi, S. Shimada, K. Masuya, K. Ibe, H. Tsunakawa, and M. Kuwabara, *J. Appl. Phys.* 98, 094305 (2005).

43. T. Pauporte, D. Lincot, B. Viana, and F. Pelle, *Appl. Phys. Lett.* 89, 233112 (2006).

44. L.K. van Vugt, S. Ruhle, and D. Vanmaekelbergh, *Nano Lett.* 6, 2707 (2006).

45. J.C. Johnson, H.J. Choi, K.P. Knutsen, R.D. Schaller, P. Yang, and R.J. Saykally, *Nat. Mater.* 1, 106 (2002).

46. S. Gradecak, F. Qian, Y. Li, H.G. Park, and C.M. Lieber, *Appl. Phys. Lett.* 87, 173111 (2005).

47. Y. Jiang, W.J. Zhang, J.S. Jie, X.M. Meng, J.A. Zapien, and S.T. Lee, *Adv. Mater.* 18, 1527 (2006).

48. X. Duan, Y. Huang, R. Agarwal, and C.M. Lieber, *Nature* 421, 241 (2003).

49. R. Agarwal, C.J. Barrelet, and C.M. Lieber, *Nano Lett.* 5, 917 (2005).

50. M.S. Hu, G.M. Hsu, K.H. Chen, C.J. Yu, H.C. Hsu, L.C. Chen, J.S. Hwang, L.S. Hong, and Y.F. Chen, *Appl. Phys. Lett.* 90, 123109 (2007).

51. A.H. Chin, S. Vaddiraju, A.V. Maslov, C.Z. Ning, M.K. Sunkara, and M. Meyyappan, *Appl. Phys. Lett.* 88, 163115 (2006).

52. M. Law, L.E. Greene, J.C. Johnson, R. Saykally, and P. Yang, *Nat. Mater.* 4, 455 (2005).

53. J.B. Baxter, A.M. Walker, K. van Ommering, and E.S. Aydil, *Nanotechnology* 17, S304 (2006).

54. J.J. Wu, G.R. Chen, H.H. Yang, C.H. Ku, and J.Y. Lai, *Appl. Phys. Lett.* 90, 213109 (2007).

55. C.J. Barrelet, J. Bao, M. Loncar, H.G. Park, F. Capasso, and C.M. Lieber, *Nano Lett.* 6, 11 (2006).

56. A.P. Goodey, S.M. Eichfeld, K.K. Lew, J.M. Redwing, and T.E. Mallouk, *J. Am. Chem. Soc.* 129, 12344 (2007).

57. Q.Q. Wang, J.B. Han, D.L. Guo, S. Xiao, Y.B. Han, H.M. Gong, and X.W. Zou, *Nano Lett.* 7, 723 (2007).

58. A.V. Maslov and C.Z. Ning, *Appl. Phys. Lett.* 83, 1237 (2003).

59. A.V. Maslov and C.Z. Ning, *Opt. Lett.* 29, 572 (2004).

60. A.V. Maslov and C.Z. Ning, *IEEE J. Quantum Electron.* 40, 1389 (2004).

61. L. Chen and E. Towe, *Appl. Phys. Lett.* 87, 103111 (2005).

62. L. Chen and E. Towe, *J. Appl. Phys.* 100, 044305 (2006).

63. L. Chen and E. Towe, *Appl. Phys. Lett.* 89, 053125 (2006).

64. H.E. Ruda and A. Shik, *J. Appl. Phys.* 101, 034312 (2007).

14

Applications in Sensors

14.1 Introduction

The properties of many materials are influenced by their surroundings. For example, resistance, capacitance, and dielectric constant of single-walled carbon nanotubes (SWNTs) change when certain gases or vapors are adsorbed on the surface [1]. Similarly, metal oxides exhibit a change in resistance when exposed to oxidizing and reducing gases and vapors [2]. Therefore, measurement and monitoring of the change in a specific property of a material such as SWNT or an inorganic nanowire (INW) provides a basis for chemical sensing. Biomolecules can also change the conductance of carbon nanotubes and some INWs—for example, biospecies containing amino or nitro groups or deoxyribonucleic acid (DNA) adsorption—which forms the basis for biosensing. When biological materials do not directly change the properties of inorganic materials, then INWs and carbon nanotubes may be used as electrodes or conductors, which can have probe molecules attached to them either on the tip or on the surface [3]. When the target interacts or hybridizes with the probe, the generated electrical or electrochemical signal is transported through the conductor for processing. This is the so-called lock and key approach where a specific receptor is selected to strongly and selectively bind to the biospecies of interest [2]. An alternative in either case (chemical or bio) is a cantilever sensor wherein molecular adsorption of a species on a cantilever (or adsorption followed by hybridization) may result in specific deflection of the cantilever that can be used to identify the target.

The chemical and biosensors are different from analytical instruments such as mass spectrometer, gas chromatograph, and other laboratory equipment, which are also capable of identifying chemical components. Indeed, there is a miniaturization effort in progress to reduce the size and power consumption of various analytical instruments using the advances in mircofabrication, microelectromechanical system (MEMS), and nanotechnology. However, this chapter only deals with applications of INWs in chemical and biosensors. In this context, some relevant terminology is introduced below as listed in Ref. [4].

A sensor is a device that produces some form of a measurable signal in response to an external stimulus. A transducer is a device that converts signal from one form to another; it converts the sensor response into a measurable quantity. The transduction mechanism usually includes electrical, optical, acoustic, thermal, and other types of transducers. The sensor accuracy relates to how closely the sensor output matches the true value. The assessment is only possible if the sensor is calibrated against a standard or through the comparison of the sensor output against another measurement system with a known accuracy. The error is the difference between the sensor measurement and the actual value as established above. Resolution is defined as the smallest incremental change in the measured environment that can produce a detectable increment in the output signal. This is necessarily limited by the noise in the system. Noise refers to random fluctuation in the output signal even when the input, i.e., the environment being measured, is not changing. The source of noise can be external such as mechanical vibration, electromagnetic interference, thermal changes in the environment, etc., as well as internal noises such as shot noise, $1/f$ noise, etc. [4]. Drift is the gradual change over time in the sensor response even when the input environment is constant and stable. Sensitivity is the ratio of the incremental change in the sensor output signal to an incremental change in the ambient under measurement. Selectivity refers to the ability of the sensor to discriminate the one component under pursuit in the presence of several others. Stability refers to the sensor's ability to produce the same output value when the input conditions are constant over a period of time. The sensor response time is the time needed by the sensor to arrive at a stable value in response to a stimulus. In practice, when a step change is introduced in the input, the response time is taken as the time required for the measured signal to reach 95% of its final value. Conversely, the recovery time is the time required for the sensor to reach 95% of its final value when the source of the stimulus is removed. Finally, the sensor's range or span represents the range of input conditions that will yield measurable output signals. Operation outside the range may be meaningless, suffer from large inaccuracies, and may even damage the system.

14.2 Chemical Sensors

14.2.1 Sensor Requirements and the Role of Nanomaterials

Chemical sensors are needed in a variety of scenarios in chemical, security, biomedical, mining, agricultural, food processing, and other industries (see Table 14.1) [5]. Figure 14.1 shows the earliest chemical sensor in human history, which is a canary in a coal mine. A dead canary in a coal mine served as a warning signal to the miners about the dangers down the mine. Though significant progress in sensor technology has been realized in the last several decades, anecdotal evidence indicates that birds are still used for sensing in coal mines in some parts of the world and for clearing threats in abandoned

TABLE 14.1

Application Summary for Chemical Sensors

Gas/Vapor	Application
H_2	Combustion gas detection in aircraft/ spacecraft, fuel cells
NH_3	Industrial, medical
NO_2	Air quality monitoring
Ethanol	Breath analyzer, food quality, wine quality
Oxygen	Combustion
NO	Pollution monitoring
N_2O	Greenhouse gas, anesthetic use
CO	Pollution, home safety
CH_4	Coal mining
HCl, Cl_2	Chemical industry
SO_2	Chemical industry
H_2S	Chemical industry
Benzene, toluene, octane	Volatile organic compound monitoring, industrial emission
Formaldehyde	Indoor air quality, industrial emission
Nitrotoluene	Bomb detection
Nerve gas	Security, threat detection

buildings in war zones; the likely reason for this sad approach still in use is because such a "live" sensor may be more reliable and inexpensive than what modern technology can offer. Chemical sensor systems in recent years have been developed using a variety of base technologies such as cantilever systems, conductive polymer sensors, metal oxide thin-film sensors, surface acoustic wave sensors, quartz crystal microbalance systems, etc. Besides the applications listed in Table 14.1, these sensors have been deployed in numerous other scenarios as discussed by Albert et al. [2] and the references therein: monitoring of fish spoilage, coffee bean roasting, classifying bananas according to ripeness, sorting teas according to quality, establishing roasting times for almonds, wine identification, beer quality monitoring, air quality monitoring in agricultural storage, identification of appropriate period for artificial insemination of cows by monitoring estrus, and fluid leaks in cooling systems.

Regardless of the base technology, what is needed is a system that consists of the following:

- Sensor component
- Preconcentrator (almost always needed)
- Microfan or pump

FIGURE 14.1
Canary used as a sensor in coal mines. (Courtesy of M. Meyyappan.)

- Heater or cooler, if necessary
- Temperature and humidity sensors
- Signal-processing chip
- Data acquisition, storage
- Interface control, input/output, readout unit
- Integration of the above (nano–micro–macro)

Given a wide range of base technologies, what are the criteria for performance and selection?

- Sensitivity, as warranted by the application (ppm down to ppb)
- Absolute discrimination, selectivity
- Rapid response
- Fast recovery
- Reliability
- Long-term stability
- Small package (size, mass) in most applications these days
- Low power consumption
- A technology that is amenable for sensor network or sensor web
- A technology that is adaptable to different platforms (portable versus stationary for example)
- A technology that can take advantage of large-scale manufacturing practices to drive the cost down

Nanotechnology offers advantages with respect to some of the criteria listed above. First, nanomaterials possess a large surface-to-volume ratio, leading to large adsorption rates for gases and vapors. Properties at the nanoscale are also different from their bulk counterparts, for example, increased reactivity. These attributes of nanomaterials aid with the criteria of increased sensitivity and response, relative to thin-film sensors. Certainly, a reduction in size, weight, and power consumption is possible due to the small amount of active material used in the sensor. Selectivity may not directly benefit from the virtues of nanotechnology. Historically, absolute discrimination has been attempted using selective coatings, filters, dopings, and cycling of temperature, voltage, or similar parameters. All of these can be equally applied with the nanomaterial-based sensors as well. In any case, these approaches are not universal for all analytes and a vast amount of work is needed to come up with a coating or specific filter for each case. A universal approach to selectivity has been intelligent pattern recognition or finger printing with the use of a sensor array (instead of a single sensor) in the so-called electronic nose mode [2,6–9], which is discussed in Section 14.2.4. Nanotechnology will enable fabrication of a large number of sensors in a small chip, thus facilitating an effective discrimination procedure through an electronic nose concept with the use of powerful pattern recognition algorithms.

14.2.2 Nanowires in Sensor Fabrication

As seen in Table 14.2, most of the sensor studies in the literature have used metal oxide nanowires such as ZnO [10–29], tin oxide [30–44], In_2O_3 [45–50], V_2O_5 [51–53], and other oxides [54–62]. Few studies on the use of Pd [63,64], GaN [65], and silicon [66–69] nanowires have also appeared. In all cases, the

TABLE 14.2

Nanowire-Based Chemical Sensors Reported in the Literature

Ref.	Nanowire Material	Construction	Gas/Vapor	Background	Sensitivity	Selectivity
Wan et al. [10]	ZnO NW	Pt IDE, heater, ZnO NW paste	Ethanol	Air, 300°C	1 ppm	NR
Li et al. [11]	ZnO NW	Backgate FET, SW	O_2	Pure, RT	NR	NR
Xue et al. [14]	$ZnSnO_3$ NW	Diode, MW paste	Ethanol	Air, 300°C	1 ppm	NR
Wang et al. [15]	ZnO nanorod with Pd cluster	Diode, MW	H_2	Air, RT	10 ppm	NR
Wang et al. [16]	ZnO nanorod with Pd, Pt, Au, Ni, Al, Ti	Diode, NW	H_2	Air, RT	10 ppm	NR
Tien et al. [17]	ZnO nanorod	Diode, MW	H_2	Air, RT	10 ppm	NR
Fan and Lu [19]	ZnO NW	Backgate FET, SW	NO_2, NH_3	Air, RT	10 ppm (NO_2) 0.5% (NH_3)	NR
Fan and Lu [20]	ZnO NW	Backgate FET, SW	O_2, NH_3, CO, NO_2	Ar, RT, 500 K	20 ppm (NO_2), 1% NH_3, 0.5% CO	NR
Sun et al. [21]	ZnO nanorod Ag doped	Diode, MW	Ethanol, H_2, LPG, NH_3	Air, 300°C	100 ppm	NR
Rout et al. [24]	ZnO NW, Pt doped	Diode, MW	H_2, ethanol	Air, 150°C	10, 1000 ppm	NR
Rout et al. [25]	In_2O_3 NW, WO_3 NW, ZnO NW	Diode, MW paste	NO_2 NO, N_2O	Air, 150°C	10 ppm	NR
Rout et al. [26]	ZnO NW, TiO_2 NW, WO_3 NW	SW, atomic force microscopy (AFM)	H_2, LPG	Air	100–1000 ppm	NR
Rout et al. [27]	ZnO NW, In_2O_3 NW, SnO_2 NW	Diode, MW paste	NH_3	Air, 300°C	800 ppm	NR
Hsueh et al. [28]	ZnO NW with Pd	Diode, MW	Ethanol	Air, RT	500 ppm	NR
Li et al. [29]	ZnO NW In-doping	Diode, MW paste	Ethanol	Air, RT	100 ppm	NR

Reference	Material	Device	Analyte	Conditions	Concentration	Notes
Comini et al. [30]	SnO_2 nanobelts	Diode, MW	CO, NO_2, ethanol	Air, 400°C	250 ppm (CO), 250 ppm (ethanol), 0.5 ppm (NO_2)	NR
Law et al. [31]	SnO_2 nanoribbon	Diode, MW	NO_2	Air, RT	10 ppm	NR
Ying et al. [34]	SnO_2 nanowhisker	Diode, MW	Ethanol	Air, 300°C	50 ppm	NR
Kolmakov et al. [36]	SnO_2 NW, nanobelt with Pd NP	Diode, SW	O_2, H_2	Pure, 200°C–270°C	NR	NR
Kalinin et al. [37]	SnO_2 NW	SW with AFM	O_2	Pure, RT	NR	NR
Sysoev et al. [38]	SnO_2 NW, In_2O_3 NW, TiO_2 NW	MW diode	H_2, CO	O_2, 350°C		3-sensor array e-nose
Sysoev et al. [39]	SnO_2 NW	MW diode	Propanol ethanol, CO	Air, 600°C	500 ppb	Multisensor array e-nose
Hernandez-Ramirez et al. [40]	SnO_2 NW	SW diode	Water vapor	N_2, air, 180°C–360°C	1500 ppm	NR
Kuang et al. [41]	SnO_2 NW	SW backgate FET	Humidity	Air, 30°C	5%	NR
Chen et al. [42]	SnO_2 nanorods	MW, diode	Ethanol	Air, 300°C	10 ppm	NR
Wan et al. [44]	SnO_2 NW	MW film, diode	Ethanol	Air, 300°C	1 ppm	NR
Xiangfeng et al. [45]	In_2O_3 NW	MW film, diode	Ethanol	Air, 370°C	1000 ppm	NR
Li et al. [46]	In_2O_3 NW	SW, backgate FET	NO_2, NH_3	Ar, air, RT	100 ppm (NO_2), 0.02% (NH_3)	NR
Zhang et al. [48]	In_2O_3 NW	SW, backgate FET	NO_2	Ar, air, RT	5 ppb	NR
Ryu et al. [49]	In_2O_3 NW	SW, diode	Ethanol, CO, H_2	Air, RT-300°C	1 ppm (ethanol), 10 ppm (CO), 50 ppm (H_2)	NR
Yu et al. [51]	V_2O_5 NW	MW, diode	He	Pure, RT	NR	NR
Raible et al. [52]	V_2O_5 NW	MW, IDE	L-Butylamine, NH_3, toulene, L-propanal	Air, RT	10 ppm, L-butylamine	NR

(continued)

TABLE 14.2 (continued)

Nanowire-Based Chemical Sensors Reported in the Literature

Ref.	Nanowire Material	Construction	Gas/Vapor	Background	Sensitivity	Selectivity
Liu et al. [53]	V_2O_5 nanobelt	MW, diode	Ethanol	Air, 200°C	5 ppm	NR
Wang et al. [54]	CuO nanorod	MW, diode	Ethanol	Air, 300°C	1 ppm	NR
Francioso et al. [55]	TiO_2 NW	MW, diode	Ethanol	Air, 500°C	1200 ppm	NR
Fu et al. [56]	CeO_2 NW	MW, IDE	Humidity	Air, RT	NR	NR
Liu et al. [57]	TeO_2 NW	MW, IDE	NO_2, NH_3, H_2S	Air, RT	10 ppm (NO_2, H_2S) 100 ppm (NH_3)	NR
Kim et al. [58]	WO_3 NW	IDE, ME	NO_2, NH_3, ethanol	Air, 20°C–250°C	3–600 ppm	NR
Ponzoni et al. [59]	WO_{3-x} NW	MW, IDE	NO_2, H_2S, CO, NH_3	Air, 200°C–500°C	50 ppb (NO_2), 10 ppm (H_2S)	NR
Polleux et al. [60]	Tungsten oxide NW	MW, diode	NO_2	Air, 150°C	50 ppb	NR
Deb et al. [61]	Tungsten oxide NW mat	MW, diode	N_2O	N_2, 400°C	1 ppm	NR
Atashbar et al. [63]	Pd NW	MW, diode	H_2	Air, RT	3%	NR
Im et al. [64]	Pd NW	SW, diode	H_2	N_2, RT	0.02%	NR
Dobrokhotor et al. [65]	GaN NW with Au NP	MW, diode	Methane	Pure, RT	NR	NR
Zhou et al. [66]	SiNW	Diode, MW	NH_3	N_2, air, RT	0.1%	NR
Talin et al. [67]	SiNW	Backgate FET array, MW	Nitrobenzene, phenol	Cyclohexane, RT	0.001–0.25M solution	4-sensor array e-nose
McAlpine et al. [68]	SiNW coated with silanes	TFT on plastic, MW film	NO_2	N_2, RT	20 ppb	NR
Chen et al. [69]	SiNW with Pd NP	Diode, MW	H_2	Ar, RT	5%	NR

NW, nanowire; NP, nanoparticle; NR, not reported; FET, field effect transistor; TFT, thin-film transistor; IDE, interdigitated electrode; ppm, parts per million; ppb, parts per billion; M, molar; MW, multiwire (like a mat) SW, singlewire; RT, room temperature; LPG, liquified petroleum gas.

as-grown nanowires dispersed in a suspension are deposited on a substrate, dried, and annealed; or alternatively nanowires in a paste are applied to the substrate followed by drying and annealing. An ageing step of several hours is also included occasionally. Contacts are deposited by thermal or e-beam evaporation, either selectively over a single nanowire in a low-density sample or simply over multiple nanowires. This diode or chemiresistor configuration is used for measuring a change in conductance upon exposure to a specified amount of sample gas or vapor mixed with argon or nitrogen or air. Alternatively, field effect transistor (FET)-like devices use the silicon substrate as a backgate, again with a single nanowire or multiple wires bridging the source and the drain. The test chamber typically consists of gas flow arrangements, heater if desired, and electronics to monitor current–voltage characteristics. The conductance or resistance as a function of time is measured in response to a known concentration of the sample admitted for a specified time duration (pulse). The sample flow is turned off at the end of the pulse and recovery characteristics are recorded with or without air/inert gas purging. This cycle may be repeated for different sample concentrations (such as 5, 50, 100 ppm, etc.). The results are reported in terms of raw current data or normalized conductance $\Delta G/G_0$ or as a ratio R_a/R_g where R_a is resistance in air and R_g is the resistance when exposed to the analyte flow. Table 14.2 does not include response times or recovery times since most studies do not report these important informations and when reported, there is no common definition of these parameters to make comparison and evaluation meaningful here. Selectivity is a subject that is generally omitted, as the goal of most studies in the literature is to show that the chosen nanowire responds with a change in conductivity upon exposure to an analyte. Selectivity has been considered only in a few cases in Table 14.2 when a small array of sensors is fabricated and used like an electronic nose [38,39,67]; this will be addressed in Section 14.2.4. A few cases from Table 14.2 are discussed below in some detail to illustrate the basic behavior of metal oxide nanowire-based devices when exposed to gases/vapors such as NO_2, H_2, CO, NH_3, ethanol, etc.

Figure 14.2 shows a silicon-based membrane that houses Pt interdigitated electrodes (IDEs) and a Pt heater [10]. The silicon substrate consists of two sputtered layers of Si_3N_4 and SiO_2 followed by a 1 μm thick low-stress SiON film deposited by plasma-enhanced chemical vapor deposition (PECVD). The Pt IDE is sputtered onto the SiON layer and a window of 1.4 mm² is formed by backside etching of silicon with KOH solution. The as-prepared ZnO nanowires (average diameter of 25 nm) using a VLS approach are dispersed on the silicon-based membrane by spin-coating followed by drying for 1 h at 400°C. Figure 14.3 shows resistance as a function of time where the above device is exposed to ethanol at concentrations of 1–200 ppm when the heater maintains 300°C. The ratio R_a/R_g increases with ethanol concentration (not shown here) with typical values of 47 for 200 ppm, 31 for 100 ppm, and 1.9 for 1 ppm. These numbers are claimed to be higher than for ZnO thin-film

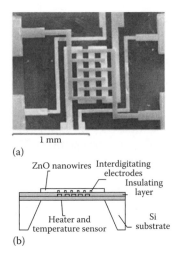

(a)

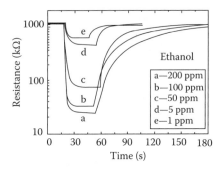

(b)

FIGURE 14.2

(a) Scanning electron microscope (SEM) image of a Pt IDE and Pt heater and (b) schematic of the device in (a). (Reprinted from Wan, Q. et al., *Appl. Phys. Lett.*, 84, 3654, 2004. With permission.)

FIGURE 14.3

Resistance change upon exposure to various concentrations of ethanol for the device in Figure 14.2. (Reprinted from Wan, Q. et al., *Appl. Phys. Lett.*, 84, 3654, 2004. With permission.)

ceramic sensors [10] which can be attributed to the higher surface-to-volume ratio of nanowires. The ratio R_a/R_g is proportional to P_g^β where P_g is the partial pressure of the analyte and β is an exponent. In the case of ethanol sensing here, β is unity in the concentration range of 1–100 ppm, and R_a/R_g appears to saturate beyond 200 ppm.

Figure 14.4 shows the resistance versus time characteristics of a device using ZnO nanorods loaded with Pd clusters and prepared by molecular beams epitaxy (MBE) [16]. A 20 nm Au thin film as catalyst, Zn metal, and O_2/O_3 plasma discharge to deliver atomic oxygen are used in the nanorod growth

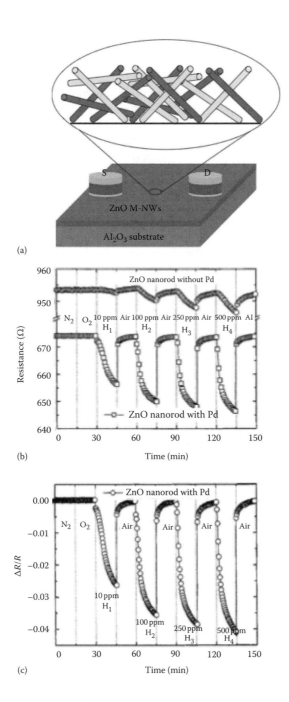

FIGURE 14.4
(a) Schematic of a ZnO nanorod hydrogen sensor. (b) Resistance versus time in response to various concentrations of hydrogen with and without Pd doping. (c) Results replotted in terms of $\Delta R/R$. (Reprinted from Wang, H.T. et al., *Appl. Phys. A*, 81, 1117, 2005; Wang, H.T. et al., *Appl. Phys. Lett.*, 86, 243503, 2005. With permission.)

on an Al_2O_3 substrate. The nanorods (30–150 nm diameter and 2–10 μm long) are contacted by Al/Ti/Au electrodes on two ends separated by 30 μm (see Figure 14.4). A 10 nm thick film of Pd is sputtered onto the nanorod surface, which forms clusters with a ~70% surface, coverage and rms roughness of ~8 nm. The device is exposed to 10–500 ppm of H_2 in air at room temperature. Several interesting features are evident from Figure 14.4. First, the addition of Pd clusters increases the response of ZnO nanorods to hydrogen pulses and there is a factor of 5 difference in response between samples with and without Pd loading. Pd may dissociate H_2 effectively and deliver atomic hydrogen to the ZnO surface, as will be discussed in Section 14.2.3. Second, the relative response of the device depends on the H_2 concentration and appears to increase with concentration. Finally, as seen in Figure 14.4, the response to O_2 exposure is negligible with or without Pd clusters. Interestingly, clusters of Pt, Au, Ag, Ni, and Ti have either little or no effect on the sensor response [15]. In a comparison with a ZnO thin-film device, the above nanorod device with Pd doping shows a larger response by a factor of 3 for 500 ppm of H_2 in air, again possibly due to higher surface-to-volume ratio of nanorods over thin films. This enhancement may not be large enough to warrant the use of ZnO nanowires if it turns out that nanowire-based devices are more expensive to manufacture than the established thin-film counterparts; however, there is no information on or even possible speculation of relative costs at present. In any case, there are certainly possibilities to optimize the nanowire sensor technology including the diameter and length of nanowires, preparation technique, Pd loading, and other relevant factors to obtain at least an order of magnitude larger response than from the ZnO thin film. Interestingly, when the hydrogen source flow is turned off, the recovery for ZnO nanorods is slower than for the thin film, which may be because the nanorods have more hydrogen per area to begin with [16].

Pd loading of ZnO nanowires also seems to have a slightly positive effect on ethanol sensing [28]. The resistance ratio R_a/R_g increases from 1.2 to 1.8 at 170°C and from 1.6 to 2.6 at 230°C with Pd loading for 500 ppm of ethanol in air. In this case, it is instructive to note that the absolute sensitivity values are rather poor compared to the results in Figure 14.3. A direct comparison to draw meaningful conclusions is not possible since ZnO preparation techniques, testing conditions etc are different between the two studies; the only obvious difference here is that the ZnO nanowire diameter is 50 nm compared to 25 nm in Figures 14.2 and 14.3. The inverse dependence of sensitivity on diameter will be discussed later in Section 14.2.3. Indium doping of ZnO has also been pursued for ethanol sensing [29] but it is not clear if there is any real advantage in spite of a band gap narrowing of 80 meV achieved by the addition of indium. The ratio R_a/R_g is 3 for 1 ppm, 27 for 100 ppm, and 37 for 300 ppm of ethanol. In contrast, the results in Figure 14.3 are 1.9 for 1 ppm, 32 for 100 ppm, and 47 for 200 ppm [10] without any doping. At first glance, In doping does not appear to be useful in enhancing the sensitivity. Considering that both results are from the same group, there are no

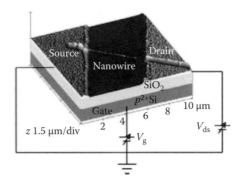

FIGURE 14.5

AFM image of a backgated FET with a ZnO nanowire. (Reprinted from Fan, Z. and Lu, J.G., *IEEE Trans. Nanotechnol.*, 5, 393, 2006. With permission.)

significant variations in nanowire preparation or device testing between the two studies. But the ZnO nanowires with In doping are 60 nm in diameter compared to the 25 nm in Figure 14.3. So, enhancement by In doping, if any, is likely to be masked by the inherently worse base performance by thicker ZnO wires.

Figure 14.5 shows an AFM image of a FET with a single ZnO nanowire as the conducting channel [20]. The nanowire is of n-type due to oxygen vacancies and Zn interstitials. ZnO nanowires suspended in alcohol are deposited on a p^{2+} silicon substrate capped with a SiO_2 layer, and e-beam lithography is used to define the electrodes in a low-density area to contact a single wire. Three-terminal measurements are made through the back-gated silicon substrate. Target gases such as NO_2, NH_3, and CO mixed with argon are tested separately for their impact on the FET's electrical characteristics. Figure 14.6 shows a decrease in conductance upon exposure to NO_2 and a stronger response with an increase in concentration from 1 to 20 ppm. Interestingly, a 50% reduction in conductance can be caused by just 2 ppm of NO_2 while it takes a concentration of 1% NH_3 to cause the same reduction. This and other unique phenomena related to NH_3 response will be discussed later under mechanisms. As seen in Figure 14.6b, applying a strong negative gate bias (30 V) aids the desorption process once the ammonia source is turned off.

Another oxide nanowire that has been prominent in chemical sensing is SnO_2, naturally due to the historical use of SnO_2 thin films in commercial sensors [2]. A SnO_2 nanowire device fabricated and tested in a generic diode configuration shows a sensitivity factor R_a/R_g of 23 for 50 ppm ethanol at 300°C [34] (for 50–200 nm diameter wires), which is higher than the value of 15 for ZnO nanowires under the same conditions in Figure 14.3 (for 25 nm diameter wires). While certainly this comparison can lead to a hasty conclusion, indicating superiority of SnO_2 nanowire over ZnO nanowire, the results for SnO_2 nanowire from the same group that produced the ZnO results in

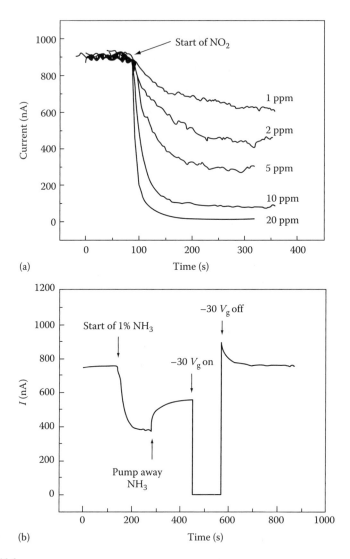

(a)

(b)

FIGURE 14.6
Sensing response of the device in Figure 14.5 for various concentrations of NO$_2$ (a) and 1% of NH$_3$ (b). (Reprinted from Fan, Z. and Lu, J.G., *IEEE Trans. Nanotechnol.*, 5, 393, 2006. With permission.)

Figure 14.3 are worse than for ZnO nanowire. For example, sensitivity to ethanol at 300°C is 13.9 for 100 ppm [42] using 4–15 nm diameter SnO$_2$ nanowires; with 3–12 nm diameter wires, the sensitivity is 30.7 for 100 ppm [43]. In contrast, the ZnO nanowire sensitivity for diameters twice as big (25 nm) from the same group is 31 for 100 ppm at 300°C (Figure 14.3). Controlled studies comparing performances of different materials under similar or identical conditions are not available.

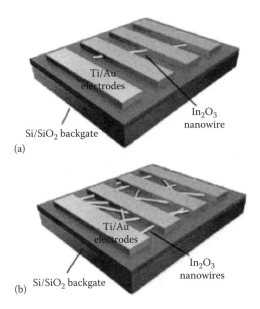

FIGURE 14.7
Backgated FETs with In_2O_3 nanowires. (a) Single wire device. (b) Device with multiple wires. (Reprinted from Zhang, D. et al., *Nano Lett.*, 4, 1919, 2004. With permission.)

Besides zinc and tin oxides, other oxides have been studied for their potential in sensor applications, and prominent among them is indium oxide, In_2O_3. Figure 14.7 shows backgated FETs with a single In_2O_3 as well as multiple nanowires as conducting channel [48]. The nanowires are prepared by laser ablation of an In_2O_3 target, which yields about 5 μm long wires with a diameter of ~10 nm. The as-fabricated devices show n-type transistor characteristics. Figure 14.8 shows the normalized conductance $\Delta G/G_0$ as a function of time when a single wire FET is exposed to pulses of 20 ppb to 1 ppm of NO_2. The device recovery part of the cycle is aided by UV light illumination using a 254 nm wavelength source. The detection limit for NO_2 is found to be lowered to 5 ppb with a multiple nanowire FET. In_2O_3 nanowires have also been investigated for their potential in ethanol sensing. A simple chemiresistor configuration with In_2O_3 wires (60–160 nm diameter) show [45] R_a/R_g of only 2 for 100 ppm of ethanol at 370°C, which is not as good as the ZnO and SnO_2 response discussed earlier. Ryu et al. [49] also reported only R_a/R_g of 2.5 for 100 ppm ethanol at 300°C for In_2O_3 nanowires (~10 nm diameter). The R_a/R_g for CuO (80 nm diameter nanorods) also is low relative to ZnO and Sn_2O_3 at about 2 for 200 ppm ethanol at 300°C [54].

14.2.3 Sensing Mechanisms

Under ambient conditions, oxygen molecules get readily adsorbed onto the oxide nanowire surface and form surface O_2^- and O^- species:

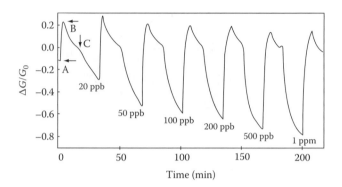

FIGURE 14.8

Response of a single wire FET from Figure 14.7 for various concentrations of NO_2. (Reprinted from Zhang, D. et al., *Nano Lett.*, 4, 1919, 2004. With permission.)

$$O_2(g) + e \rightarrow O_2^-(ads) \tag{14.1}$$

$$O_2^-(ads) + e \rightarrow 2O^-(ads) \tag{14.2}$$

This depletes the electrons near the surface of an n-type nanowire, leading to a higher resistance state. Now when this surface layer is exposed to a reducing gas environment as in the following cases, electrons are released from the respective surface reactions, which transfer back into the conductance band and increase the conductivity of the nanowires:

$$2H_2 + O_2^-(ads) \rightarrow 2H_2O + e \tag{14.3}$$

$$2C_2H_5OH^- + O_2^-(ads) \rightarrow 2CH_3CHO + 2H_2O + e \tag{14.4}$$

$$2NH_3 + 3O^-(ads) \rightarrow N_2 + 3H_2O + 3e \tag{14.5}$$

$$CO + O^-(ads) \rightarrow CO_2 + e \tag{14.6}$$

When the sensor is exposed to an oxidizing species such as NO_2, the resistance increases since NO_2 adsorbs and forms negatively charged surface species, NO_2.

The results discussed in the previous section for various nanowire structures are consistent with the above picture for both reducing and oxidizing environments. NH_3 sensing however shows interesting variations depending on temperature and doping [20,47]. The electron-donating reaction (Equation 14.5) and the resulting increasing in conductance is observed in ZnO nanowire FET only at elevated temperatures of 500°K. At room temperature, NH_3 adsorption actually decreases the conductance, indicating an electron-withdrawing nature. Fan and Lu [20] attributed this reversal in behavior to shift in Fermi energy, E_F, at elevated temperatures. Upon adsorption of gas molecules onto the nanowire surface, electron transfer occurs from the system at higher chemical potential to that at lower chemical potential until reaching equilibrium. At room temperature, E_F for ZnO is above the chemical potential of ammonia, thus leading to an electron transfer out of the nanowire and consequent decrease in conductance. When the temperature is increased, both the nanowire Fermi level and ammonia's chemical potential are lowered but the reduction of the former is larger, thus leading to a state wherein E_F lies below the chemical potential of NH_3. Then the electron-donating situation depicted in reaction (Equation 14.5) happens, resulting in an increase in conductance. The Fermi-level shift has also been shown to occur with different doping concentrations in ZnO nanowires, thus exhibiting both oxidizing and reducing behavior in ammonia sensing [47].

It was mentioned earlier that Pd clusters deposited upon the ZnO and SnO_2 nanowires increase the sensitivity to H_2 sensing [16,36]. There is a similar increase observed in sensing O_2 and ethanol as well with Pd clusters [28,36]. It is believed that Pd effectively dissociates H_2 and O_2 and delivers reactive atomic H and O to the nanowire surface, which enhance the electron-transfer reactions discussed above [36]. The conclusions on the effectiveness of other metal clusters in enhancing sensitivity appear to be inconsistent in various reports. Wang et al. [15] indicated ZnO nanorods coated with Pt, Au, Ag, Ni, and Ti thin films (all sputtered) do not appear to show any enhancement in sensitivity for H_2 over the uncoated specimens. Their samples are typically 30–150 nm in diameter prepared by MBE under a VLS mechanism. In contrast, Rout et al. [24] showed that 1% Pt loading increases ZnO nanowire sensitivity to 56 from 32 at 200°C for 1000 ppm H_2 in air. The response time (to reach the 90% of the equilibrium value according to their definition) is 18 s for Pt-coated sample versus 52 s for the uncoated sample. Similarly, Sun et al. [21] showed that 0.5 wt% loading with Ag increases the sensitivity of ZnO nanowires (10–30 nm diameter) for ethanol, H_2, NH_3, and liquefied petroleum gas (LPG). They believe that Ag clusters act as electron acceptors and enlarge the surface space charge layer, which results in electron depletion near the surface. Under a reducing environment, Ag clusters donate the electrons back to the nanowire conduction band, thus enhancing conductivity.

Wherever the recovery time appears to be slow, UV illumination of nanowires [46] and carbon nanotubes [1,70] aids in the desorption of species and speeds up the recovery. UV exposure generates electron–hole pairs in the nanowires, which interact with the surface-charged species appropriately; for example, in the case of NO_2 sensing, NO_2 combines with a hole and releases NO_2. The UV exposure has been shown to work effectively for NH_3, oxygen, and water vapor as well.

Based on the description of the sensing mechanisms above, the rate of change of electrical resistance should be proportional to the steady-state adsorbate concentration on the nanowire surface. Assuming a first-order kinetics for adsorption and desorption, a balance equation can be written as [61]:

$$\frac{d[\theta]}{dt} = K_a[C] - K_d[\theta] \qquad (14.7)$$

where
    $[\theta]$ is the steady-state adsorption coverage
    $[C]$ is the gas phase concentration
    K_a and K_d are adsorption and desorption rate constants, respectively

For a given gas concentration $[C]$, the solution to Equation 14.7 is

$$[\theta] = a - be^{-ct} \qquad (14.8)$$

where a, b, and c are constants that depend on the adsorption and desorption rates. The constant, a, is proportional to the adsorption rate and thus depends on gas phase concentration. The constant, c, represents the rate constant for the desorption process and hence the time constant in the response curve. Equation 14.8 describes the rise in response immediately following the exposure to the analyte, which has been seen in all the sensor characteristics as discussed in the chapter thus far. Indeed, fit of experimental data in each case would yield a, b, and c for a given concentration and facilitate direct comparison of various materials and sensor constructions. During recovery when the analyte flow is turned off as in all the experiments described here, the second term in Equation 14.7 representing the rate of desorption dominates and the solution to Equation 14.7 then would be given by

$$[\theta] = \theta_0 + \theta_1 e^{-ct} \qquad (14.9)$$

where θ_0 and θ_1 are constants. This solution represents the decay in resistance during the recovery when the sensor proceeds to return to the base line. Again, the fit of experimental data to the above form would yield the two constants θ_0 and θ_1 [61].

The advantage of using nanowires over the corresponding oxide thin film is the enhanced surface to volume ratio as mentioned earlier. Then sensitivity should depend on the nanowire radius and improve with a decrease in size. A controlled study using the experimental setup in Figure 14.5 indeed proves this radius dependence to be true [20]. Figure 14.9 shows the ZnO nanowire sensitivity represented by $\Delta G/G_0$ as a function of radius for O_2. This behavior is explained by noting that the conductance G of a nanowire exposed to a gas is given by [20]

$$G = \frac{\pi r^2}{\ell} n_e e \mu_e \qquad (14.10)$$

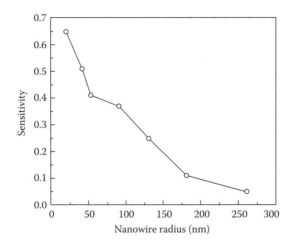

FIGURE 14.9
Dependence of sensor response $\Delta G/G_0$ as a function of nanowire radius. Results are for ZnO nanowire FET from Figure 14.5 in response to oxygen exposure. (Reprinted from Fan, Z. and Lu, J.G., *IEEE Trans. Nanotechnol.*, 5, 393, 2006. With permission.)

where

   r and ℓ are radius and length of the nanowire
   μ_e is the electron mobility
   n_e depends on the surface density of the chemisorbed species, N_s and
      charge transfer coefficient α:

$$n_e = n_0 - \frac{2\alpha N_s}{r} \tag{14.11}$$

where n_0 is the electron density prior to gas exposure. This yields

$$\frac{\Delta G}{G_0} = \frac{1}{r}\frac{2\alpha N_s}{n_0} \tag{14.12}$$

showing clearly the inverse dependence of sensitivity on the nanowire radius.

14.2.4 Selectivity and Electronic Nose

A sensor fabricated by any of the approaches described earlier is not inherently selective for any analyte, regardless of the morphology of the sensor material (nanowire, nanoparticle, continuous thin film, etc.). Any gas or vapor can change the conductivity of an oxide and one cannot specify what causes such a response just by looking at the response curve. It is possible to use selective filters that allow only the species of interest to reach the sensor. The nanowires may be selectively doped with a metal that can give dominant response only to selected species. Alternatively, the nanowires may be coated with a polymer material selective to a particular species. All such approaches may work in some isolated cases, but have little utility in a broad sense to develop sensors that can selectively discriminate the species of interest. The only general approach for selective discrimination appears to be the so-called electronic nose [2,6,7], which attempts to mimic the mammalian olfactory system. The latter consists of hundreds to thousands of different olfactory receptor genes and when stimulated by an odor, responses from various receptors are sent to the olfactory bulb and then to the olfactory cortex for processing and recognition [2]. The receptors are not highly selective to specific odorants. Indeed, each receptor responds to several odor-causing agents and multiple receptors may respond to the same stimulant. Mimicking this system, the electronic nose uses an array of sensors instead of a single sensor. Typically, all sensors in the array are not the same and it is important to introduce some variations. There may be physical differences between different sensors in the array but more importantly some chemical or material differences including coatings and dopants. Each sensor, physically and/or chemically different from its neighbor in the array, provides a specific interaction with gases and vapors in varying degrees. A number of different sensors can provide a pattern for a specific target and each target will have its own pattern from the same set of sensors. Using various advanced pattern recognition techniques, the targets can be distinguished from one another. Other signal-processing approaches include statistically based chemometric methods and neural networks [2]. Nanotechnology enables fabrication of a large array of sensors in a small area chip. Multiple sensors are ideal when complex mixtures are involved, background is unknown, or constantly changing background is encountered. The large array is also helpful to reduce ambiguity, allow redundancy, and improve signal-to-noise ratio [2].

As seen from Table 14.2, the use of an array and e-nose approach is not common for nanowire-based sensing in the literature except for Refs. [38,39], demonstrating the principle with three sensors. A carbon nanotube sensor array from the literature [70,71] is described below as an example for sensitive and selective gas discrimination though the carbon nanotubes (CNTs) can be simply replaced with any of the nanowires in Table 14.2 or better yet, the array can have individual sensors using CNTs and different oxide nanowires. Figure 14.10 shows a sensor chip containing 32 sensor elements on a 1 cm × 1 cm area. The microfabrication approach enables to reduce the chip size and/or increase the number of sensor elements as desired and 64- and 96-sensor element chips are already available [72]. Also, since this is wafer fabrication using standard semiconductor processing techniques, scale-up is straightforward from the 4 in. wafer in Ref. [71] to larger wafers. The chip is mounted on a ceramic carrier and consists of a heater, thermistor, a temperature sensor, and a humidity sensor.

Figure 14.10 also shows an individual sensor consisting of an IDE system. A p-type boron-doped silicon (100) wafer with a resistivity of 0.006–0.01 Ωcm

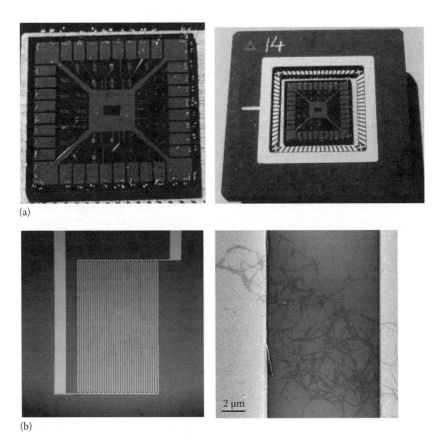

(a)

(b)

2 µm

FIGURE 14.10

(a) A 32-sensor array chip mounted on a ceramic chip carrier. (b) An individual sensor consisting of an IDE array pattern and SWNTs bridging adjacent fingers. (Courtesy of J. Li.)

and thickness of $500 \pm 25\,\mu m$ is used as the substrate. A $0.5\,\mu m$ layer of SiO_2 is thermally grown on the substrate and then 200 nm thick patterned Pt lines are deposited on top of the oxide for electrical heating and resistive temperature detection (RTD). This is followed by a $1\,\mu m$ silicon nitride to serve as an insulating layer between the heater and the IDE above. A layer of 200 nm Pt on top of 20 nm Ti is deposited on top of the nitride layer onto the designed finger patterns. The IDE pattern consists of 4, 8, 12, and $50\,\mu m$ finger gaps with 10 and $20\,\mu m$ finger widths.

Purified SWNTs are used as the sensor medium here; however as mentioned earlier, any of the oxide nanowires can be used as well. In situ CVD to deposit the conducting material on the sensor array may be complicated and expensive, especially if different materials were to be used in different sensors across the array. A simple solution casting or ink-jetting approach would be easier and less expensive. SWNTs are dispersed in dimethyl formamide

(DMF) to form a suspension. A small amount (0.05 μl) of this dispersion (concentration 3 mg/L) is drop-deposited onto the IDEs of each sensor. The evaporation of DMF leaves a network of nanotubes across the fingers and a drying step is used to get rid of the DMF residue. The density of the SWNTs across the electrodes of any one sensor can be adjusted by varying either the quantity (number of drops) or concentration of the SWNT–DMF solution. There are several variations introduced in the sensor array besides any physical (finger width, gap) or SWNT concentration differences. A chlorosulfonated polyethylene is used in a couple of sensors as a coating for SWNT to facilitate detection of chlorine. Hydroxypropyl cellulose coating of nanotubes is used in some sensors for HCl detection. In each case, the above materials mixed with the appropriate solvents are drop-deposited in small quantities over SWNTs in some of the sensor elements. These two particular sensor types will give stronger responses to chlorine and HCl, respectively, than the rest in the arrays. On another sensor element, a solution of monolayer-protected clusters of gold is cast over the nanotubes. Finally, in one of the sensors, the SWNTs in solution already have Pd-sputtered onto the nanotubes prior to dispersion. Pd is ideal for hydrogen and methane detection. It is important to note that signal processing is not possible if there is no detectable signal for a given analyte, and therefore the first priority is to elicit signals from some of the sensors in the array. This is where dopings, coatings, metal loadings, etc., can be useful.

The sensor array chip was "trained" by exposing to NO_2, HCN, HCl, Cl_2, acetone, and benzene at ppm concentration levels in air using a computerized multicomponent gas blending and dilution system [71]. The resistance was measured under various conditions and pattern recognition was done by principal component analysis (PCA). The sensor response is represented in terms of $(R_t - R_0)/R_0$ where R_0 is the initial resistance and R_t is the resistance during the exposure. Figure 14.11 shows unique fingerprints for NO_2 and Cl_2 from the 32 sensor elements in the electronic nose. It is clear that each sensor in the array gives different response to either NO_2 or chlorine. The same sensor also gives distinctly different response for the two analytes. Since the response is a function of concentration, temperature, and other influencing external factors, it is a common practice to normalize the data for meaningful comparison. Then, the lengths of all the data vectors are forced to be the same by equating the sum of the squares to a constant:

$$\sum_{k=1}^{N} x_{ik}^2 = C_i \tag{14.13}$$

where
  k denotes the sensor number
  i denotes the gas/vapor analyte

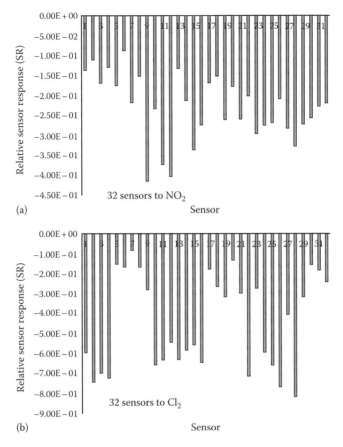

FIGURE 14.11
Distinct response patterns from the 32 sensors in Figure 14.10 in response to NO₂ (a) and chlorine (b). (Courtesy of J. Li.)

N is the total number of sensors (=32 in Figure 14.11)
C_i is any real number

The data is also autoscaled by

$$x'_{ik} = \frac{x_{ik} - \overline{x}_k}{S_k} \tag{14.14}$$

where
x_{ik} is the autoscaled response
x_k is the mean value of the normalized response
S_k is the standard deviation

S_k is given by

$$S_k = \left[\frac{1}{N-1} \sum_{i=1}^{N} (x_{ik} - \bar{x}_k)^2 \right]^{0.5}$$

(14.15)

The purpose of autoscaling is to remove inadvertent weighting that may arise from arbitrary units.

Next, a pattern recognition algorithm using PCA is applied to the data set generated above. PCA expresses the information contained in the variables $X = \{x_k, k = 1, 2, ..., K\}$ into a lower number of variables $Y = \{y, y_2, ..., y_z\}$ where $z < K$. Y is then the principal components of X. Figure 14.12 shows the first three principal components for the data from $N = 32$ sensors for

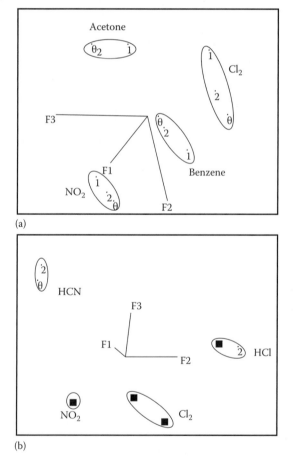

(a)

(b)

FIGURE 14.12
PCA plot showing the discriminating ability of the sensor array. (Courtesy of J. Li.)

gases NO_2, HCl, Cl_2, HCN, acetone, benzene, and other tested gases. The principal components are labeled F_1, F_2, and F_3. It is clear that all the analytes are distinctly and completely separated in the principal component space, demonstrating that this 32-sensor array is capable of discriminating these gases and vapors. Further details on the working principles and nuances of the electronic nose can be found in Refs. [6,7,71]. It is emphasized again that though the description above relates to a CNT sensor array as this was one of the first and rare efforts to date to apply e-nose principles to nanosensors, the applicability is equally valid for nanowire sensor arrays. Indeed, a sensor array with independent CNT sensors and SnO_2, gallium oxide nanowire sensors have been demonstrated [72] where the use of nanowires and SWNTs provide sufficient orthogonality to produce distinct patterns.

14.3 Biosensors

The use of INWs for constructing biosensors has not been as extensive [73–92] as for chemical sensors described in Table 14.2 and previous section. This may be likely due to the reason that the oxide thin films, especially tin oxide, have been used successfully in commercial chemical sensors for quite some time and the use of nanowires is an obvious extension to advance sensitivity levels while reducing the size and power consumption by taking advantage of the large surface area. Neither that kind of history nor an obvious extension using nanowires is the case with biosensors. The advantages and limitations of using nanowires to design a biosensor can be seen by an analysis similar to the balance Equation 14.7 presented earlier [93]:

$$\frac{dN}{dt} = k_f \left(N_0 - N \right) \rho_s - k_r^N \qquad (14.16)$$

where
  N is the number density of probe or receptor molecules bound to the target
  N_0 is the initial density of probes on the nanowire surface
  k_f and k_r are the rate constants for attachment and detachment, respectively
  ρ_s is the density of the targets at the nanowire surface

The first term on the right-hand side of Equation 14.16 represents the target–probe conjugation and the second term stands for detachment events. The target density itself is given by solving the corresponding transport equation:

$$\frac{d\rho}{dt} + V \cdot \nabla \rho = D \nabla^2 \rho \qquad (14.17)$$

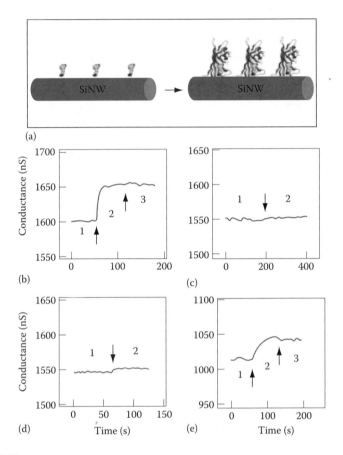

FIGURE 14.13
Detection of protein binding using a silicon nanowire device. (a) SiNW modified with biotin
(left) and after streptavidin binding (right). (b) Conductance versus time for biotin-modified
SiNW. 1, 2, and 3 correspond to regions of buffer solution, addition of 250 nM streptavidin, and
pure buffer solution, respectively. (c) Conductance versus time for unmodified nanowire. 1 and
2 have the same meaning as before. (d) Conductance versus time for a biotin-modified nanowire. Region 1 is same as before and region 2 represents addition of 250 nM streptavidin prein-
cubated with biotin. (e) Conductance versus time for a biotin-modified nanowire. Region 1 as
before, region 2 represents 25 pM streptavidin, and region 3 pure buffer solution. (Reproduced
from Cui, Y. et al., *Science*, 293, 1289, 2001. With permission.)

where the second term on the left denotes convection of the target to the
surface due to flow and the right-hand side term represents delivery of the
target by diffusion. D is the corresponding diffusion coefficient. Analyzing
biosensors constructed using thin films, nanowires, and particles, Nair and
Alam [93] reach several conclusions. First, a nanowire-based sensor can give
three to four orders of magnitude higher detection limit than a planar thin-
film sensor. Second, going further down to spherical geometry offers no special advantage. Finally, there is a tradeoff between the response time and

minimum detectable concentration. Their analysis shows that femtomolar detection would take an incubation time of hours to days. There are however several approaches to improve the response time, which include decreasing the size of the nanowire, increasing the diffusion by raising the temperature, and by using convection or flow effectively.

Biosensors have been fabricated mostly using silicon nanowires [73–77, 83,84] and unlike the case of chemical sensors, oxide nanowires are not prominent [78–80]. Gold and platinum nanowires have also been used successfully [87–92]. The demonstrated applications include detection of DNA and sequence variations, virus, cholesterol, prostate-specific antigen (PSA), glucose, cytokeratin in cancer diagnostics, and foodborne pathogens.

A FET using a nanowire can be used to detect conjugation between a receptor and a target by monitoring the conductance change of the nanowire. Cui et al. [73] fabricated a SiNW FET with a backgate similar to the description in Chapter 6 and modified the surface with biotin (see Figure 14.13). The conductance of these modified nanowires increases rapidly when a 250 nM streptavidin solution is added. This conductance value is maintained even with the addition of a pure buffer solution in sequence. The streptavidin is a negatively charged species and when it binds to a p-type nanowire surface, the conductance increases. When there is no biotin, addition of streptavidin does not change the nanowire conductance, confirming that the origin of conductance change is the specific binding of streptavidin to biotin. Silicon nanowire devices have also been used for detecting DNA sequence variations [74], viruses [75], and PSA [76]. Figure 14.14 shows the principle behind the virus detection wherein two nanowire devices have their nanowire surface modified with antibodies [75]. When a virus binds to the antibody receptor on one device, the conductance of this device changes and after unbinding, the conductance returns to the baseline value. The second device, which does not have the conjugation process, acts as an internal control. This technique has been shown [75] to work with influenza A virus.

Indium oxide has been used as an alternative to SiNW for biosensors [78–80] and unlike silicon, these nanowires do not have a native oxide coating between the nanowire surface and the receptor. A backgated FET with a single In_2O_3 nanowire as in Figure 14.7 has been used to detect biospecies containing amino or nitro groups [78]. Figure 14.15 shows results for the detection of low-density lipid (LDL) cholesterol—which has a protein containing positively charged amino groups surrounding a hydrophobic core as shown in the inset of Figure 14.15c. When a drop of a pure ethanol is added to the nanowire surface, the device current increases momentarily and then decreases quickly following the evaporation of ethanol. This can be repeated many times. When a drop of LDL in ethanol suspension is added, the current increases again but stabilizes at an intermediate level even after ethanol evaporates. Here the mechanism is similar to the description in Section 14.2.3. The amino groups in the LDL proteins act as a reducing agent after the binding and donate electrons to the In_2O_3 nanowires, thus increasing

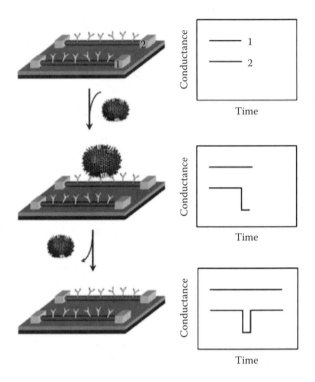

FIGURE 14.14

Virus detection using SiNWs. Two nanowire devices, 1 and 2 have different antibody receptors functionalized on the surface. Virus–receptor conjugation on device 2 changes its conductance (right) and when virus detaches from the surface, conductance returns to its original value. (Reproduced from Patolsky, F. et al., *PNAS*, 101, 14017, 2004. With permission.)

the current. The same transistor configuration with a single In_2O_3 nanowire has been used to detect PSA [80], which is a marker for prostate cancer. Figure 14.16 shows the current versus time history of the nanowire transistor during the detection event. When a buffer solution is first added, the current does not change from the background value, which is indicative of the device stability. When a nontarget bovine serum albumin (BSA) is added (100 nM), the current still does not change, indicating a negative response to a nonspecific binding. Finally, when exposed to 0.14 nM PSA, the current increases dramatically above the background value. Early results indicate a PSA detection limit of 250 pg/mL.

14.3.1 Nanoelectrode Arrays

The examples described in the previous section are single devices demonstrating the utility of nanowires in fabricating the biosensor and the detection capabilities. In practice, it would be ideal to have an array of sensors,

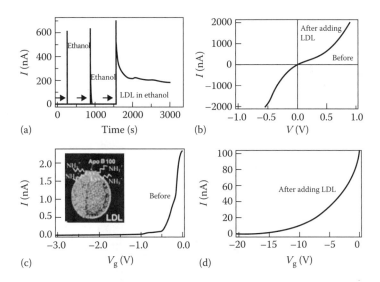

FIGURE 14.15
LDL cholesterol detection with an In$_2$O$_3$ nanowire transistor. (a) Current versus time after exposure to pure ethanol pulses followed by LDL in ethanol. (b) I–V results before and after exposure to LDL in ethanol. (c, d) I–V_g results before and after exposure to LDL in ethanol. (Reproduced from Li, C. et al., *Appl. Phys. Lett.*, 83, 4014, 2003. With permission.)

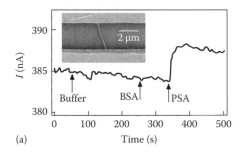

FIGURE 14.16
Detection of PSA using an In$_2$O$_3$ nanowire device similar to that in Figure 14.7 (inset). Current signal recorded over time after sequential exposure to buffer, BSA, and PSA. (Reproduced from Li, C. et al., *J. Am. Chem. Soc.*, 127, 12484, 2005. With permission.)

which can be used to simultaneously detect several targets with multiplexing capabilities. Since this type of biosensing works based on the "lock and key" mechanism, i.e., selectivity being guaranteed by specific probe–target interaction (see Figure 14.17), such nanoelectrode arrays (NEAs) can be successfully used in any of the applications described in the previous section. They are amenable to both electrical and electrochemical transductions.

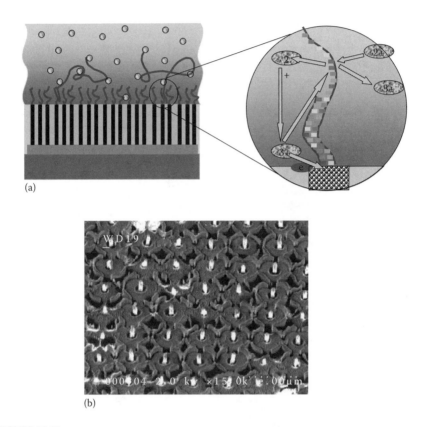

(a)

(b)

FIGURE 14.17

(a) Schematic of a NEA functionalized with a receptor. After binding with a target DNA sequence, electrochemical detection is done with the aid of a mediator Ru(bpy)$_3^{2+}$. (b) SEM image of a NEA after SiO$_2$ deposition and CMP. (Courtesy of J. Li and P. Arumugam.)

Construction of an NEA for biosensing is described below [3,94,95] using carbon nanofibers (CNFs) although the process equally applies to the use of any conductive nanowires that can be grown vertically on a substrate in a controllable manner.

PECVD is used to grow individual, freestanding, vertically aligned CNFs. Thermal CVD is not suitable since it will yield tower-like multiwalled nanotubes (MWNT) with a large number of tubes bound together by van der Waals force. Individual NEAs need to be apart from each other by a certain distance to avoid overlap of the radial diffusion layers of neighboring electrodes; otherwise, the electrode behavior will be that of a bigger micro- or macroelectrode. Such will be the case with thermally grown MWNT towers.

A silicon wafer with a 500 nm thick thermal oxide is used as substrate and a thin layer of Pt or Cr is deposited to serve as electrical contact connecting the CNFs to the measuring circuit. This is followed by sputtering of a 10–30 nm iron or nickel as catalyst layer. Acetylene is used as source gas at a pressure of 1–3 Torr and growth temperatures of 500°C–600°C for the PECVD of CNFs. When a thin film is used as catalyst, as in the case of nanowire growth described in this book, the film breaks up into droplets of various sizes distributed about a mean at random locations on the wafer. The thickness of the film can be varied to change the NEA density and diameter distribution but not the location. Also, a distribution in diameter always leads to a difference in CNF heights as well. Therefore, NEA construction ideally requires patterning of the catalyst in prespecified locations in order to obtain a tight control on diameter, height, and position (see Figure 14.17). This description equally applies to an NEA of silicon or oxide nanowires.

In an NEA serving as biosensor, neighboring electrodes need to be isolated from each other to avoid cross-talk and the possibility of behaving like a big micro- or macroelectrode. Mechanical stability of NEA is also critical when dipping into a fluidic environment during sensing. Both these goals can be achieved by intercalating the NEA with a dielectric such as SiO_2. This is readily done using thermal CVD with a popular precursor such as tetraethoxy silane, which covers the spacing between the vertical fibers with SiO_2. Then chemical mechanical polishing (CMP) is used to obtain a smooth top surface of the oxide with ends of CNFs poking out as seen in Figure 14.17. These tips now can be functionalized with probe molecules. Li and coworkers [3,94,95] demonstrated functionalizing this NEA with an oligonucleotide probe with a sequence related to the wild-type allele of cancer-related BRCA 1 gene and hybridizing with a complementary sequence. AC voltammetry results were generated using $Ru(bpy)_3^{2+}$ as an electrochemical mediator where the guanine in the incoming target oxidizes the intermediate from the 2+ to 3+ state, thus releasing an electron. The mediators transfer electrons efficiently from the guanine bases to the electrode even when they are not in direct contact, which allows detection of hybridization of less than a few attomoles of oligonucleotide targets with a $20 \mu m^2$ electrode. The ability to detect in the range of 100 molecules using this NEA approaches the detection limit of laser fluorescence techniques in DNA microarrays.

The above CNF biosensor has not yet demonstrated the detection limits of some of the FET-based devices described in the previous section; nevertheless it is plenty sufficient to meet numerous needs. More importantly, this NEA is amenable to large-scale fabrication (see Figure 14.18) much more easily than the nanowire FETs (or even nanotube FETs) discussed earlier. The latter are more cumbersome to fabricate compared to simple NEAs. The most challenging aspect of an FET is the horizontal placement of a nanowire or nanotube across every pair of source and drain across an entire wafer. In situ growth of nanowires or carbon nanotubes is too difficult since they may not naturally grow horizontally to bridge a specified gap. The alternative, which is "pick and place," may not turn out to be a manufacturable process.

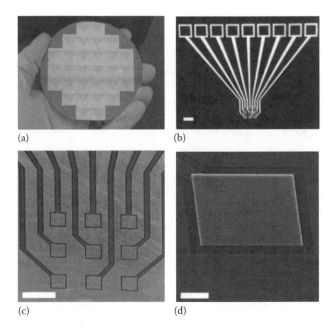

FIGURE 14.18
Wafer fabrication of NEAs for biosensing. (a) Optical microscopy image of a 4 in. wafer with 30 dies. (b) Single die consisting of 3 × 3 array of electrode pads individually wired to nine 1 mm × 1 mm contact pads. (c) The layout of the 3 × 3 electrode array which is 20 μm × 20 μm. (d) One of the electrodes in (a). (Courtesy of J. Li.)

References

1. J. Li, Chemical and physical sensors, in *Carbon Nanotubes: Science and Applications*, M. Meyyappan, Editor, CRC Press, Boca Raton, FL (2004).
2. K.J. Albert, N.S. Lewis, C.L. Schauer, G.A. Sotzing, S.E. Stitzel, T.P. Vaid, and D.R. Walt, *Chem. Rev.* 100, 2595 (2000).
3. J. Li, J.E. Koehne, A.M. Cassell, H. Chen, H.T. Ng, Q. Ye, W. Fan, J. Han, and M. Meyyappan, *Electroanalysis* 17, 15 (2005).
4. K. Kalantar-Zadeh and B. Fry, *Nanotechnology-Enabled Sensors*, Springer, New York (2008).
5. J. Janata, *Principles of Chemical Sensors*, Plenum Press, New York (1989).
6. J. Gardner and P.N. Bartlett, *Electronic Nose: Principles and Applications*, Oxford University Press, New York (1999).
7. J. Li, *Sensors* 17, 56 (2000).
8. K.C. Persaud and G.H. Dodd, *Nature* 299, 352 (1982).
9. T.C. Pearce, S.S. Schiffman, H.T. Nagle, and J.W. Gardner, eds., *Handbook of Machine Olfaction: Electronic Nose Technology*, Wiley-VCH, Weinheim (2003).
10. Q. Wan, Q.H. Li, Y.J. Chen, T.H. Wang, X.L. He, J.P. Li, and C.L. Lin, *Appl. Phys. Lett.* 84, 3654 (2004).

11. Q.H. Li, Y.X. Liang, Q. Wan, and T.H. Wang, *Appl. Phys. Lett.* 85, 6389 (2004).

12. P. Feng, Q. Wan, and T.H. Wang, *Appl. Phys. Lett.* 87, 213111 (2005).

13. T. Gao and T.H. Wang, *Appl. Phys. A* 80, 1451 (2005).

14. X.Y. Xue, Y.J. Chen, Y.G. Wang, and T.H. Wang, *Appl. Phys. Lett.* 86, 233101 (2005).

15. H.T. Wang, B.S. Kang, F. Ren, L.C. Tien, P.W. Sadik, D.P. Norton, S.J. Pearton, and J. Lin, *Appl. Phys. A* 81, 1117 (2005).

16. H.T. Wang, B.S. Kang, F. Ren, L.C. Tien, P.W. Sadik, D.P. Norton, S.J. Pearton, and J. Lin, *Appl. Phys. Lett.* 86, 243503 (2005).

17. L.C. Tien, P.W. Sadik, D.P. Norton, L.F. Voss, S.J. Pearton, H.T. Wang, B.S. Kang, F. Ren, J. Jun, and J. Lin, *Appl. Phys. Lett.* 87, 222106 (2005).

18. Z. Fan, D. Wang, P.C. Chang, W.Y. Tseng, and J.G. Lu, *Appl. Phys. Lett.* 85, 5923 (2004).

19. Z. Fan and J.G. Lu, *Appl. Phys. Lett.* 86, 123510 (2005).

20. Z. Fan and J.G. Lu, *IEEE Trans. Nanotechnol.* 5, 393 (2006).

21. Z.P. Sun, L. Liu, L. Zhang, and D.Z. Jia, *Nanotechnology* 17, 2266 (2006).

22. Y. Chen, C.L. Zhu, and G. Xiao, *Nanotechnology* 17, 4537 (2006).

23. J.X. Wang, X.W. Sun, Y. Yang, H. Huang, Y.C. Lee, O.K. Tan, and L. Vayssieres, *Nanotechnology* 17, 4995 (2006).

24. C.S. Rout, S.H. Krishna, S.R.C. Vivekchand, A. Govindaraj, and C.N.R. Rao, *Chem. Phys. Lett.* 418, 586 (2006).

25. C.S. Rout, K. Ganesh, A. Govindaraj, and C.N.R. Rao, *Appl. Phys. A* 85, 241 (2006).

26. C.S. Rout, G.U. Kulkarni, and C.N.R. Rao, *Nanotechnol. J. Phys. D, Appl. Phys.* 40, 2777 (2007).

27. C.S. Rout, M. Hegde, A. Gorindaraj, and C.N.R. Rao, *Nanotechnology* 18, 205504 (2007).

28. T.J. Hsueh, S.J. Chang, C.L. Hsu, Y.R. Lin, and I.C. Chern, *Appl. Phys. Lett.* 91, 053111 (2007).

29. L.M. Li, C.C. Li, J. Zhang, Z.F. Du, B.S. Zou, H.C. Yu, Y.G. Wang, and T.H. Wang, *Nanotechnology* 18, 225504 (2007).

30. E. Comini, G. Faglia, G. Sberveglieri, Z. Pan, and Z.L. Wang, *Appl. Phys. Lett.* 81, 1869 (2002).

31. M. Law, H. Kind, B. Messer, F. Kim and P. Yang, *Angew. Chem. Int. Ed.* 41, 2405 (2002).

32. A. Maiti, J.A. Rodriguez, M. Law, P. Kung, J.R. McKinney, and P. Yang, *Nano Lett.* 3, 1025 (2003).

33. Y. Wang, X. Jiang, and Y. Xia, *J. Am. Chem. Soc.* 125, 16176 (2003).

34. Z. Ying, Q. Wan, Z.T. Song, and S.L. Feng, *Nanotechnology* 15, 1682 (2004).

35. A. Kolmakov, Y. Zhang, G. Cheng, and M. Moskovits, *Adv. Mater.* 15, 997 (2003).

36. A. Kolmakov, D.O. Klenov, Y. Lilach, S. Stemmer, and M. Moskovits, *Nano Lett.* 5, 667 (2005).

37. S.V. Kalinin, J. Shin, S. Jesse, D. Geohegan, A.B. Baddorf, Y. Lilach, M. Moskovits, and A. Kolmakov, *J. Appl. Phys.* 98, 044503 (2005).

38. V.V. Sysoev, B.K. Button, K. Wepsiec, S. Dmitriev, and A. Kolmakov, *Nano Lett.* 6, 1584 (2006).

39. V.V. Sysoev, J. Goschnick, T. Schneider, E. Strelcov, and A. Kolmakov, *Nano Lett.* 7, 3182 (2007).

40. F. Hernandez-Ramirez, S. Barth, A. Tarancon, O. Casals, E. Pellicer, J. Rodriguez, A. Ramano-Rodriguez, J.R. Morante, and S. Mathur, *Nanotechnology* 18, 424016 (2007).
41. Q. Kuang, C. Lao, Z.L. Wang, Z. Xie, and L. Zheng, *J. Am. Chem. Soc.* 129, 6070 (2007).
42. Y.J. Chen, X.Y. Xue, Y.G. Wang, and T.H. Wang, *Appl. Phys. Lett.* 87, 233503 (2005).
43. Y.J. Chen, L. Nie, X.Y. Xue, Y.G. Wang, and T.H. Wang, *Appl. Phys. Lett.* 88, 083105 (2006).
44. Q. Wan, J. Huang, Z. Xie, T.H. Wang, E.N. Dattoli, and W. Lu, *Appl. Phys. Lett.* 92, 102101 (2008).
45. C. Xiangfeng, W. Caihong, J. Dongli, and Z. Chenmou, *Chem. Phys. Lett.* 399, 461 (2004).
46. C. Li, D. Zhang, X. Liu, S. Han, T. Tang, J. Han, and C. Zhou, *Appl. Phys. Lett.* 82, 1613 (2003).
47. D. Zhang, C. Li, X. Liu, S. Han, T. Tang, and C. Zhou, *Appl. Phys. Lett.* 83, 1845 (2003).
48. D. Zhang, Z. Liu, C. Li, T. Tang, X. Liu, S. Han, B. Lei, and C. Zhou, *Nano Lett.* 4, 1919 (2004).
49. K. Ryu, D. Zhang, and C. Zhou, *Appl. Phys. Lett.* 92, 093111 (2008).
50. P. Feng, X.Y. Xue, Y.G. Liu, and T.H. Wang, *Appl. Phys. Lett.* 89, 243514 (2006).
51. H.Y. Yu, B.H. Kang, U.H. Pi, C.W. Park, S.Y. Choi, and G.T. Kim, *Appl. Phys. Lett.* 86, 253102 (2005).
52. I. Raible, M. Burghard, V. Schlecht, A. Yasuda, and T. Vossmeyer, *Sens. Actuators B* 106, 730 (2005).
53. J. Liu, X. Wang, Q. Peng, and Y. Li, *Adv. Mater.* 17, 764 (2005).
54. C. Wang, X.Q. Fu, X.Y. Xue, Y.G. Wang, and T.H. Wang, *Nanotechnology* 18, 145506 (2007).
55. L. Francioso, A.M. Taurino, A. Forleo, and P. Siciliano, *Sens. Actuators B* 130, 70 (2008).
56. Y.Q. Fu, C. Wang, H.C. Yu, Y.G. Wang, and T.H. Wang, *Nanotechnology* 18, 145503 (2007).
57. Z. Liu, T. Yamazaki, Y. Shen, T. Kikuta, N. Nakatani, and T. Kawabata, *Appl. Phys. Lett.* 90, 173119 (2007).
58. Y.S. Kim, S.C. Ha, K. Kim, H. Yang, S.Y. Choi, Y.T. Kim, J.T. Park, C.H. Lee, J. Choi, J. Paek, and K. Lee, *Appl. Phys. Lett.* 86, 213105 (2005).
59. A. Ponzoni, E. Comini, G. Sbervegileri, J. Zhou, S.I. Deng, N.S. Xu, Y. Ding, and Z.L. Wang, *Appl. Phys. Lett.* 88, 203101 (2006).
60. J. Polleux, A. Gurlo, N. Bursan, V. Weimar, M. Antonietti, and M. Niederberger, *Angew. Chem. Int. Ed.* 45, 2610020 (2006).
61. B. Deb, S. Desai, G.U. Sumanasekera, and M.K. Sunkara, *Nanotechnology* 18, 285501 (2007).
62. P. Feng, X.Y. Xue, Y.G. Liu, Q. Wan, and T.H. Wang, *Appl. Phys. Lett.* 89, 112114 (2006).
63. M.Z. Atashbar, D. Banerji, and S. Singamaneii, *JEEE Sens. J.* 5, 792 (2005).
64. Y. Im, C. Lee, R.P. Vasquez, M.A. Bangar, N.V. Myung, E.J. Menke, R.M. Penner, and M. Yun, *Small* 2, 356 (2006).
65. V. Doborokhotor, D.N. McIlroy, M.G. Norton, A. Abuzir, W.J. Yeh, I. Stevenson, R. Pouy, J. Bochenek, M. Cartwright, L. Wang, J. Dawson, M. Beaux, and C. Berven, *J. Appl. Phys.* 99, 104302 (2006).

66. X.T. Zhou, J.Q. Hu, C.P. Li, D.D.D. Ma, C.S. Lee, and S.T. Lee, *Chem. Phys. Lett.* 369, 220 (2003).
67. A.A. Talin, L.L. Hunter, F. Leonard, and B. Rokad, *Appl. Phys. Lett.* 89, 153102 (2006).
68. M.C. McAlpine, H. Ahmad, D. Wang, and J.R. Heath, *Nature Mater.* 6, 379 (2007).
69. Z.H. Chen, J.S. Jie, L.B. Luo, H. Wang, C.S. Lee, and S.T. Lee, *Nanotechnology* 18, 345502 (2007).
70. J. Li, Y. Lu, Q. Ye, M. Cinke, J. Han, and M. Meyyappan, *Nano Lett.* 3, 929 (2003).
71. Y. Lu, C. Partridge, M. Meyyappan, and J. Li, *J. Electro Anal. Chem.* 593, 105 (2006).
72. J. Li, unpublished observations.
73. Y. Cui, Q. Wei, H. Park, and C.M. Lieber, *Science* 293, 1289 (2001).
74. J.I. Hahm and C.M. Lieber, *Nano Lett.* 4151 (2004).
75. F. Patolsky, G. Zheng, O. Hayden, M. Lakamamyali, X. Zhuang, and C.M. Lieber, *PNAS* 101, 14017 (2004).
76. G. Zheng, F. Patolsky, Y. Cui, W.V. Wang, and C.M. Lieber, *Nature Biotechnol.* 23, 1294 (2005).
77. W. Wang, C. Chen, K.H. Lin, Y. Fang, and C.M. Lieber, *PNAS* 102, 3208 (2005).
78. C. Li, B. Lei, D. Zhang, X. Liu, S. Han, T. Tang, M. Rouhanizadeh, T. Hsiai, and C. Zhou, *Appl. Phys. Lett.* 83, 4014 (2003).
79. M. Curreli, C. Li, Y. Sun, B. Lei, M.A. Gundersen, M.E. Thomson, and C. Zhou, *J. Am. Chem. Soc.* 127, 6922 (2005).
80. C. Li, M. Curreli, H. Lin, B. Lei, F.N. Ishikawa, R. Datar, R.J. Cote, M.E. Thompson, and C. Zhou, *J. Am. Chem. Soc.* 127, 12484 (2005).
81. Z. Li, Y. Chen, X. Li, T.I. Kamins, K. Nauka, and R.S. Williams, *Nano Lett.* 4, 245 (2004).
82. I. Park, Z. Li, X. Li, A.P. Pisano, and R.S. Williams, *Biosens. Bioelectron.* 22, 2065 (2007).
83. W. Chen, H. Yao, C.H. Tzang, J. Zhu, M. Yang, and S.T. Lee, *Appl. Phys. Lett.* 88, 213104 (2006).
84. K. Yang, H. Wang, K. Zou, and X. Zhang, *Nanotechnology* 17, 5276 (2006).
85. A.K. Wanekaya, W. Chen, N.V. Myung, and A. Mulchandani, *Electroanalysis* 18, 533 (2006).
86. A. Vaseashta and J. Irudayaraj, *J. Optoelectron. Adv. Mater.* 7, 35 (2005).
87. M. Yang, F. Qu, Y. Lu, Y. He, G. Shen, and R. Yu, *Biomaterials* 27, 5944 (2006).
88. M. Yang, F. Qu, Y. Li, Y. He, G. Shen, and R. Yu, *Biosens. Bioelectron.* 23, 414 (2007).
89. J. Wang, N.V. Myung, M. Yun, and H.G. Monbouquette, *J. Electroanal. Chem.* 575, 139 (2005).
90. A. Cusma, A. Curulli, D. Zane, S. Kaciulis, and G. Padeletti, *Mater. Sci. Eng. C* 27, 1158 (2007).
91. S. Aravamudhan, N.S. Ramgir, and S. Bhangali, *Sens. Actuat. B* 127, 29 (2007).
92. S.J. Patil, A. Zajac, T. Zhukov, and S. Bhansali, *Sens. Actuat. B* 129, 859 (2008).
93. P.R. Nair and M.A. Alam, *Appl. Phys. Lett.* 88, 233120 (2006).
94. J. Li, Biosensors, in *Carbon Nanotubes: Science and Applications*, CRC Press, Boca Raton, FL (2004).
95. J. Li, H.T. Ng, A. Cassell, W. Fan, H. Chen, Q. Ye, J. Koehne, J. Han, and M. Meyyappan, *Nano Lett.* 3, 597 (2003).

15

Applications in the Energy Sector

15.1 Introduction

With the increasing demand for energy and the detrimental effect of fossil fuels on the environment, renewable energy sources are fast emerging as an important alternative to traditional energy sources. Among the alternatives, solar energy holds great potential due to the enormous amount of solar radiation the earth receives. It is estimated that the earth receives about 162,000 TW of solar energy every hour, which is more than 10,000 times the total worldwide energy consumption in 2005. Even if 0.1% of the land area is covered with 10% efficient solar cells, we can meet our energy demands [1]. In addition to energy conversion, energy storage devices are also gaining increased importance due to the emergence of portable electronic devices and hybrid vehicles. Some popular energy storage devices include lithium ion batteries, electrochemical capacitors, and hydrogen storage materials.

Inorganic materials are an important component in most of the systems for energy conversion and storage. These materials are used either as the active materials for light absorption or as charge transport media in solar cells and energy storage in Li-ion batteries. The most suitable materials for these applications require that they have appropriate optical and electronic properties, high strain tolerance, and high surface areas. Most of the current devices use sintered nanoparticle networks for these applications. Their drawback is that electron transport is inefficient through them due to the grain boundaries present at the interparticle contacts. In contrast, the one-dimensional (1-D) structures such as nanowires (NWs) and nanotubes (NTs) provide direct channels for charge and mass transport to the back contact, minimizing efficiency losses (Figure 15.1). Also, several different architectures and hybrid structures based on 1-D structures and nanoparticles can further improve the charge collection properties. In addition to better charge transport characteristics, nanowires exhibit better light harvesting properties and can provide better strain relaxation during the Li-ion battery operation for retaining high charge capacities. In this chapter, concepts involving the use of nanowires in solar electric, solar chemical, electrochromic, and Li-ion battery applications will be illustrated.

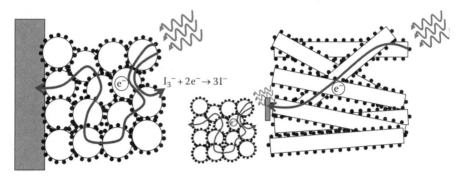

FIGURE 15.1
Electron transport through a network of (a) nanoparticles and (b) nanowires.

15.2 Solar Cells

Solar cells are classified into three broad generations. The first generation solar cells consist of large area Si single crystal p–n junctions. The second generation solar cells consist of thin epitaxial semiconductor layers such as amorphous silicon, CdTe, and CuInSe/S on lattice-matched wafers. Third generation solar cells do not achieve charge separation by utilizing a p–n junction. Instead charge separation in these cells depends on the preferential charge injection into one side of the material interface. Solar cells of this type include dye-sensitized solar cells (DSSCs), organic solar cells, photoelectrochemical (PEC) cells, nanocrystalline solar cells, and those made with organic–inorganic blends. Nanomaterials hold tremendous promise to reduce the cost of the second and third generation cells.

15.2.1 Dye-Sensitized Solar Cells

DSSCs have attracted great interest in the last few years as low-cost alternatives to the traditional p–n junction solar cells. This is due to the use of inexpensive materials and processing requirements. The DSSC is a special class of excitonic solar cells in which charge generation and charge separation take place in two different media. The DSSC solar cell utilizes wide band gap semiconductor nanostructures sensitized with a monolayer of visible light absorbing dye that is in contact with a redox electrolyte or a hole conductor. Figure 15.2 shows the various processes taking place in a DSSC solar cell. Upon visible light excitation, the excited dye injects electrons into the conduction band of the semiconductor nanostructure. The electrons percolate through the network of nanoparticles and are extracted at the back contact, typically a fluorinated tin oxide (FTO) substrate and to the external circuit. The dye is regenerated by electron donation from the I^-/I_3^- electrolyte, which

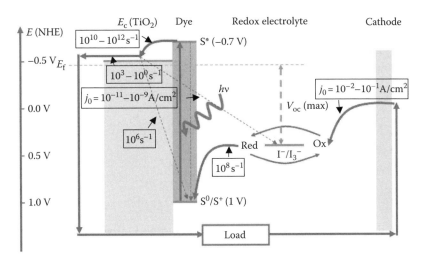

FIGURE 15.2
Schematic showing the various processes taking place in a DSSC and the time scales in which they occur.

in turn is regenerated by the reduction of triiodide at the counter electrode. The open circuit voltage is the potential difference between the quasi-Fermi level in the semiconductor under illumination and the redox potential of I^-/I_3^- [2].

The construction of a DSSC is simple. A paste/dispersion of the nanoparticles is deposited on an FTO glass plate and sintered at 450°C–500°C. The prepared electrode is sensitized with the dye (for example, Ru 535 or N719, Solaronix) by soaking in the dye solution overnight. The platinum counter electrode is prepared either by evaporation or by heating the FTO substrates coated with a solution of chloro-platinic acid (Plastisol, Solaronix). The cells are assembled by heating them with an insulating plastic spacer between them with a small hole to fill the electrolyte. The electrolyte is filled in the space between the electrodes and then the cell is completely sealed. The efficiency of the solar cell is determined from the *I–V* characteristics of the cell by the expression:

$$\eta = \frac{I_{sc} \times V_{oc} \times FF}{P} \times 100 \qquad (15.1)$$

where
I_{sc} is the short circuit current
V_{oc} is the open circuit potential
FF is the fill factor
P is the incident power

These measurements can be done using a potentiostat–galvanostat set up under an AM 1.5 100 mW/cm^2 light source. The short circuit current density (J_{sc}) in a DSSC is given by

$$J_{sc} = \int IPCE \, d\lambda \qquad (15.2)$$

where IPCE is the incident photon to current conversion efficiency. The IPCE is calculated from the expression,

$$IPCE(\%) = \frac{1240 \times I_{sc}(A/cm^2)}{\lambda(nm) \times I_{inc}(W/cm^2)} \qquad (15.3)$$

The different wavelengths of light can be obtained either with the use of a monochromator or with a tunable laser. The IPCE is given by

$$IPCE = LHE \varphi_{inj} \varphi_{coll} \qquad (15.4)$$

where
  LHE is the light harvesting efficiency
  φ_{inj} is the efficiency of electron injection from the dye molecules into the semiconductor
  $\varphi_{coll.}$ is the electron collection efficiency at the back contact

LHE depends on the morphology of the semiconductor and the amount of dye loading on the semiconductor. φ_{inj} depends on the semiconductor–dye interface and the overlap of the conduction band in the semiconductor and the lowest unoccupied molecular orbital (LUMO) of the dye. The $\varphi_{coll.}$ of the electrons is defined as the ratio of the τ_r and τ_c (τ_r is the electron recombination time constant; τ_c is the electron transport time constant). From these expressions, it is clear that LHE, φ_{inj}, and $\varphi_{coll.}$ in the films should be high in order to obtain high efficiencies.

Figure 15.2 also shows the timescales associated with the various processes that take place within a DSSC. Charge injection from the dye into the semiconductors has been found to be in the femto–pico second timescales, which is much faster than the electron relaxation within the dye molecule, occurring in the timescale of pico–nano seconds. The electron transfer from the semiconductor conduction band to the highest occupied molecular orbital (HOMO) and the electrolyte occurs in the micromilli second timescale. Charge transport by the electrolyte occurs in the pores of the semiconductor film to the counter electrode and that of injected electrons within the nanocrystalline film to the back contact should be fast enough to compete efficiently with the electron recapture reaction.

Nanoparticle-based films used in DSSCs are highly disordered structures, which can slow down the transport dynamics. It has been shown that electron transport in TiO_2 nanoparticle networks is two to three orders of magnitude slower than that in single-crystal TiO_2 and gets slower with increasing porosity due to higher number of interparticle contacts. Longer diffusion lengths imply that the electrons spend longer times within the films before reaching the back contact. In order to achieve high charge collection efficiencies, the electron transport should be faster than electron recombination with the electrolyte and dye [3]. Nanowires have the potential to improve the transport characteristics of these films since they have fewer defects and grain boundaries acting as electron traps at interparticle contacts. This means that they can be laid as thicker films on substrates, extending the absorption of light, thus improving the overall solar cell efficiency.

The use of 1-D materials for the anode has resulted in significant improvement in the transport and recombination characteristics compared to nanoparticle electrodes, but the overall efficiencies of most of the reported cells still remain less than the nanoparticle cells. The main reason for this has been the low surface area of the nanowire-based electrodes. A few different 1-D materials and their solar cell performance are discussed next. Growth and characterization of these materials have been discussed under various chapters in this book and therefore only their solar cell performance is discussed below.

15.2.1.1 Titania Nanowire–Based DSSCs

TiO_2 is an interesting material for DSSC applications because of its electronic structure and surface properties, which enable high dye adsorption and high efficiency. Since nanowires can further improve the performance of DSSCs due to their high charge collection efficiencies, TiO_2 nanowires is an interesting system for these applications. TiO_2 nanowires with different aspect ratios and slight variations in morphology have been synthesized using different synthesis routes (see Section 9.2). One of the earliest reports on the use of TiO_2 NWs in DSSCs used nanowires synthesized by low-temperature surfactant-assisted self-assembly [4]. Although, these wires did not have uniform diameters along the axis, the anatase TiO_2 nanoparticles oriented themselves such that all the lattice planes of the attached particles were perfectly aligned, exposing mainly the (1 0 1) surfaces, which have high dye adsorption properties [5,6]. Efficiencies as high as 9.2% were achieved using these nanowires. The J_{sc} in these cells, compared to similarly prepared cells using P-25 TiO_2 was found to be higher, which was attributed to the high dye adsorption on the (1 0 1) surfaces. Higher efficiencies as summarized in Table 15.1 were also observed in DSSCs fabricated using TiO_2 nanorods synthesized by hydrothermal method [7]. In addition to higher currents, the photocurrents increased faster and linearly with film thickness in both the above reports, compared to nanoparticle-based films, in which the photocurrents saturated after a certain thickness. This indicated a more efficient transport of electrons through the films. In an alternative approach TiO_2

TABLE 15.1

Summary of Several Studies on Nanowire-Based Materials for Different Solar Cells: DSSCs; Direct Absorption PEC Solar Cells; and p–n Junction Solar Cells

Materials System	Diameter/Length	Specific Comments	V_{oc}	I_{sc}	FF	η (%)	Reference
DSSC							
	10–30 nm/2–5 nm	Oriented attachment of particles. (1 0 1) planes exposed	720	19.2	0.68	9.2	[4]
	20–30 nm/100–400 nm		748	13.4	0.7	7	[7]
TiO$_2$ nanowires	12–28/4–10 μm	Polycrystalline grown directly on Ti foils. (1 0 1) growth direction	590	4.21	0.6	1.5	[8]
	50 nm/several μm	Brookite phase, nanorods	610	2.29	0.64	1.8–50 mW/cm²	[9]
	30–40 nm/100–300 nm	Sensitized with CdS	440	1.31	0.43	0.8–32 mW/cm²	[28]
TiO$_2$ nanoribbons	50–130 nm wide/few mm	Brookite phase nanoribbons	630	3.55	0.49	2.2–50 mW/cm²	[9]
Multicore TiO$_2$ nanofibers	200 nm tubes filled with 24 nm fibers	Multicore TiO$_2$ nanofibers	730	16.09	0.49	5.77	[10]
ZnO nanowires	130–200 nm/18–24 μm	Vertically oriented	710	5.85	0.37	1.5	[11]
	100 nm/several microns + branching on them	Four generations of dendritic growth	740	1.62	0.38	0.5	[12]
	50–100 nm/5–6 μm	Branched nanowires, vertically oriented, straight	690	1.26	0.39	0.34	[13]
	50–100 nm/5–6 μm	Branched nanowires, vertically oriented, branched	620	1.84	0.40	0.46	[13]
	120–150 nm/15 μm		540	6.79	0.5	1.7	[14]
	200 nm, 4.5 μm		460	1.81	0.6	0.5	[14]
	190 nm, 7 μm	Vertically oriented on flexible substrates	600	3.5	0.42	0.9	[15]

Material	Dimensions	Description					Ref.
	200 nm tips	Flower-like	650	5.5	0.53	1.9	[16]
	75–125 nm, 12 µm	Sensitized with CdSe QD's of 3–4 nm	500–600 nm	1–2	0.3	0.4	[29]
SnO_2 nanowires	20–200 nm/several µm		520–560	5.7	0.5	2.1–70 mW/cm²	[17]
ZnO NW/P3HT	60–100/0.5–1.0 µm		250	3	0.27	0.2	[30]
	40–50 nm/500 nm		440	2.17	0.56	0.53	[31]
TiO_2 nanotubes	9 nm/several 100's of nm		720	16.6	0.65	7.83	[19]
	9 nm/several 100's of nm	Blended with 2% TiO_2 NP's	720	18.1	0.64	8.43	[19]
	10 nm/100's of nm		790	13.3	0.68	7.1	[20]
	80 nm/360 nm	Pore size of 46 nm, vertically oriented	750	7.87	0.49	2.97	[21]
	50 nm/5.7 µm	Pore size of 30 nm. 30 nm crystallite size, vertically oriented	610	9	0.55	3.0	[22]
	130 nm/14.4 µm	Pore size of 90 nm, vertically oriented. Donor-antenna dye	723	13.44	0.63	6.1	[23]
	58 nm/6.3 µm	Pore diameters of 38 nm, structural disorder removed by supercritical CO_2 drying	600	4.9	0.53	1.6	[3]
	58 nm/6.3 µm		580	5.7	0.56	1.9	[3]
TiO_2 nanotubes	115 nm, 4.5 µm	TiO_2/P3HT/redox couple	700	5.5	0.55	2.1	[24]
	115 nm, 4.5 µm	TiO_2/P3HT/PEDOT-PSS	~320	2.0	~0.25	0.16	[24]
	115 nm, 4.5 µm	TiO_2/P3HT-PCBM/PEDOT-PSS	450	6.5	0.34	1.0	[24]
ZnO nanotubes	250 nm, 60 nm	Pore size of 210 nm, grain size of 20 nm	739	3.3	0.64	1.6	[26]
ZnO NW–ZnO NP	200 nm, 10 µm	Second generation dendritic nanowires with sparse NPs attached to nanowire's	840	1.7	~0.52	~0.75	[32]

(continued)

TABLE 15.1 (continued)

Summary of Several Studies on Nanowire-Based Materials for Different Solar Cells: DSSCs; Direct Absorption PEC Solar Cells; and p–n Junction Solar Cells

Materials System	Diameter/Length	Specific Comments	V_{oc}	I_{sc}	FF	η (%)	Reference
	200nm, 10μm	Second generation dendritic nanowires with dense nanoparticles filling the nanowire network	700	3.5	0.53	1.3	[32]
	200nm, 10μm	Second generation dendritic nanowires with nanoparticles sitting on top of nanowire network	750	4.0	~0.42	~1.26	[32]
TiO$_2$ NW–TiO$_2$ NP	30–80nm/several μm	20wt.% nanowires	600	26	0.55	8.6	[33]
ZnO NW–TiO$_2$ shell	120–180nm/13–17μm	20nm shell thickness	800	4.8	0.56	2.27	[34]
SnO$_2$ NW–TiO$_2$ nanoparticles	20–200nm/several μm	SnO$_2$ nanowire's coated with 5–10nm TiO$_2$ nanoparticles	686	8.56	0.49	4.1–70mW/cm^2	[17]
Direct Absorption PEC Cells							
Si NWs	2μm/20μm	Arrays	407	1.57			[38]
	20–300nm/20μm	Arrays	660–730	0.65–1.36			[39]
Cd(Se,Te) NW	200nm, 2–3μm	Photoetched	230–290	4.4–6.8	0.44	0.4–0.8	[40]
CdSe NW	7nm/60nm	Blended with P3HT	700	6	0.4	1.7	[41]
p–n Junction Solar Cells							
Si NWs	100nm/10μm	Vertically aligned, p–n junction	549	26.06	0.65	9.31	[42]
	100nm/10μm	Slantingly aligned	580	27.14	0.72	11.37	[43]
Si NWs	200nm to 1.5μm/30–50μm	Single nanowire solar cell	190	5	0.4	0.46	[44]
	300nm/10μm	Radial p-i-n junction	260	24	0.55	3.5	[45]

nanowires were directly grown on Ti substrates; $Na_2Ti_2O_5 \cdot H_2O$ nanotubes were first grown by their hydrothermal treatment in a basic solution, followed by their transformation into $H_2Ti_2O_5 \cdot H_2O$ by ion exchange in HCl and then by annealing these nanotubes at 500°C to form TiO_2 nanowires [8]. The nanowires are 12–28 nm in diameter and 4–10 μm long. They are polycrystalline with (1 0 1) planes of the individual grains nearly perpendicular to the local wire axis. Brookite phase TiO_2 nanorods and nanoribbons were also synthesized and were used to fabricate DSSCs. The photovoltaic measurements showed that TiO_2 nanoribbons were found to perform better than the nanorods [9]. X-ray photoelectron spectroscopy (XPS) analysis revealed that in the prepared samples, TiO_2 nanorods had Ti(III) defect states on their surface, which were acting as electron traps and thus resulted in low efficiency. Other structures such as multicore cable like TiO_2 nanofibrous membranes have also been used as the anode material for solar cells. These structures were synthesized by the calcinations of electrospun composite PVAc/titania membranes, giving efficiencies of 5.7% (Table 15.1) [10]. Current research efforts include synthesis of single crystal vertical nanowire arrays and highly crystalline nanowires in bulk quantities using cost-effective means that can boost the efficiencies to greater than 10%.

15.2.1.2 ZnO Nanowire–Based DSSCs

ZnO is another interesting material system for DSSCs for several reasons: (a) it is easily produced both by wet chemical and vapor phase techniques and (b) it can produce voltages in excess of 700 mV as its work function values are similar to those for titania. The first reports on ZnO nanowire-based DSSCs used solution growth methods to grow vertically aligned nanowires [11] and metal organic chemical vapor deposition (MOCVD) to grow dendritic nanowires [12]. The efficiencies obtained using these nanowires were less than 1.5%, attributed to the insufficient dye absorption due to the low surface area of the nanowires. In contrast, branched and high aspect ratio nanowires performed better than plain nanowires due to the increase in surface area [13,14]. In addition to the improved transport properties evident from the linear photocurrent rise with nanowire length, femtosecond transient absorption spectroscopy measurements showed that the charge transfer is faster from the dye to nanowires compared to charge transfer to nanoparticles [11]. This was attributed to the differences in the surfaces of nanoparticles and nanowires. The ZnO nanowires synthesized had mostly the (1 0 0) planes exposed. Other variations such as nanowire DSSCs on flexible substrates [15] and DSSCs with flower-like nanowires [16] have also been demonstrated with better performance. The flower-like nanowire DSSCs showed high light scattering efficiency than the vertically oriented nanowires due to better light scattering. The nanowire DSSCs on flexible substrates showed no degradation in the performance and better crack resistance, while the performance of nanoparticle-based cells degraded after cyclic bending.

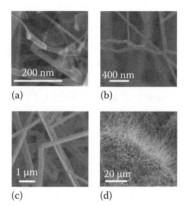

(a)   (b)

(c)   (d)

FIGURE 15.3
SnO_2 nanowires synthesized by reactive vapor transport. (From Gubbala, S. et al., *Adv. Funct. Mater.*, 18, 2411, 2008. With permission.)

15.2.1.3 SnO₂ NW–Based DSSCs

Another material of interest for DSSCs is SnO_2 but not much attention has been paid to the nanowire form until recently [17]. Several different morphologies of nanowires were obtained using the reactive vapor transport of Sn on quartz substrates in H_2 and O_2 atmospheres as seen in Figure 15.3. The cells made using the SnO_2 nanowires showed higher open circuit voltage ($V_{oc} = 520$–$560\,mV$) and fill factor (FF = 0.5) than the corresponding nanoparticle electrodes ($V_{oc} = 350\,mV$ and FF = 0.36). The currents however are less than for the nanoparticle electrodes due to their lower surface area (5.7 mA/cm² compared to 11 mA/cm² for nanoparticle electrodes). Higher efficiencies were observed for electrodes made using branched and highly interconnected nanowire samples (2.1%) compared to the electrodes using a powder containing mainly individual nanowires (0.58%). V_{oc} values measured at high light intensities are 250 mV higher than for the nanoparticle electrodes. These differences were attributed to the differences in the work functions of these two structures. Since nanowires typically have only one or two types of facets, the surface of all nanowires represents that of a single crystal surface. Nanoparticles on the other hand have mixed facets and a large number of edge sites, leading to a high degree of surface polycrystallinity. In a different configuration, indium tin oxide (ITO) nanowires were also tested by coating them with TiO_2 and sensitizing them with dye [18].

15.2.1.4 Inorganic Nanotubes, Polymers, and Nb₂O₅ Nanowires for DSSCs

Inorganic nanotubular structures also present an interesting morphology for electrochemical energy conversion devices. These structures can potentially offer up to about twice the surface area than nanowires for electrolyte access due to their hollowness. The first few reports on DSSCs based on TiO_2

NTs were made with randomly oriented TiO_2 NTs on FTO substrates using the surfactant-assisted templating method [19,20]. The nanotube-based films showed improved crack resistance when small amounts of P-25 particles were added to these pastes, resulting in efficiencies of 8.43%. The V_{oc} of the nanotube-based devices was also found to be ~40–50 mV more than for the P-25 particles, which was attributed to the more negative Fermi level in the TiO_2 nanotubes due to higher electron density [20].

Vertically aligned TiO_2 NTs have also been used as the anode material in DSSCs [21–23]. The efficiencies of the nanotube-based DSSCs were improved to 6.1% from when donor-antenna dyes with high molar extinction coefficients were used [23]. An improvement in the performance of the solar cells was also observed with the removal of structural disorder (removing bundling of nanotubes and cracks in the films) from these films [3]. The structural disorder was removed in the nanotubes using a supercritical CO_2 drying technique. Such oriented nanotubes gave J_{sc} of 5.7 mA/cm², V_{oc} of 0.58 V, and a fill factor of 0.56. This is in contrast to the performance of disordered nanotubes, which gave J_{sc} of 4.9 mA/cm², V_{oc} of 0.6 V, and FF of 0.53, resulting in an overall conversion efficiency of 1.6%.

Polymer-based DSSCs have an advantage over the liquid electrolyte solar cells in terms of cost and scalability. Some of the configurations that have been tried with polymer as either the photosensitizer or the hole transport medium are described here. TiO_2 nanotubes were sensitized with a self-assembled hybrid polymer (P3HT). The open circuit potentials of these cells were more than those of cells made with TiO_2 nanoparticles. The nanotubes were also used to make an all solid state solar cell wherein poly(3,4-ethylenedioxythiopene) poly(styrenesulfonate) (PEDOT-PSS) was used as the hole conducting medium. In another configuration, a double heterojunction solar cell was made using the TiO_2 nanotubes with a 1:1 mixture of P3HT/[6,6]-phenyl-C_{61}-butyric acid methyl ester (PCBM) as hole-conducting medium [24]. In addition to the above studies, flexible DSSCs with TiO_2 nanotubes have also been demonstrated [25] using TiO_2 nanotubes grown on Ti foils and ITO/polyethylene napthalate (PEN) coated with Pt as the counter electrode. The cells gave an overall efficiency of 3.6%, which was more than that of DSSCs fabricated with FTO glass substrates giving 3.3%. The higher current densities in the flexible cells were attributed to reduced optical loss through the electrolyte as the counter electrode was placed directly on top of the TiO_2 NT layer. The low V_{oc} and FF in the flexible cells were attributed to high series resistance in the nanotubes and low activity of Pt on Pt/ITO–PEN as indicated by the impedance measurements. In addition to TiO_2 nanotubes, ZnO nanotubes grown by atomic layer deposition (ALD) on anodic aluminium oxide (AAO) membranes have also been used as anode material for DSSCs [26].

In addition to the above-mentioned materials, Nb_2O_5 is also interesting for DSSC applications, due to the position of its conduction band edge. DSSCs were made using the Nb_2O_5 nanowire arrays grown directly on Nb foils using plasma oxidation (Figure 15.4) [27]. Efficiencies of ~0.7% were achieved with these 7 µm long nano-wires. The solar cells gave J_{sc} of 1.1 mA/cm², V_{oc} of 560 mV, and fill factors of 0.4 at 35 mW/cm² light intensity.

FIGURE 15.4
Nb$_2$O$_5$ NW array synthesized on Nb foils by plasma oxidation.

15.2.1.5 Quantum Dot Sensitizers for Nanowire-Based Solar Cells

One of the variations of the nanowire DSSC is the quantum dot-sensitized solar cell [28,29]. For example, CdSe semiconductor nanocrystals were used as the sensitizer for single-crystal ZnO nanowires [29]. CdSe quantum dots, capped with mercaptopropionic acid, are attached to the surface of the nano-wires. These solar cells exhibit short circuit currents ranging from 1 to 2 mA/cm² and open-circuit voltages of 0.5–0.6 V, resulting in an overall energy conversion efficiency of 0.4%. The main advantage of using quantum dots as sensitizers is that the optical absorption of the quantum dots can be tuned by changing the diameter of the quantum dots and secondly, higher efficiencies are expected due to the multiple exciton generation from a single photon. However, it has not been proven that such an effect has shown any beneficial effects for DSSCs.

In a different configuration, the dye-sensitized electrode is in contact with a polymer hole conducting medium. Inorganic nanowire–polymer DSSCs have also shown a two orders-of-magnitude reduction in the recombination kinetics with nanorods compared to nanoparticles. In the ZnO nanorod–P3HT system, the V_{oc} and J_{sc} of the cells are higher than for the nanoparticle cells, resulting in a fourfold improvement in the overall efficiency of the device [30]. The performance of the cells is found to be dependent on the density and orientation of the nanowires. For example, in the case of vertically aligned NW–P3HT system, the performance decreases with increasing density; this is due to the poor wetting by P3HT of the total layer of the nanowires with a (0 0 1) surface, which is polar and (1 0 0) side facets that are nonpolar [31].

15.2.1.6 Hybrid/Composite Structures

The performance of DSSCs based on 1-D structures can be optimized by designing new architectures, which may increase the available surface area for dye absorption and reduce electron recombination due to improved charge transport through the nanowires and hence the light harvesting efficiency of the solar cell. Different strategies have been tried in this regard including nanowire–nanoparticle composites involving the same or different materials, and nanowires coated with a thin layer of another material. When using different materials for composites, it is important to design such systems based on the appropriate band edge position to help with both charge separation at the nanowire/nanoparticle interface and strategies to reduce the recombination reaction.

FIGURE 15.5
Hybrid nanowire structures consisting of TiO₂ NW-NP. (From Tan, B. and Wu, Y., *J. Phys. Chem. B*, 110, 15932, 2006. With permission.)

Different configurations of ZnO nanowire/nanoparticle composites have been compared: ZnO nanowires sparsely coated with ZnO nanoparticles; the voids in ZnO nanowires films completely filled with ZnO nanoparticles; and ZnO nanowires coated with ZnO nanoparticles [32]. The configuration with a high density of nanoparticles combined with nanowires exhibits the best performance. The sparsely covered nanowires showed only a small increase in photocurrent due to a small increase in the surface area of these electrodes, while nanoparticles deposited over nanowires as a separate layer gave lower currents than the nanoparticle electrodes. This is probably due to the bottleneck for electron transfer by nanowires due to low density, which further leads to high electron recombination and low V_{oc}. In another study, DSSCs made using single crystal TiO₂ nanowire–nanoparticle composites showed an improvement in performance compared to the nanoparticle cells (Figure 15.5) [33]. The nanowires were anatase phase, 30–80 nm in diameter, and several microns long. These composites with 5%–20% nanowires showed higher J_{sc} than pure TiO₂ nanoparticle cells in spite of a decrease in the total internal surface area of the composite structure due to more efficient charge collection at high nanowire loadings, the J_{sc} suffered due to low surface area. However, the photocurrent continued to rise with increasing film thickness up to 17 μm in 77% nanowire-loaded films, compared to 10 μm in the 5%–2% nanowire-loaded films. This again implied better charge collection. The improved J_{sc} value for the nanowire-loaded films was also attributed to the high light scattering within the films. Among all the cells, the best efficiency was 8.6%, achieved from films with 20% nanowires with a film thickness of 14 μm, compared to 6.7% from pure nanoparticle cells.

In order to reduce electron recombination, barrier layers, in the form of core–shell nanowires have also been used, in which the shell layer can retard electron recombination. For example, ZnO nanowires were coated with TiO₂ layers using ALD to form core–shell structures (Figure 15.6a) [34]. At shell

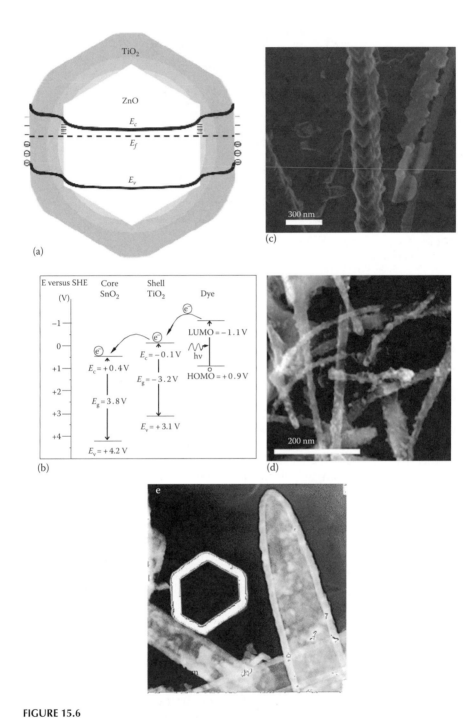

FIGURE 15.6
Hybrid structures of (a) ZnO NW–TiO$_2$ NP. (From Law, M. et al., *J. Phys. Chem. B*, 110, 22652, 2006.) (b) SnO$_2$ NW–TiO$_2$ NP hybrid structures and their corresponding electronic band diagrams. (From Gubbala, S. et al., *Adv. Funct. Mater.*, 18, 2411, 2008. With permission.)

thicknesses below 5 nm, the shells were amorphous, but transformed into polycrystalline shells for thicknesses above 5 nm. The V_{oc} of solar cells made using the core–shell (ZnO–TiO$_2$) nanowires was considerably higher by ~250 mV compared to pure ZnO NW array-based solar cells. Dark current measurements showed a decrease with increasing shell thickness, decreasing electron recombination, and resulting in higher electron density reflected in higher open circuit potentials. This was due to the formation of an n–n$^+$ layer with a more negative TiO$_2$ conduction band edge. The J_{sc} decreased with shell thickness until 5 nm and increased to the original value with further increase in shell thickness. The initial decrease in J_{sc} was due to the poor electron injection into the amorphous TiO$_2$ caused by the higher density of electron traps. As the thickness of the amorphous layer increases, the surface area of amorphous TiO$_2$ also increases, thus leveling off the J_{sc}. Overall, the cell efficiency doubled in response to TiO$_2$ shells 10–35 nm thick, jumping from 0.85% to 1.7%–2.1%. Further increase in the TiO$_2$ thickness layer results in amorphous to crystalline transition which improves both, the electron injection and transport, thus improving the charge collection.

SnO$_2$ nanowire-based hybrid structures (Figure 15.6b) showed higher efficiencies compared to either pure SnO$_2$ nanowire or TiO$_2$ nanoparticle electrodes [17]. In this configuration, the SnO$_2$ nanowires were coated with a thin layer of TiO$_2$ nanoparticles by dip-coating them in an aqueous TiCl$_4$ solution. As illustrated in the band diagram Figure 15.6b, the electrons injected into TiO$_2$ from the excited dye are expected to be efficiently injected into SnO$_2$. The excellent transport properties of interconnected nanowires should allow fast transport of electrons to the back contact, thus lowering the recombination rate. X-ray diffraction (XRD) analysis revealed that the TiO$_2$ nanoparticles were rutile phase. The TiO$_2$–SnO$_2$ hybrid structures showed the highest efficiencies (4.1%), at 70 mW/cm^2 light intensity with a high short circuit current (8.56 mA/cm^2) and open circuit voltage (686 mV). The current densities of the TiO$_2$–SnO$_2$ hybrid electrodes was found to depend on the morphology of SnO$_2$ nanowires and typical values for samples made in this work ranged from 3 to 9 mA/cm^2. It was also seen that the onset of recombination current occurs at higher potential for SnO$_2$ nanowire/TiO$_2$ nanoparticle hybrid electrodes when compared to the SnO$_2$ nanoparticle and nanowire electrodes, which indicates retardation in the electron recombination reaction in the SnO$_2$ nanowire–TiO$_2$ nanoparticle hybrid electrode. Due to its recombination and transport properties, the SnO$_2$ nanowires could serve as good underlying matrix for many other semiconductor materials in DSSCs. Further optimization of hybrid structures with different material systems can greatly enhance the overall performance of the solar cells.

15.2.1.7 Transport and Recombination

The transport and recombination times are determined by either photocurrent or photovoltage decay measurements or optical impedance measurements. In a photocurrent decay measurement, the cells are probed with a weak laser pulse in the presence of background illumination. The intensity of background illumination varies using neutral density filters. The decay is recorded and the time constant of the decay is measured by fitting the decay to $e^{-(t/\tau_c)}$. From the time constant τ_c, the diffusion constant of the electrons in the film can be determined from the expression, $D = d^2/2.35\tau_c$, where d is the film thickness and D is the diffusion coefficient. Open circuit photovoltage decay measurements are done to determine the electron recombination time constants. In this technique, the photovoltage decay is recorded at open circuit conditions. The lifetime of the electron in the film is determined using the expression $\tau_r = -\left(kT/e\right)\left(dV_{oc}/dt\right)^{-1}$.

In addition to these techniques, optical impedance measurements can also be performed to get the same information. If the current is measured as a response to the modulation, the technique is called intensity-modulated photocurrent spectroscopy (IMPS); if the voltage is measured, the technique is called intensity-modulated photovoltage spectroscopy (IMVS). The current and voltage response are measured using a lock-in amplifier. The time constants are extracted from the Nyquist or Bode representations of the solar cell response.

One-dimensional structures offer a great improvement in the charge collection efficiencies of DSSCs. Many studies have been performed to investigate the transport time constant and the recombination lifetime of electrons in these cells and compared them with the nanoparticle-based cells. In one such study, transient photocurrent measurements on MOCVD grown ZnO nanowires (see Figure 15.7a) showed that electron transport within the nanorod arrays is two orders of magnitude faster (30 μs) than in nanoparticles (10 ms). However, the recombination kinetics within nanowire-based cells is similar to that in nanoparticle cells [35]. IMPS and IMVS measurements performed on ZnO nanorod arrays grown in solution also reveal that the transport in nanowires is faster (by 10–100 times) than in nanoparticles, but

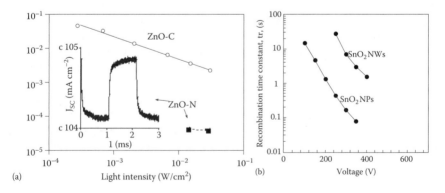

FIGURE 15.7
DSSCs behaviors: (a) ZnO transport. (From Galoppini, E. et al., *J. Phys. Chem. B*, 110, 16159, 2006.) (b) SnO₂ recombination of nanoparticles and nanowires. (From Gubbala, S. et al., *Adv. Funct. Mater.*, 18, 2411, 2008. With permission.)

the electron recombination kinetics in the nanowire and nanoparticles are similar [36]. The transport time constants in the case of nanowire electrodes are independent of the light intensity, in contrast to the observation for nanoparticle films, where the transport time constants decrease with increasing light intensity due to the progressive filling up of traps. The fact that diffusion coefficient is constant with increasing light intensity in nanowire electrodes suggests that either all the traps are filled even at low light intensity or that the traps are all concentrated around one energy.

Interestingly, in a different study, it was found that the electron transport time in polycrystalline TiO_2 nanowires was similar to that of TiO_2 nanoparticle networks, but the recombination was found to be much slower in nanowires [8]. This resulted in charge collection efficiencies of 150 in nanowires compared to 50 and 10 in nanotubes and nanoparticles. The slower recombination in nanowires was attributed to the large diameter for nanowires that could support a radial electric field, keeping the electrons away from the surface. In contrast, the nanoparticles have a different spatial distribution of traps. For example, in nanowires, the majority of the traps may be at the internal grain boundaries, where they are not exposed to the electrolyte. Similar results were seen in other reports, with recombination lifetimes, about one order of magnitude longer for nanowire films. The measurements showed that the slopes of the plots for the diffusion constants and the J_{sc} are parallel for nanowires and nanoparticles, indicating that the two materials have similar distribution of traps.

In the case of tin oxide nanowires, transport time constants of various nanowire samples were different depending on how well they are interconnected and branched [17]. These values ranged from 4.5 ms for high-aspect ratio branched nanowires to 30 ms for low-aspect ratio loosely connected wires. Also, the electron transport in the nanowire electrode was faster than the nanoparticle electrode. Figure 15.7b shows the electron recombination lifetime for different electrodes. The SnO_2 nanowire DSSCs show about two orders of magnitude slower recombination rate than the SnO_2 nanoparticles over the entire range of V_{oc} values.

Transport and recombination measurements were also made for nanotube structures. The transport time constants in TiO_2 nanotubes were similar to those of nanoparticles due to similar crystallite sizes in nanoparticles and in these nanotubes [3]. However, the different slopes of the transport time constants with photon flux indicate that the density and distribution of traps in nanoparticles and nanotubes were different. The recombination time constants for nanotubes were an order of magnitude longer than the nanoparticle films indicating fewer electron recombination sites [21]. At the maximum power, the charge collection efficiency of nanotube films was 25% larger than the nanoparticle films. In addition to improved charge collection efficiency, these films also show better light harvesting efficiencies. In addition, the transport and recombination characteristics are significantly improved when structural disorders are removed from these materials [3].

15.2.2 Direct Absorption PEC Cells

PEC cells based on low-bandgap inorganic materials have also been investigated for solar energy conversion. Si is an excellent material for this purpose due to its band gap of 1.12 eV. Si was used to make the first nanowire-based PEC cell [37], in which the p-type nanowires were grown using the vapor–liquid–solid method. Cyclic voltammogram of a Si NW array electrode under illumination in $[Ru(bpy)_3]^{2+}$ (bpy = 2,2′-bipyridyl)/acetonitrile solution revealed that the reduction of $Ru(bpy)_3^{2+}$ occurred at potentials of about 220 mV more positive than that using the Pt electrode. In the dark, the Si NW array exhibited negligible cathodic current in the same potential range. The photocurrents were about twice compared to currents obtained using planar Si wafers. However, the photovoltages were 500 mV for planar Si electrodes. The lower photovoltages for Si NW electrodes was attributed to excessive doping of the nanowires, which lead to the formation of an insufficiently thick space charge region, creating a leaky diode junction. In a similar study, micron-thick n-Si nanowire arrays were used as PEC cells [38] with an electrolyte of 1,1′-dimethylferrocene $(Me_2Fc)^{+/0}$ redox system in CH_3OH. The nanowire array electrodes exhibited about 350–400 mV and J_{sc} of ~1.5 mA/cm². These high V_{oc} values suggest a very low recombination reaction with the $(Me_2Fc)^{+/0}$–CH_3OH interface. The Si nanowires were also grown by electroless etching of Si wafers and their photovoltaic properties were measured in a 40% HBr and 3% Br solution [39]. These structures gave voltages between 660 and 730 mV and current densities from 0.65 to 1.36 mA/cm². In addition to the photovoltaic measurements, these nanowire arrays showed excellent antireflectivity.

In Cd(Se,Te) nanowire arrays grown using anodic alumina template on Ti substrates exhibited lower V_{oc} and J_{sc} values compared to planar electrodes. The spectral response of the nanowire electrodes showed enhanced charge collection efficiencies of low-energy photons absorbed far from the front surface as shown in Figure 15.8 [40]. In a different study, the CdSe nanorods with different aspect ratios were fabricated into solar cells using P3HT polymer. The nanorod radii were varied to tune the absorption into the material. Although, the onset of the external quantum efficiency of CdSe nanorods with diameters of 7 nm is broader than the one with diameters of 3 nm by ~70 nm (wavelength) in the red region, the relative improvement in the performance of the solar cell was very low. In contrast, the nanowires with constant diameters showed a large improvement in the efficiency with increasing lengths due to better transport properties as shown in Figure 15.9 with the highest efficiency of 1.7% [41].

15.2.3 p–n Junction Solar Cells

The p–n junction solar cells based on nanowires have been demonstrated in two different configurations (Figure 15.10). In the first case, the junction is radial, while in the other, the junction is planar, but with the nanowire arrays acting

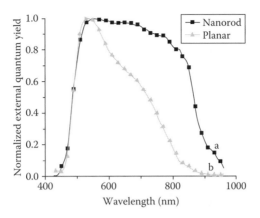

FIGURE 15.8
Quantum efficiency of Cd(SeTe) nanowires and nanoparticles showing an enhancement in the efficiency for nanorod electrodes. (From Spurgeon, J.M. et al., *J. Phys. Chem. C*, 112, 6186, 2008. With permission.)

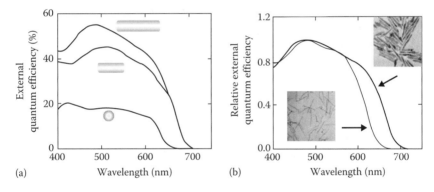

FIGURE 15.9
Quantum efficiencies for CdSe nanorods of (a) different lengths and (b) different diameters. (From Huynh, W.U. et al., *Science*, 295, 2425, 2002. With permission.)

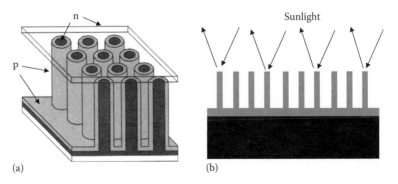

FIGURE 15.10
p–n junction solar cells (a) radial p–n junction. (From Brendan, M.K. et al., *J. Appl. Phys.*, 97, 114302, 2005.) (b) Nanowires acting as an antireflecting coating in a p–n junction solar cell.

as an antireflecting material. Vertically aligned single crystal Si nanowires were grown on a p-type Si wafer and the excellent antireflection properties of these nanowires gave conversion efficiencies of up to 9.31% [42]. Further improvement in the efficiencies to 11.37% were obtained for Si NW-based p–n junction solar cells by improving their antireflection properties and better electrical contact of the cells by employing slanted nanowire structure [43]. These efficiencies are, however, lower than the single-crystal Si solar cells in spite of higher antireflectivity due to increased carrier recombination.

In addition to solar cells made with arrays of nanowires, studies on single nanowire solar cells have given insight into the photovoltaic performance of nanowires [44]. The Si nanowire gave a J_{sc} of $5\,mA/cm^2$, V_{oc} of $190\,mV$, and a fill factor of 0.4, giving an overall efficiency of 0.46%. The agreement between the peak positions for coherent thin film absorption and the measured quantum efficiency implied interference between the front and back surfaces of the nanowires. Also, the external quantum efficiency and the absorption of the nanowires showed a similar trend, indicating constant internal quantum efficiency throughout the length of the nanowire, indicating no electron recombination loss. In the solar cells made with radial p-i-n junctions of Si, efficiencies of up to 3.5% have been observed with the following characteristics: $J_{sc} \sim 24\,mA/cm^2$, V_{oc} of $260\,mV$, and FF of 0.55. Here the J_{sc} scales linearly with the nanowire length, indicating that the photogenerated carriers are collected uniformly all along the length of the nanowires [45].

15.2.4 PEC Cells for Chemical Conversion

Photolysis is the process by which a chemical compound is decomposed with the assistance of a photoactive catalyst or electrode in the presence of light. Photolysis is employed in a number of applications, such as large-scale purification of waste water. A significant amount of research in this area is devoted to PEC electrolysis of water for the production of hydrogen. The hydrogen can then be used in a fuel cell stack to produce electricity for consumption. The goal is clean, sustainable, and secure yet inexpensive energy generation and storage.

It is possible for both the anode and cathode to be semiconductors, but the most common setup involves a photoanode (an n-type semiconductor) and a metal cathode (such as platinum). As Figure 15.11 illustrates, upon illumination by sunlight, the generated electrons flow from the bulk of the photoanode to the cathode, whereupon they participate in the evolution of hydrogen. The positively charged holes oxidize water at the surface of the photoanode to produce oxygen and positive ions.

There are a number of challenges associated with searching for a suitable and stable material for performing photolysis. The criteria are

1. Band gap energy between 1.8 and 3.1 eV (roughly).
2. Appreciable reaction kinetics and charge carrier conductivity.

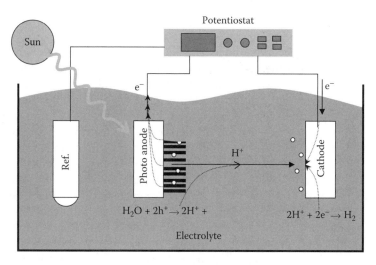

FIGURE 15.11
An illustration of a PEC water splitting system in which the photoanode is the semiconductor nanowire array. Upon illumination, hydrogen and oxygen are evolved on opposite sides.

3. The energies of the HOMOs and LUMOs "straddle" the redox potentials for water decomposition.

4. Stability in aqueous environments.

5. Earth-abundant and mass-producible.

The minimum band gap requirement is dictated by the minimum energy needed to split water (1.23 eV), thermodynamic losses, and overpotentials needed to increase the reaction kinetics. The upper band gap requirement ensures that the semiconductor absorbs a significant portion of the terrestrial solar spectrum. It is not enough that the semiconductor absorb the photons, however, because it also must be able to separate and conduct the electrons and holes. As for stability, the semiconductor electrode may undergo corrosion. The third requirement refers to the relative energy band positions of the semiconductor. Essentially, the energy of the valence band should be greater than the water oxidation potential (1.23 eV versus normal hydrogen electrode [NHE]) and the energy of the conduction band should be less than the water reduction potential (0 eV versus NHE). The kinetics of charge transfer from the surface to the electrolyte must be faster than the decomposition reaction. For example, TiO_2 has excellent stability in a wide range of pH levels whereas the corrosion of ZnO by oxidation is preferred over oxidation of water by the holes [46]. ZnO is stable as a photocathode (p-type), however, because the water reduction potential is greater than the reduction corrosion potential [46]. Finally, the material cannot be composed of precious elements or else its application would be limited from the beginning. Similarly, there should exist, or be developed, a method for inexpensive mass-production in order to economically feasible.

There are many approaches to meet these criteria, from finding new materials to manipulating well-known materials. In this regard, the use of nanowires can be advantageous. Nanowires offer a high surface-to-volume ratio, which means that using nanowires rather than a flat surface increases the reaction surface area, leading to greater current. Nanowires can have fewer (or no) grain boundaries [47,48], which are a serious problem in polycrystalline films because they act as electron–hole trap sites [49]. Furthermore, nanowires are an excellent platform for doping and alloying studies. Hematite (alpha-Fe_2O_3), for example, is a promising material for PEC because it has an optimal band gap of 2 eV and is electrochemically stable in moderate pH environments [50,51], but it is plagued by slow water oxidation kinetics at the interface, which favors charge recombination. One way to increase the efficiency of hematite PEC electrodes is to minimize the distance through which the minority carriers must diffuse [52], thus reducing the probability of recombination losses. The geometry of nanorods/nanowires achieves just this, and in addition the perpendicular orientation better facilitates charge transport to the back-contact without loss due to random pathways such as in nanoparticle systems. Reducing the feature size, to sub-5 nm dimensions may lead to quantum confinement [53], which may actually be beneficial if it leads to a decrease (upward shift) in the conduction band energy without a significant increase in the total band gap. The ideal characteristics of an n-type nanowire array photoelectrode are illustrated in Figure 15.12. 1-D nanostructures are relatively new in this field, but several studies have already demonstrated that they perform better than particles or thin films.

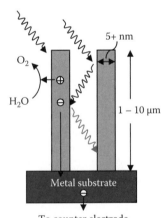

FIGURE 15.12

Illustration of n-type nanowire array electrode. As photoanodes, the nanowires oxidize water and the electrons proceed to the counter electrode for the reduction reaction. A long height increases light absorption and the small diameter matches the short hole diffusion length.

Probably the most widely studied material for PEC water splitting is TiO_2. Although it is a wide band gap semiconductor, and thus requires UV light for efficient electrolysis, it is stable and relatively inexpensive. Some groups synthesized TiO_2 nanowires and nanotubes and demonstrated improved performance compared to films or particles [54–56]. One study in particular found that the nanowires achieve a twofold increase in photoconversion efficiency over a single-layer film [54], and another study demonstrated a 20-fold increase in photocurrent over a nanoparticulate film [56]. These studies have attributed the increased activity to increased surface area and reduced grain boundaries. These observations are not limited to TiO_2, however. Other materials, such as hematite—as discussed above—and ZnO have been shown to benefit from 1-D geometries for PEC water-splitting. An array of ZnO nanorods exhibited a photocurrent twice that of a ZnO thin film in the same study [57]. Synthesis of hematite nanowire or nanorod architectures has been demonstrated using different methods [27,48,53,58–65]. In one particular study, RF plasma was used to synthesis nanowire arrays of several oxides in a matter of seconds [48]. Subsequent studies showed that the hematite nanowires were single crystalline and exhibited a superstructure of repeating oxygen vacancy planes parallel to the growth direction [48,66]. Oxygen vacancies in hematite are traditionally responsible for significant losses in charge carrier conductivity, so the potential advantage of these nanowires for PEC electrolysis is that the carriers may conduct to back contact without encountering vacancy trap sites. Discoveries such as these continuously demonstrate the benefits of nanowire architectures for PEC electrolysis.

To date, the highest measured PEC efficiency is 12.4%, which was achieved using a $GaInP_2/GaAs$ monolithic tandem cell [67]. However, this cell exhibited significant degradation. The photoconversion efficiency is calculated using Equation 15.5 [68]:

$$\varepsilon = \frac{1.229 I_P}{P_t} \tag{15.5}$$

where
the constant 1.229 is expressed in volts
I_P is photocurrent in A/m^2
P_t is the input power (or light irradiance) in W/m^2

The constant comes from dividing the standard Gibbs energy for hydrogen evolution at 25°C and 1 bar (237.2 kJ/mol) by number of moles of electrons used to generate 1 mol of H_2 (2), and by the Faraday constant (96,485 C). The numerator and denominator are both in terms of power units, and therefore efficiency is the ratio of power output to power input. This method assumes standard temperature and pressure conditions and that all the carriers are utilized for evolving hydrogen and oxygen. The latter assumption is contradicted if the power output comes from the photoelectrode corrosion. In order to use Equation 15.5 to calculate efficiency, the photocurrent (current difference between dark and light conditions) must be generated without an external bias.

Although it is desirable for a photoelectrode to split water spontaneously upon illumination, some externally applied bias is most often necessary to encourage photocurrent. In such a case, Equation 15.5 must be modified as follows [68]:

$$\varepsilon = \frac{(1.229 - V_{bias})I_P}{P_t} \tag{15.6}$$

where

V_{bias} is the potential difference between the working electrode (photoelectrode) and the counter electrode.

It has been observed that nanowires are attracted to each other during PEC such that they can be seen to clump together at the tips. Scanning electron microscope (SEM) images of WO_3, W_2N, and alpha-Fe_2O_3 nanowire before and after electrochemistry show this effect in Figure 15.13. The suspected reason for this phenomenon is the existence of electric fields due to charge build up at the nanowire surface. Carbon nanotubes have been widely demonstrated to align to a positive electric field, and may permanently deform if the field is high enough [69–72].

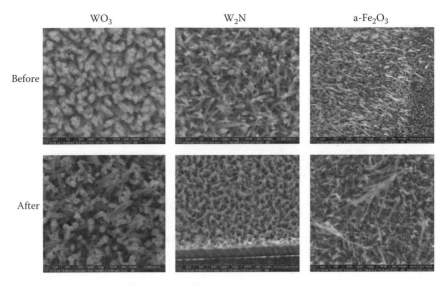

FIGURE 15.13
Nanowires imaged before and after PEC characterization were found to clump together due to apparent attraction at the tips.

15.3 Electrochromic Devices

Electrochromism (EC) is the phenomenon of inducing a reversible change in the optical properties of a material by the application of a small electric field. Transmission modulation in these films is achieved by varying the oxidation state of the electrochromic material by an electric field–assisted insertion and extraction of small alkali metal ions, like lithium (Li^+) as shown in Figure 15.14. In general, the oxide materials used are in the form of nanoparticles or thin films. The use of nanowires with faster charge and mass transport properties are expected to enable faster switching between the colored and bleached states, and yield higher color contrasts.

Much of the early work on EC devices using nanowires has focused on the use of WO_3 nanowires synthesized using a hot-wire assisted chemical vapor deposition (CVD) [73]. The synthesis process resulted in 40–60 nm individual WO_3 nanowires and also bundles of nanowires. The bundles of nanowires typically contained six nanowires with an overall diameter of ~300 nm. Two device configurations, vertically aligned nanowires and mat-like nanowires were used to fabricate the devices (see Figure 15.15) [74]. For the electrochromic effect, coloration, and bleaching potentials of −3.5 and 2.5 V were applied. In the devices made with vertical arrays, the transmission modulation in these films was between 0% and 76%, with the coloration talking place almost instantaneously, but bleaching took almost 30 min to reach ~71%. The coloration process followed a biexponential function with time constants of 38 and 1.2 s while bleaching followed a single time constant of 138 s. The two time constants observed during coloration were attributed to the diameter distribution in these nanowires. The faster time constants are observed from

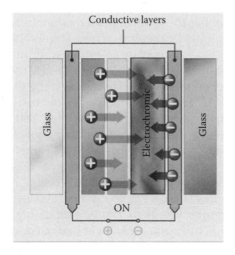

FIGURE 15.14
Schematic showing the working of an EC device.

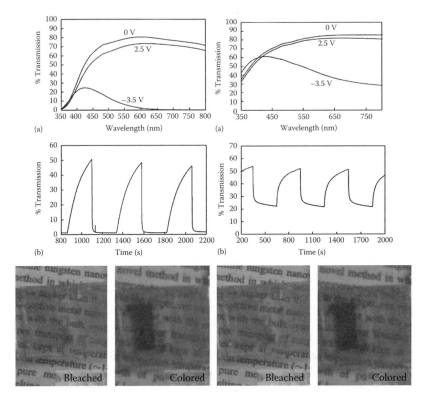

FIGURE 15.15
EC device performance of electrodes based on WO_3 nanowires in two configurations: (a) arrays and (b) mats. (From Gubbala, S. et al., *Sol. Energy Mater. Sol. Cells*, 91, 813, 2007. With permission.)

the rapid coloration of the 40–60 nm diameter nanowires and the slower time constants from the bundled nanowires. These values are consistent with the time constants obtained theoretically assuming a diffusion constant of 10^{-11} cm^2/s for Li^+ ion inside WO_3 solid [75]. The characteristic timescale can be estimated from the expression $\tau = l^2/D$, where l is the diameter of the nanowire and D is the diffusion constant of Li^+ ions in WO_3. In nanowires, there are typically two characteristic lengths, diameter, and length. The majority of the intercalation and deintercalation in nanowires occurs through the walls of the nanowires as shown in Figure 15.16. Therefore, the characteristic length in these structures for the intercalation and deintercalation is the diameter of the wires. Using this expression, the time constant for individual and bundled nanowires is found to be 2.5 and 90 s, respectively [76]. There is another possible reason why there can be a much higher time constant at high coloration. Li^+ ion diffusion constant at high coloration can be an order of magnitude lower at Li^+ ion fraction reaching 0.3 in Li_xWO_3.

The continued application of coloration potential and subsequent intercalation of Li^+ into the interior parts of the thick WO_3 nanowire arrays and

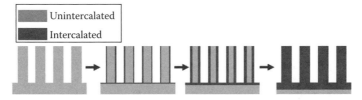

FIGURE 15.16
Schematic showing the Li$^+$ ion intercalation into WO$_3$.

the film does not seem to affect the optical contrast significantly. The rapid coloration of these films due to the surfaces of the nanowires is further confirmed by the XPS spectra of these films taken after 15 s of coloration. The bleaching process in these films is found to be very slow. The time constant for bleaching in the vertically oriented nanowires was found to be 138 s. This can be attributed to the slower rates of deintercalation from the bundles and the wires. The deintercalation from the thin nanowires does not seem to affect the optical contrast in the films. The slow bleaching is characteristic of the continued bleaching taking place from the thick 300 nm nanowires. In addition to this process, the intercalation and deintercalation kinetics are also affected by the Li$^+$ ion concentration in the films.

The electrochromic performance of mat electrodes was also measured in a similar fashion. The transmission at 700 nm for these films at 0 bias is about 85%, which falls down to 32% upon Li$^+$ intercalation. The transmission goes back to 80% upon bleaching. The high transmission through these films in the colored state is due to the low density of these wires on the FTO substrate, which allows the direct passage of light through the FTO, without interacting with the nanowires. The time constants for coloration are 83 and 3.3 s, while for bleaching are 153 and 14 s. These differences arise due to the two sets of wires present (bundled and single nanowires). Larger time constants in these films compared to the vertically aligned films may be due to the high wire to wire resistances. Since the films in this case are of low density, the optical contrast from both the individual and bundled nanowires can be clearly observed.

Other studies were performed using W$_{18}$O$_{49}$ nanowire arrays grown using thermal CVD. The nanowires are 39–65 nm in diameter and 5 μm long and the EC devices made using these nanowire arrays were tested for coloring and bleaching with applied potentials at −3.5 and 2 V, respectively. The devices showed a change of 34.5% in the transmission upon coloration. The coloration and bleaching times (defined as the time to reach two-thirds of the total change in the transmission) was found to be less than 2 s [77]. The EC devices were also made using W$_{18}$O$_{49}$ nanowire bundles synthesized using solvothermal method, each bundle consisting of 15–20 nanowires of 6 nm in diameter and 600 nm long. These nanowire bundles showed coloration and bleaching times of 3.5 and 1.1 s. The coloration efficiency of these nanowires was 55 cm^2/C compared to 32 cm^2/C for sputter-deposited films [78].

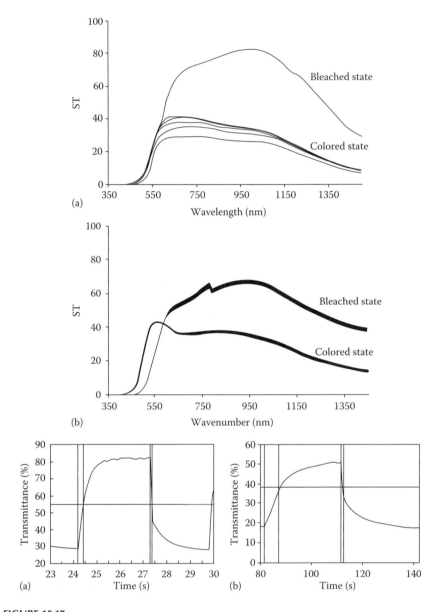

FIGURE 15.17
EC devices based on (a) vanadium oxide nanowires and (b) silver vanadium oxide nanowires. (From Xiong, C. et al., *ACS Nano*, 2, 293, 2008. With permission.)

In addition to tungsten oxide, V_2O_5 nanowires were also grown using thermal CVD [79]. The resulting nanowires were 10–20 nm in diameter and have lengths of 250–450 nm. The devices showed a 40% change in transmission at ~400 nm. The switching times in these devices were ~6 s [80]. In a different study, V_2O_5 nanowires with lengths of 30 μm and 10–20 nm in diameter were synthesized using a hydrothermal method. The EC devices made with these nanowires showed a transmission change of 33.7% at 1005 nm. The switching time, defined here as the time for 50% change in the total color change, was 6 and 1 s. In the same study, silver vanadium oxide nanowires were also synthesized, with similar dimensions. The transmission change in these EC devices was 60% at 1020 nm and the switching times were 0.2 and 0.1 s for coloration and bleaching, respectively (see Figure 15.17). The faster switching times for the SVO films were explained by the higher diffusion coefficients of SVO films (5.3×10^{-10} cm^2/C) versus 7.5×10^{-11} cm^2/C.

In summary, the electrochromic devices made using nanowires show promising results in terms of both color contrast and the switching speeds. At present, there is not much information on long-term stability, which requires further study.

15.4 Li-Ion Batteries

The need for clean and efficient energy storage is important in consumer electronics, medical devices, automotive, and other applications. In the context of energy storage, Li-ion batteries have gained much interest. They have a high energy density and capacity, which makes them the automatic choice for cell phones, laptops, digital cameras, camcorders, power tools, and hybrid vehicles. At present, the worldwide market for Li-ion batteries is estimated to be around $10 billion [81].

A Li-ion battery comprises of a negative electrode (anode), which is typically graphite, a positive electrode (cathode, usually $LiCoO_2$ or $LiCoMnO_2$), and a nonaqueous liquid electrolyte. During the charge process, the Li$^+$ ions are deintercalated from the layered structure of $LiCoO_2$, diffuse through the liquid electrolyte, and are then intercalated into the graphite. During discharge, the whole process is reversed. The rate of discharge and charge process is measured in terms of a metric called C. A rate of $C/2$ indicates that the half cycle of the process takes 2 h. Similarly, a rate of $C/10$ indicates that the half cycle of the process takes 10 h. The electrons travel through the external circuit and the higher the rates, the faster the electrons are driven into the external circuit (see Figure 15.18). If the negative electrode is lithium metal, extremely high capacities of 4000 mA h/g are possible, but the repeated cycling of the electrode leads to a dendritic growth due to the replating of lithium after each cycle, which eventually leads to explosion hazards.

The Li-ion battery tests are generally carried out in a three electrode cell or a coin-type cell. In a three-electrode setup, the working electrode, which is the material being tested, is soldered to the gauze and Li metal is attached to the other two gauzes, which act as a reference and an auxiliary electrode.

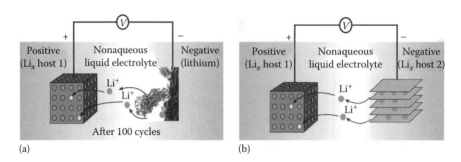

FIGURE 15.18
Schematic showing the concept of a lithium ion battery with (a) Li metal as anode and (b) a graphite as anode. (From Tarascon, J.M. and Armand, M., *Nature*, 414, 359, 2001. With permission.)

The fact that materials are hardpressed onto buttons makes the coin cell arrangement different from the wire setup. The handling of the Li metal is carried out in an oxygen-free glove box to prevent the oxidation of lithium. In most cases, the charge–discharge cycles are carried out in a glove box although an easier option would be to fabricate the cell inside the glove box and then run the electrochemical tests outside the glove box. The charge–discharge measurements are performed using a potentiostat/galvanostat.

15.4.1 Challenges with Anode Materials

There has been a widespread search for anode materials in lithium ion batteries. Conventional graphite and carbon electrodes are the preferred materials but the storage capacity extracted from the carbons is low. Reversible capacities close to 450 mA h/g can be achieved using chemically or physically modified carbon electrodes. But, as evident from Figure 15.19, other materials with higher capacity and a little higher voltage (versus lithium) than carbons are needed to improve the energy density of the batteries. Research efforts have led to the emergence of Li transition-metal nitrides ($Li_{3-x}Co_xN$) as potential candidates for anodes with a stable reversible capacity (600 mA h/g), but their capabilities are overshadowed by the manufacturing efforts required to handle these moisture-sensitive materials. The other interesting classes of materials are metal oxides, intermetallic alloys, and semiconductors, which have higher specific capacity compared to carbons and have a low voltage (versus lithium). But these favorable properties come at the expense of enormous volume changes associated with lithium alloying of metals and semiconductors (Si, Ge, Sn, and Al) compared to carbons, as a result of which the electrode integrity is lost. In particular, the stress and strain induced by volume expansion leads to the cracking of the electrode, which causes the material to lose its ability to store charge, and a rapid capacity loss is observed with repeated cycling. This problem of electrode degradation with cycling can be overcome by having shorter Li-ion path lengths and better

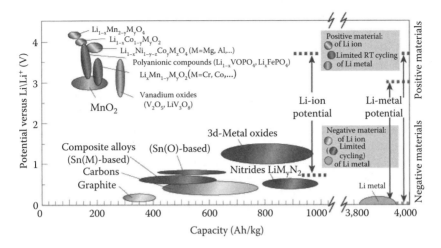

FIGURE 15.19

A comprehensive plot showing the capacities and potentials for different materials as positive and negative electrode materials when used in a Li-ion battery. (From Tarascon, J.M. and Armand, M., *Nature*, 414, 359, 2001. With permission.)

strain accommodation in the materials, which is feasible when using nanoscale materials, especially 1-D nanowires. Nanowires also offer higher surface area per unit volume, which enunciates Li-ion alloying/dealloying as well as sustains higher charge and discharge rates. In addition, electron path lengths are shorter along with the Li-ion path lengths, which significantly lower the electron/Li$^+$ conductivities contingent for Li-ion battery operation. But the major drawbacks of using nanowires as negative electrodes stems from the fact that very little is known about the undesirable electrode/electrolyte reactions, which arise due to the higher available surface area in nanowire electrodes. The packing density obtained using nanowires is significantly lower than that of the bulk material electrodes, which curtails the overall volumetric energy density. From a practical standpoint, the viability of bulk synthesis of the nanowires is an open question and even if it is, the production cost is relatively higher than the commercially used carbon electrodes. The limited availability of the data on the calendar (shelf) life of the batteries made of these materials calls for an extensive and in-depth analysis of this phenomenon to make these batteries a reality [82].

15.4.2 Challenges Facing Cathode Materials

In the case of Li-ion batteries, the positive electrode has lithium because the negative electrode is devoid of lithium in almost all the cases. However, Li-solid polymer electrolyte (SPE) cells use Li-free V_2O_5 or its derivatives as positive electrodes. $LiCoO_2$ is still the most widely used positive electrode which operates at 4 V. Although $LiNiO_2$ has better specific capacity compared to that of $LiCoO_2$, safety concerns arise due to the exothermic oxidation of the electrolyte during the collapse of the delithiated Li_xNiO_2 structure. Addition of an electrochemically inert di-, tri-, or tetravalent substitute for Ni or Co (Ga, Mg, or Al) has shown to stabilize the structural framework of the material along with high-capacity retention. Other important cathode materials under consideration include $Li(Mn/Fe/Cr)O_2$, $LiFePO_4$, $LiMn_{2-x}Al_xO_{4-y}F_y$, and Li_xVO_y compounds. The performance of these is rather close, which indicates that one of the ways to enhance the performance is to tune the morphology and the texture of the electrode. In this context, nanowires because of their enhanced properties as discussed before can serve the purpose. However, the use of the nanowire-based cathodes has been studied to a lesser extent compared to anodes. Nanowire-based cathodes can lead to more reactions at the interface of the electrolyte/electrode, leading to safety concerns, much more likely at the higher temperatures that they are operated at. Other disadvantages are quite similar as the ones described for the nanowire anode electrodes [82].

15.4.3 One-Dimensional Materials for Anodes

15.4.3.1 Carbon Nanotubes (CNTs)

Carbon nanotubes (single-walled nanotubes [SWNTs] and multiwalled carbon nanotubes [MWNTs]) have gained considerable interest because of their high stability as anodes. Lithiated carbons (Li_xC_6) are the most widely used anodes because of their higher redox potential versus the cathode materials as well as their superior cycling performance than polymers, metal oxides, metals, and semiconductors. Lithiated carbons are considerably safe and their reactivity is high. The general electrochemical reaction scheme of lithium ions into carbonaceous materials is as follows:

$$Li_xC_n \leftrightarrow xLi^+ + xe^- + C_n \qquad (15.7)$$

During the electrochemical reduction, i.e., the charge process, the Li ions move from the cathode through the electrolyte to the carbon, forming a lithiated carbon Li_xC_n. Subsequent oxidation reaction takes out the Li ions from the carbon matrix, which is termed as the discharge process. The first nanoscale carbons used as negative electrodes are the MWNTs, which show reversible capacities in the range of 100–400 mA h/g depending on the processing conditions of the nanotubes. The SWNT synthesized using laser ablation of Ni/Co catalyst containing graphite targets at a temperature of 1150°C are shown in Figure 15.20a. The pure SWNTs showed a reversible capacity of 681 mA h/g, which is equivalent to $Li_{1.8}C_6$ [83]. Similarly, etching of the SWNTs increased the reversible capacity to 1000 mA h/g ($Li_{2.7}C_6$) with further cycling of the electrode, resulting in a very little loss of capacity [83]. Other studies have shown that the defective and shorter length, MWNTs as shown

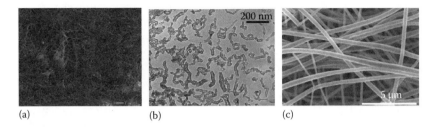

(a)   (b)   (c)

FIGURE 15.20
SEM images of various carbon tubular structures investigated as anode materials in Li-ion battery: (a) MWNTs. (From Gao, B. et al., *Chem. Phys. Lett.*, 307, 153, 1999. With permission.) (b) Short and broken MWNTs. (From Wang, X.X. et al., *Adv. Funct. Mater.*, 17, 3613, 2007. With permission.) (c) Carbon fibers made by electrospinning. (From Kim, C. et al., *Adv. Funct. Mater.*, 16, 2393, 2006. With permission.)

in Figure 15.20b show higher capacity of 593 mA h/g at a current density of 25 mA/g [84]. Even the electrospun, long carbon fibers (Figure 15.20c) exhibited a reversible capacity of about 500 mA h/g after 50 cycles [85].

15.4.3.2 Metal/Metal Oxide Nanowires

The metal/metal oxides and semiconductors form alloys with Li, which give rise to much higher theoretical capacities than that with carbon as summarized in Table 15.2. Soon after Dey demonstrated the electrochemical formation of lithium alloys in 1971 [86], metallic lithium was replaced by lithium alloys. The reversibility of the reaction is as follows:

$$Li_xM \leftrightarrow xLi^+ + xe^- + M \qquad (15.8)$$

The alloy formation is in general very complex involving the formation of several intermetallic compounds and different phases of Li_xM. Alloy formation with many metals like Al, Si, Sn, In, Co, Fe, Bi, and Sb have been studied in detail. However, these metals undergo large volume changes, resulting in severe structural deformities and successive formation of several phases. Large structural deformities cause large morphological changes in the materials and reduce the available surface area for Li alloying during subsequent cycles.

On the other hand, the interaction of Li with metal oxides occurs through the following overall, irreversible reaction by the formation of amorphous Li_2O and metal.

$$MO_x + 2xLi \rightarrow xLi_2O + M \qquad (15.9)$$

The metal oxides reduce to metal irreversibly with the formation of Li_2O phase. Bulk materials suffer from cyclability and capacity fading due to slow reactivity and cracking caused by the mechanical strain associated with large volume changes. On the other hand, the nanoscale materials offer

TABLE 15.2

Summary of Studies Dealing with Nanowire-Based Materials as Anodes in Lithium Ion Batteries

Materials	Starting Materials	Theoretical Capacity (mA h/g)	Retained Capacity (mA h/g)	Discharge Rate	Comments on Retained Morphology	References
Co_3O_4	Nanoparticles CoO nanoparticles	890	700 @ 100 cycles < 800 @ 100 cycles	C/5	Initial crystalline nanoparticles break down to smaller nanoparticles dispersed in Li_2O matrix	[125]
	Nanotubes, nanowires, nanoparticles	890	520 (nanotubes), 500 (nanowires), 475 (nanoparticles), @100 cycles	50 mA/g	XRD indicates the presence of Co phase supporting the reversibility of Li_2O	[87]
	Nanowires; gold nanoparticle-loaded nanowires	890	700 @ 10 cycles; 1000 @ 10 cycles	C/26.5	No postlithiation characterization	[89]
	Nanowire arrays on Ti foils		>300 @ 20 cycles >500 @ 20 cycles >700 @ 20 cycles <400 @ 20 cycles (broken nanowires) >100 @ 20 cycles (powders)	50C 20C 1C 1C 1C	Nanowire arrays remain intact as shown by SEM and TEM	[90]
Fe_2O_3	Copper nanorod array coated with Fe_2O_3; Fe_2O_3 nanorods; nanotubes		<900 @ 50 cycles; 600 @ 1 cycle; 510 @ 100 cycles	8C; 0.2 mA/cm; 100 mA/g	Stable electrode structure after 100 cycles as indicated by the SEM	[94–96]
	Nanowires	781	<700 @ 15 cycles	0.5 mA/cm	Nanowires remain intact with lots of nanoparticles on the periphery of the nanowire as confirmed by the TEM	[93]

Material	Morphology	Capacity @ cycles	Current rate	Characterization	Reference
SnO_2	Nanowires; nanotubes; nanoparticles	>300 @ 50 cycles; <300 @ 50 cycles; <150 @ 50 cycles	100 mA/g	No postlithiation characterization	[91,92]
	Nanorods (15 nm diameter)	700 @ 60 cycles; 400 @ 60 cycles	(5 mV to 1 V); (5 mV to 2 V)	Formation of smaller nanorods and nanoparticles after electrochemical testing	[126]
SnO_2/In_2O_3	Hollow nanostructures	<500 @ 40 cycles	0.2C	No postlithiation characterization	[127]
	Pure phase SnO_2 nanowires; SnO_2 nanowires coated with In_2O_3	400 @ 10 cycles; >700 @ 10 cycles	C/5; C/5	No postlithiation characterization	[128]
a-MoO_3	Nanobelts; lithiated nanobelts	<175 @ 15 cycles; ~225 @ 15 cycles	30 mA/g	No postlithiation characterization	[119]
MnO_2	Nanowires annealed at 100, 300, 500C	<50 @ 100 cycles; >600 @ 100 cycles; <800 @ 100 cycles		Morphology of the sample changes after the first cycle showing that the Li intercalation into the material is not similar to that of graphite	[129]
NiO	Nanoplates	350 @ 20 cycles	111 mA/g	No postlithiation characterization	[130]
CdO	Nanofibers	70 @ 10 cycles	111 mA/g	No postlithiation characterization	[130]
TiO_2	Anatase nanowires	>200 @ 100 cycles	C/2	No postlithiation characterization	[99]
	Nanotubes, nanowires	170 @ 50 cycles (nanotubes) 140 @ 50 cycles (nanowires)	110 mA/g	No postlithiation characterization	[100]
	Ordered tubular array	162 @ 100 cycles; 120 @ 100 cycles; 105 @ 100 cycles	1 mA/g; 10 mA/g; 40 mA/g		[97]
Hydrogen titanates	Nanowires	~200 @ 20 cycles; 166.1 @ 20 cycles; 132.2 @ 20 cycles	300 mA/g; 1250 mA/g; 2500 mA/g	No postlithiation characterization	[131]

high surface area for faster kinetics with Li intercalation/deintercalation and better strain accommodation. Also, at the nanoscale, it has been proven that the above reaction can be reversed. It has also been suggested that the presence of metal nanoparticles can help with the decomposition of otherwise electrochemically inactive Li_2O, making the above reaction reversible. An example reaction scheme is presented below with CoO, which shows the reversible electrochemical mechanism of Li with CoO:

$$CoO + 2Li^+ + 2e^- \leftrightarrow Li_2O + Co$$

$$2Li \leftrightarrow 2Li^+ + 2e^-$$

$$CoO + 2Li \leftrightarrow Li_2O + Co \qquad (15.10)$$

So, there has been a considerable effort with investigating various nanoscale architectures of metal oxides as anode materials. In particular, a number of studies have specifically focused on studying the behavior of nanowire-based metal oxide materials. Of several metal oxide material systems, the nanowires of cobalt oxide (Co_3O_4) [87–90], tin oxide (SnO_2) [13,91–93], iron oxide (Fe_2O_3/Fe_3O_4) [94–96], nickel oxide (NiO), and titania (TiO_2) [97–100] have been widely studied, many of which are summarized in Table 15.2. TiO_2 nanotubular architectures that were investigated as anode materials are shown in Figure 15.21. As indicated in Table 15.2, in all metal oxides, the nanowires/nanotubes seem to hold promise over their micro/nanoparticle counterparts in terms of reversible capacities and cyclability. However, many studies reporting different 1-D architectures for the same metal oxide system have shown a range of values and contradictory reports on the stability. Many of the results shown for various metal oxide systems can be understood with a rational design concept for metal nanocluster covered metal oxide nanowires described in Section 15.4.4.1. Here, various metal oxide nanowires/nanotubes results are discussed.

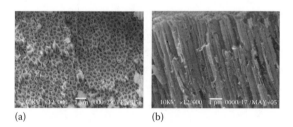

(a)   (b)

FIGURE 15.21
SEM images of (a) TiO_2 nanotubes synthesized using 0.2 µm membranes. (b) Side view of the nanotubes used in the Li-ion battery. (From Wang, K.X. et al., *Adv. Mater.*, 19, 3016, 2007. With permission.)

The Co_3O_4 nanowire/nanotubular structures seem to yield better capacity retention than the nanoparticle counterparts. The reported, stable capacities spanned over a range of values from 400–800 mA h/g [87]. The presence of Au nanoparticles on Co_3O_4 nanowires exhibited even higher capacity values of more than 1000 mA h/g [89]. The Au clusters do not by themselves contribute to the observed increase in capacity but may have played a catalytic role in decomposing Li_2O as well as in increasing the conductivity. The Co_3O_4 nanowire arrays grown on conducting substrates exhibited high stability with cycling while retaining a capacity of over 700 mA h/g even after 50 cycles. The transmission electron microscope (TEM) observations showed a high-density Co metal nanoclusters being present within amorphous matrix that could have helped with the morphological stability during Li alloying/dealloying. In addition, the postlithiation characterization using XRD indicated the presence of CoO, Co, and Li_2O, indicating that the formation reaction of Li_2O was reversible [90].

In the case of SnO_2, several studies investigated the use of SnO_2 nanowires but reported severe capacity fading [91,93]. In almost all the cases using nanowires, the initial charge/discharge capacity was reported to be much higher than 1100 mA h/g, exceeding the theoretical capacity values of both metal and metal oxide [91,93]. Even though it is not completely understood, higher capacity in the initial stages may be due to high available surface area of the nanowires and higher lithium incorporation through the presence of oxygen vacancies. The conventional electrochemical mechanism of Li alloying with SnO_2 is as below [101]:

$$SnO_2 + 4Li^+ + 4e^- \rightarrow 2Li_2O + Sn \qquad (15.11)$$

$$Sn + xLi^+ + xe^- \leftrightarrow Li_xSn \quad (0 \leq x \leq 4.4) \qquad (15.12)$$

The discrepancies in the values reported by different studies on any type of metal oxide nanowires irrespective of minor changes in their morphology and structure can be understood by considering the variations in the presence of unreacted metal. For example, the studies using pure phase SnO_2 nanowires showed severe capacity fading. In this experiment, the SnO_2 nanowires were oxidized again to ensure oxidation of any unreacted metal [102]. The SEM image in Figure 15.22a shows the SnO_2 nanowires. After several cycles of lithiation, the SnO_2 nanowires got converted to become Sn and Li_2O globules and lost the capacity irreversibly to values below 200 mA h/g after 15 cycles as shown in Figure 15.22. This result suggests that pure metal oxide nanowires may not be useful for attaining reversibility and cyclability. As described later, it is shown that the presence of metal and metal clusters helps attain reversibility at high capacities.

New innovative architectures based on nanowire/nanotube arrays are also possible to engineer lower solid state diffusion lengths for Li ion to improve the capacity of metal oxide-based electrodes. For example, Fe_3O_4-coated copper nanorod array on current collector plate exhibited high-capacity

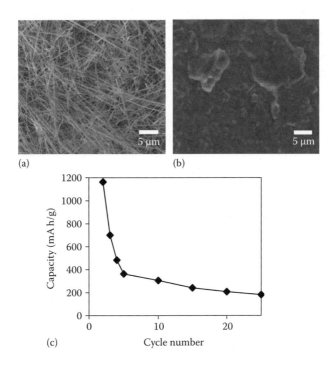

FIGURE 15.22
SEM images in (a) and (b) show SnO_2 nanowire sample before and after lithiation, respectively. (c) Capacity as a function of discharge cycles. (From Meduri, P. et al., *Nano Lett.*, 9, 612, 2009. With permission.)

retention of over 1000 mA h/g over 100 cycles [96]. In addition to solid state diffusion length scales for Li ion, one can also lower the conduction and Li ion diffusion through electrolyte using nanotube-based electrodes. The TiO_2 nanotube arrays coated with carbon black allowed for stable operation with fast charge rates due to improved Li ion diffusion and electronic conduction through carbon black [97]. The performance using CuO and NiO nanowires over their bulk material counterparts has also been reported [103]. However, it is imperative that one needs to develop a more rational design concept for all metal/metal oxide-based materials systems. This is explained later in Section 15.4.4.

15.4.3.3 Nanowires of Silicon and Related Materials

In the case of elemental semiconductors and their compounds, the volume changes associated with Li alloying can be as high as 400% (Si). Si can form a number of intermetallic phases with Li with composition as high as $Li_{22}Si_5$. But, the reactivity of bulk Si with Li is limited to high temperatures. On the other hand, the use of nanocrystalline Si can provide the reactivity with Li at room temperature [104]. Similarly, nanocrystalline Ge can form alloys with Li at room temperature. In addition, the room-temperature diffusivity of Li in Ge is 400 times higher than that of Si, which makes it an attractive material for high power-rate Li-ion batteries [105]. Both Si and Ge exhibit low potential with respect to Li^+/Li, making them attractive candidates for anode applications. Hence, Si, Ge, and their compounds can provide high lithium storage capacities if they can be modified either physically or chemically, to sustain the strain associated with volume changes and decrease the lithium diffusion lengths. In this regard, a number of studies have focused on making nanowires made of Si, Ge, and their compounds with other metals such as tin for lithium ion battery applications. Earlier results using Si nanowires resulted in a large irreversible loss of capacity with cycling owing to the possible presence of oxide sheath. Later studies as explained in Section 15.4.4.2 using Si and Ge arrays grown directly on iron collector substrates have shown reversible capacities as high as 80% of their theoretical values.

In addition to pure Si and Ge nanowires, highly branched, $Sn_{78}Ge_{22}$/carbon core–shell nanowires were produced by thermally annealing butyl capped $Sn_{78}Ge_{22}$ clusters [106]. The electrochemical reactivity of the coin-type cells made of this material at different rates over a range from 0.3C to 8C showed good stability and capacity ranging from 1100 to 1034 mA h/g. At a rate of 0.3C, the capacity of >950 mA h/g was obtained up to 45 cycles with very low-capacity fading. The increased capacity and rate capability has been attributed to that increased surface contact area and reduced lithium diffusion lengths because of the lower diameters on the nanowires, which increase the Li reactions with the nanowire. TEM images of the postlithiated nanowires show stacking faults and microtwins with their fault (1 0 1) plane parallel to the nanowire axis but retain the nanowire architecture. Pristine nanowires are devoid of these defects, which indicate that these arise from the volume changes of Sn and Ge during the lithiation and delithiation process but Sn and Ge are not segregated as evident in the line mapping of the nanowires. The nanowire arrays directly grown on current collector will allow faster carrier transport and strain relaxation during lithium intercalation and deintercalation. This is specifically true with elemental or compound nanowires where no irreversible alloys with Li can result during Li intercalation.

15.4.4 Rational Concepts for Nanowire-Based Architectures

15.4.4.1 A Concept of Nanometal Cluster-Decorated Metal Oxide Nanowires

In this concept, a simple generic design of hybrid structures involving metal nanoclusters covered metal oxide nanowires is shown to provide stable anode materials with high reversible capacity [102]. The design principle for the proposed hybrid structures is that the SnO_2 nanowires are covered with Sn nanoclusters with spacing ~1.4 times the diameter of each cluster as shown in Figure 15.23a. The spacing is necessary to accommodate the

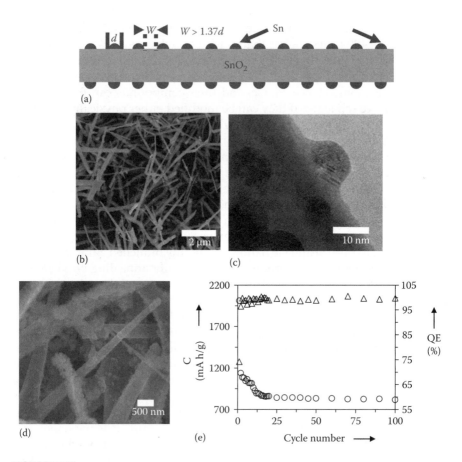

FIGURE 15.23

(a) A schematic showing metal nanocluster decorated metal oxide nanowires as stable anode materials. SEM images showing SnO$_2$ nanowires in three stages: (b) as-synthesized; (c) Sn nanocluster decorated; (d) after 100 charge–discharge cycles; and (e) discharge capacity and coulombic efficiency with cycling for the Sn-nanocluster-covered nanowires. (From Meduri, P. et al., *Nano Lett.*, 9, 612, 2009. With permission.)

volume expansion of Sn cluster during alloying, thereby preventing the Sn agglomeration. The faster electron transport through the underlying SnO$_2$ nanowires is expected to allow for efficient Li alloying and dealloying while the exposed Sn nanoclusters and SnO$_2$ nanowire surfaces serve as Li alloying sites. The SEM image in Figure 15.23b distinctly shows the as-synthesized, SnO$_2$ nanowires with diameters ranging from 50 to 200 nm and microns in length. The high-resolution TEM image in Figure 15.23c shows a SnO$_2$ nanowire covered with 15-nm sized, crystalline Sn nanoclusters evenly spaced from each other. The SEM image in Figure 15.23d distinctly shows unblemished hybrid nanowires after 100 charge–discharge cycles. Figure 15.23e depicts the discharge specific capacity with cycling at 100 mA/g current density, demonstrating that the mechanical stability of the material can be sustained for up to 100 cycles with an exceptional reversible capacity of above 800 mA h/g in a potential window of 0–2.2 V. The capacity fading at a rate of ~1.3% for the initial 15 cycles and ~0.8% after the 15th cycle is considerably lower than that reported for other nanoscale SnO_2 material systems. The hybrid nanoparticle/nanowire concept with Sn and SnO_2 material system can be extended to a wide range of metal oxides, nitrides, and other semiconductors such as Si, Ge.

The titania nanotubes (12 nm diameter) decorated with tin nanoparticles (10 nm in size) are shown to have good reversible capacity retention of 312 mA h/g after 50 cycles between 0.05 and 2.0 V [98]. In an electrochemical window of 0.05 and 1.3 V, capacity retention of 203 mA h/g for up to 50 cycles is observed. The nanotubes accommodate the tin expansion and hence the capacity of the material is believed to be retained with good stability. Of course, one can easily increase the capacity further by increasing the loading of Sn nanoclusters.

15.4.4.2 Nanowire Arrays on Conducting Substrates

The other types of high-capacity materials such as Si, Ge, and their compounds also undergo high volume changes and therefore a number of studies using these materials failed to exhibit high reversible capacity. The nanowires by themselves offer interesting possibilities and the nanowire arrays grown on conducting substrates with appropriate spacing between them seem to accommodate strain while maintaining fast charge transport for fast Li alloying and dealloying kinetics. In this regard, Si pillar arrays fabricated on Si substrates showed that the Si pillars maintain their structural integrity over 50 cycles compared to the micron-scale Si films, which exhibit cracks [107]. In another study, the Si nanowires arrays grown directly on the stainless steel substrates exhibited an initial capacity of 4277 mA h/g at a C/20 rate, equivalent to the theoretical capacity [108]. The discharge capacities remained constant at ~3100 mA h/g after the second cycle for up to 10 cycles. Electrochemical reactivity was conducted at rates of C/20, C/10, C/2, C/5, and 1C, all of which showed good stable capacities (see Figure 15.24b and c). The capacities remained at ~3500 at a rate of C/5 for 20 cycles and >2100 mA h/g at a rate of 1C. The direct contact of the nanowires with the current collector prevailed even after the lithiation and delithiation process, giving rise to minimal capacity fading with cycling. Large volume changes of 400% associated with Si lithiation is shown to be relieved by the widening and the increase in the length of the nanowires, with the nanowires remaining intact without breaking into smaller nanowires and nanoparticles. These results are still preliminary and one needs to understand many aspects about the state of nanowires systems after they are subjected to lithiation.

Similarly, the electrochemical cycling of the Ge nanowires arrays has shown a stable capacity of ~1000 mA h/g for 20 cycles at a C/20 rate [105].

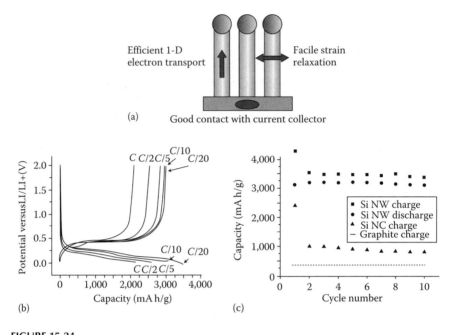

FIGURE 15.24
(a) Schematic showing the concept of Si nanowire array directly on current collector plate to accommodate strain during charge–discharge cycles; (b) first discharge capacity curves at different discharge rates; (c) capacity as a function of cycles. (From Chan, C.K. et al., *Nat. Nanotechnol.*, 3, 31, 2008. With permission.)

Good cyclability and reversibility was also reported at higher rates of C/5, C/2, 1C, and 2C. A stable capacity of ~600 mA h/g at a 2C rate indicates the good Li diffusivity in Ge. The postlithiated Ge nanowires in this particular case are amorphous unlike the previous reports, which report a series of Li–Ge phases during lithiation and delithiation. Along with being amorphous, the nanowire morphology is intact and they are well connected to the current collector, which gives rise to such a high capacity. Also, the volume changes during lithiation are accommodated by the widening and the linear expansion of the nanowire, thus further increasing the stability of the material.

15.4.4.3 Miscellaneous Concepts of 3-D Geometries

The concepts of 3-D geometries for electrode materials and their integration into Li-ion batteries are attracting attention. More recently, 3-D integrated all solid state batteries have shown promise as efficient rechargeable batteries. The 3-D fabrication of the electrode is as shown in Figure 15.25a. Si electrode covered with a solid state LiPON electrolyte showed reversible capacity retention of ~3500 mA h/g until 60 cycles with an excellent stability [109]. A similar Si electrode in a conventional organic Li-ion battery electrode

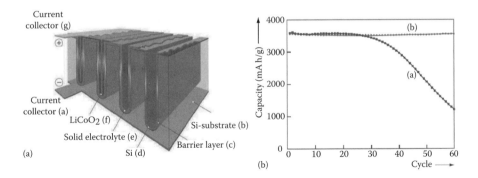

FIGURE 15.25
(a) A schematic showing the concept of 3-D architecture using trenches in Si as anodes for Li-ion battery; (b) plot showing the corresponding capacity as a function of cycles. (From Notten, P.H.L. et al., *Adv. Mater.*, 19, 4564, 2007. With permission.)

showed capacity of >1000 mA h/g for up to 60 cycles with a high capacity fading after the 30th cycle, significantly lower than the solid state battery configuration depicted above in Figure 15.25b. Hence, these concepts of solid state batteries combined along with the hybrid/pure phase nanomaterials will pave ways for significant advancement in battery technology.

15.4.5 Nanowire-Based Materials for Cathodes

There has been a slow development of nanowires/nanotube-based materials for cathodes compared to anode materials. The main motivation for using nanowires/nanotube-based materials for cathodes include the following: (a) short solid state diffusion lengths for Li-ions for higher discharge rates; (b) increased electron transport; (c) higher electrolyte/electrode interfacial area; and (d) possibility of phase stability over a large range of compositions [82]. Of course, the increased contact area and reactivity of nanowires/nanotube-based materials could pose problems in terms of high activity toward electrolyte decomposition to form solid electrolyte interface and dissolution into electrolyte. A number of studies involving VO_x-based nanotubes such as manganese vanadium oxide nanotubes [110], disordered vanadium oxide nanorolls [111], molybdenum-doped vanadium oxide nanotubes [112] and Na^+ ion exchanged vanadium oxide nanotubes [113] showed reversible capacities ranging from 100 to 170 mA h/g after 100 cycles with discharge rates of 50 mA/g. Some of these vanadium oxide nanotubes/nanorolls exhibited initial capacities as high as 250 mA h/g. All the vanadium pentoxide [114], $LiMn_2O_4$ nanotubular arrays [115], $LiCoO_2$, $LiMn_2O_4$, and $LiNi_{0.8}Co_{0.2}O_2$ nanotubes [116] grown on conducting substrates using template-assisted synthesis methods exhibit good cycling behavior with high discharge rate capabilities. In addition, $LiNi_{0.5}Mn_{0.5}O_2$ nanowire arrays using anodized alumina templates have been synthesized [117]. Similarly,

TABLE 15.3

Summary of Studies Using Nanowire-Based Materials for Development of Cathodes in Lithium Ion Batteries

Materials	Description of Starting Materials	Retained Capacity (mA h/g)	Discharge Rate	References
β-FeOOH	Nanowires	278 @ 15 cycles	0.1 mA/cm	[132]
VOOH	Hollow dandelions	125 @ 50 cycles	20 mA/g	[133]
MnO$_2$	Cr^{3+} substituted nanowires	180 @ 30 cycles	20 mA/g	[134]
	Al^{3+} substituted nanowires	>135 @ 50 cycles	20 mA/g	
	Pure phase nanowires	>135 @ 50 cycles	20 mA/g	
α-CuV$_2$O$_6$	Nanowires	>510 @ 1 cycle	20 mA/g	[135]
	Mesowires	>440 @ 1 cycle	20 mA/g	
	Microrods	>400 @ 1 cycle	20 mA/g	
	Bulk particles	>320 @ 1 cycle	20 mA/g	
V$_2$O$_5$	Nanofibers	145 @ 1 cycle	1C	[136]
		108 @ 1 cycle	10C	
		73 @ 1 cycle	40C	
	Nanotubes array	>160 @ 9 cycles		[137]
	Film	>120 @ 9 cycles		

LiCoO$_2$ nanowires were synthesized using reaction between Co$_3$O$_4$ nanowires with lithium containing precursor [118]. In addition, lithiated, a-MoO$_3$ nanobelts [119] retained capacity up to 220 mA h/g after 15 cycles with a discharge rate of 30 mA/g. The a-MoO$_3$ microrods [120] synthesized using vapor phase techniques exhibited high reversible capacities of 199 mA h/g after 100 cycles at 50 mA/g discharge rate. Other interesting material systems are presented in Table 15.3 with the corresponding capacities.

In particular, Li$_x$Mn$_{0.67}$Ni$_{0.33}$O$_2$ nanowires synthesized at low temperatures show a stable capacity of 240 mA h/g for 12 cycles at a rate of 20 mA/g and that the nanowires retain their structural integrity as confirmed by the postlithiation SEM and XRD [121]. Ni substitution increases the stabilization effect, which in turn preserves the material intact after repeated charge–discharge cycles. In another report, Li[(Ni$_{0.8}$Co$_{0.1}$Mn$_{0.1}$)$_{1-x}$]O$_2$ encapsulated by Li[(Ni$_{0.5}$Mn$_{0.5}$)$_x$]O$_2$ is used as a cathode material with x varying between 0 and 1 [122]. For a composition of $x = 0.2$, the electrode performs the best with a value of 190 mA h/g for 40 cycles (rate of 40 mA/g). Similarly layered Li$_{0.88}$[Li$_{0.18}$Co$_{0.33}$Mn$_{0.49}$]O$_2$ nanowires have shown excellent capacity retention of 220 mA h/g at a 15C rate [123]. A desert rose structured LiCoO$_2$ electrode shows a capacity of about 110 mA h/g (7C rate), and about 100 mA h/g (36C rate) compared to the commercial electrode of the same materials, which retains a capacity of 10–20 mA h/g at a 7C rate [124]. All the different material electrodes were cycled for 15 cycles.

References

1. Basic Research Needs for Solar Energy Utilization, Report of the Basic Energy Sciences Workshop on Solar Energy Utilization, April 18–21, 2005, Published by the Office of Science, US Department of Energy.
2. M. Gratzel, *Nature* 414, 338 (2001).
3. K. Zhu, T. B. Vinzant, N. R. Neale, and A. J. Frank, *Nano Lett.* 7, 3739 (2007).
4. M. Adachi, Y. Murata, J. Takao, J. Jiu, M. Sakamoto, and F. Wang, *J. Am. Chem. Soc.* 126, 14943 (2004).
5. V. Shklover, Y. E. Ovchinnikov, L. S. Braginsky, S. M. Zakeeruddin, and M. Gratzel, *Chem. Mater.* 10, 2533 (1998).
6. M. Gratzel, *Pure Appl. Chem.* 73, 459 (2110).
7. J. Jiu, F. Wang, S. Isoda, and M. Adachi, *Chem. Lett.* 34, 1506 (2005).
8. E.P. Emil, E. B. Janice, and S. A. Eray, *Appl. Phys. Lett.* 91, 123116 (2007).
9. K. Pan, Q. Zhang, Q. Wang, Z. Liu, D. Wang, J. Li, and Y. Bai, *Thin Solid Films* 515, 4085 (2007).
10. K. Hiroshi, D. Bin, N. Takayuki, T. Hiroki, and S. Seimei, *Nanotechnology* 18, 165604 (2007).
11. M. Law, L. E. Greene, J. C. Johnson, R. Saykally, and P. Yang, *Nat. Mater.* 4, 455 (2005).
12. J. B. Baxter and E. S. Aydil, *Appl. Phys. Lett.* 86, 053114-1 (2005).
13. D. I. Suh, S. Y. Lee, T. H. Kim, J. M. Chun, E. K. Suh, O. B. Yang, and S. K. Lee, *Chem. Phys. Lett.* 442, 348 (2007).
14. Y. Gao, M. Nagai, T. C. Chang, and J. J. Shyue, Cryst. *Growth Des.* 7, 2467 (2007).
15. C. Y. Jiang, X. W. Sun, K. W. Tan, G. Q. Lo, A. K. K. Kyaw, and D. L. Kwong, *Appl. Phys. Lett.* 92, 143101 (2008).
16. C. Y. Jiang, X. W. Sun, G. Q. Lo, D. L. Kwong, and J. X. Wang, *Appl. Phys. Lett.* 90, 263501 (2007).
17. S. Gubbala, V. Chakrapani, V. Kumar, and M. K. Sunkara, *Adv. Funct. Mater.* 18, 2411 (2008).
18. E. Joanni, R. Savu, M. de Sousa Góes, P. R. Bueno, J. N. de Freitas, A. F. Nogueira, E. Longo, and J. A. Varela, *Scr. Mater.* 57, 277 (2007).
19. S. Ngamsinlapasathian, S. Sakulkhaemaruethai, S. Pavasupree, A. Kitiyanan, T. Sreethawong, Y. Suzuki, and S. Yoshikawa, *J. Photochem. Photobiol., A* 164, 145 (2004).
20. Y. Ohsaki, N. Masaki, T. Kitamura, Y. Wada, T. Okamoto, T. Sekino, K. Niihara, and S. Yanagida, *Phys. Chem. Chem. Phys.* 7, 4157 (2005).
21. G. K. Mor, K. Shankar, M. Paulose, O. K. Varghese, and C. A. Grimes, *Nano Lett.* 6, 215 (2006).
22. K. Zhu, N. R. Neale, A. Miedaner, and A. J. Frank, *Nano Lett.* 7, 69 (2007).
23. K. Shankar, J. Bandara, M. Paulose, H. Wietasch, O. K. Varghese, G. K. Mor, T. J. LaTempa, M. Thelakkat, and C. A. Grimes, *Nano Lett.* 8, 1654 (2008).
24. K. Shankar, G. K. Mor, H. E. Prakasam, O. K. Varghese, and C. A. Grimes, *Langmuir* 23, 12445 (2007).
25. D. Kuang, J. Brillet, P. Chen, M. Takata, S. Uchida, H. Miura, K. Sumioka, S. M. Zakeeruddin, and M. Gratzel, *ACS Nano* 2, 1113 (2008).
26. A. B. F. Martinson, J. W. Elam, J. T. Hupp, and M. J. Pellin, *Nano Lett.* 7, 2183 (2007).

27. M. Mozetic, U. Cvelbar, M. K. Sunkara, and S. Vaddiraju, *Adv. Mater.* 17, 2138 (2005).
28. H. Jia, H. Xu, Y. Hu, Y. Tang, and L. Zhang, *Electrochem. Commun.* 9, 354 (2007).
29. K. S. Leschkies, R. Divakar, J. Basu, E. Enache-Pommer, J. E. Boercker, C. B. Carter, U. R. Kortshagen, D. J. Norris, and E. S. Aydil, *Nano Lett.* 7, 1793 (2007).
30. P. Ravirajan, A. M. Peiro, M. K. Nazeeruddin, M. Graetzel, D. D. C. Bradley, J. R. Durrant, and J. Nelson, *J. Phys. Chem. B* 110, 7635 (2006).
31. D. C. Olson, S. E. Shaheen, R. T. Collins, and D. S. Ginley, *J. Phys. Chem. C* 111, 16670 (2007).
32. J. B. Baxter and E. S. Aydil, *Sol. Energy Mater. Sol. Cells* 90, 607 (2006).
33. B. Tan and Y. Wu, *J. Phys. Chem. B* 110, 15932 (2006).
34. M. Law, L. E. Greene, A. Radenovic, T. Kuykendall, J. Liphardt, and P. Yang, *J. Phys. Chem. B* 110, 22652 (2006).
35. E. Galoppini, J. Rochford, H. Chen, G. Saraf, Y. Lu, A. Hagfeldt, and G. Boschloo, *J. Phys. Chem. B* 110, 16159 (2006).
36. A. B. F. Martinson, J. E. McGarrah, M. O. K. Parpia, and J. T. Hupp, *Phys. Chem. Chem. Phys.* 8, 4655 (2006).
37. A. P. Goodey, S. M. Eichfeld, K. K. Lew, J. M. Redwing, and T. E. Mallouk, *J. Am. Chem. Soc.* 129, 12344 (2007).
38. J. R. Maiolo, B. M. Kayes, M. A. Filler, M. C. Putnam, M. D. Kelzenberg, H. A. Atwater, and N. S. Lewis, *J. Am. Chem. Soc.* 129, 12346 (2007).
39. K. Peng, X. Wang, and S. T. Lee, *Appl. Phys. Lett.* 92, 163103 (2008).
40. J. M. Spurgeon, H. A. Atwater, and N. S. Lewis, *J. Phys. Chem. C* 112, 6186 (2008).
41. W. U. Huynh, J. J. Dittmer, and A. P. Alivisatos, *Science* 295, 2425 (2002).
42. K. Peng, Y. Xu, Y. Wu, Y. Yan, S. T. Lee, and J. Zhu, *Small* 1, 1062 (2005).
43. F. Hui, L. Xudong, S. Shuang, X. Ying, and Z. Jing, *Nanotechnology* 19, 255703 (2008).
44. M. D. Kelzenberg, D. B. Turner-Evans, B. M. Kayes, M. A. Filler, M. C. Putnam, N. S. Lewis, and H. A. Atwater, *Nano Lett.* 8, 710 (2008).
45. B. Tian, X. Zheng, T. J. Kempa, Y. Fang, N. Yu, G. Yu, J. Huang, and C. M. Lieber, *Nature* 449, 7164 (2007).
46. H. Gerischer and B. O. Serahin, Ed., Topics in *Appl. Phys.*, 31, 115 (1979).
47. N. Beermann, L. Vayssieres, S. E. Lindquist, and A. Hagfeldt, *J. Electrochem. Soc.* 147, 2456 (2000).
48. Z. Chen, U. Cvelbar, M. Mozetic, J. He, and M. K. Sunkara, *Chem. Mater.* 20, 3224 (2008).
49. U. Björksten, J. Moser, and M. Grätzel, *Chem. Mater.* 6, 858 (1994).
50. R. Shinar and J. H. Kennedy, *J. Electrochem. Soc.* 130, 860 (1983).
51. R. Shinar and J. H. Kennedy, *Sol. Energy Mater.* 6, 323 (1982).
52. T. Lindgren, H. L. Wang, N. Beermann, L. Vayssieres, A. Hagfeldt, and S. E. Lindquist, *Sol. Energy Mater. Sol. Cells* 71, 231 (2002).
53. L. Vayssieres, C. Sathe, S. M. Butorin, D. K. Shuh, J. Nordgren, and J. H. Guo, *Adv. Mater.* 17, 2320 (2005).
54. S. U. M. Khan and T. Sultana, *Sol. Energy Mater. Sol. Cells* 76, 211 (2003).
55. J. Jitputti, Y. Suzuki, and S. Yoshikawa, *Catal. Commun.* 9, 1265 (2008).
56. J. H. Park, S. Kim, and A. J. Bard, *Nano Lett.* 6, 24 (2006).
57. K. S. Ahn, S. Shet, T. Deutsch, C. S. Jiang, Y. F. Yan, M. Al-Jassim, and J. Turner, *J. Power Sources* 176, 387 (2008).

58. J. D. Holmes, K. P. Johnston, R. C. Boty, and B. A. Korgel, *Science* 287, 1471 (2000).
59. L. Vayssieres, N. Beermann, S. E. Lindquist, and A. Hagfeldt, *Chem. Mater.* 13, 233 (2001).
60. Y. Y. Fu, R. M. Wang, J. Xu, J. Chen, Y. Yan, A. Narlikar, and H. Zhang, *Chem. Phys. Lett.* 379, 373 (2003).
61. T. Yu, Y. W. Zhu, X. J. Xu, K. S. Yeong, Z. X. Shen, P. Chen, C. T. Lim, J. T. L. Thong and C. H. Sow, *Small* 2, 80 (2006).
62. C. H. Kim, H. J. Chun, D. S. Kim, S. Y. Kim, J. Park, J. Y. Moon, G. Lee, J. Yoon, Y. Jo, M. H. Jung, S. I. Jung, and C. J. Lee, *Appl. Phys. Lett.* 89, 223103 (2006).
63. R. M. Wang, Y. F. Chen, Y. Y. Fu, H. Zhang, and C. Kisielowski, *J. Phys. Chem. B* 109, 12245 (2005).
64. X. G. Wen, S. H. Wang, Y. Ding, Z. L. Wang, and S. H. Yang, *J. Phys. Chem. B* 109, 215 (2005).
65. Y. L. Chueh, M. W. Lai, J. Q. Liang, L. J. Chou, and Z. L. Wang, *Adv. Funct. Mater.* 16, 2243 (2006).
66. U. Cvelbar, Z. Q. Chen, M. K. Sunkara, and M. Mozetic, *Small* 4, 1610 (2008).
67. O. Khaselev and J. A. Turner, *Science* 280, 425 (1998).
68. O. K. Varghese and C. A. Grimes, *Sol. Energy Mater. Sol. Cells* 92, 374 (2008).
69. P. V. Kamat, K. G. Thomas, S. Barazzouk, G. Girishkumar, K. Vinodgopal, and D. Meisel, *J. Am. Chem. Soc.* 126, 10757 (2004).
70. Y. Wei, C. Xue, K. A. Dean, and B. F. Coll, *Appl. Phys. Lett.* 79, 4527 (2001).
71. Y. Avigal and R. Kalish, *Appl. Phys. Lett.* 78, 2291 (2001).
72. A. Srivastava, A. K. Srivastava, and O. N. Srivastava, *Appl. Phys. Lett.* 72, 1685 (1998).
73. J. Thangala, S. Vaddiraju, R. Bogale, R. Thurman, T. Powers, B. Deb, and M. K. Sunkara, *Small* 3, 890 (2007).
74. S. Gubbala, J. Thangala, and M. K. Sunkara, *Sol. Energy Mater. Sol. Cells* 91, 813 (2007).
75. C. G. Granqvist, *Sol. Energy Mater. Sol. Cells* 60, 201 (2000).
76. G. Suresh, PhD Dissertation, 88 (2008).
77. C. C. Liao, F. R. Chen, and J. J. Kai, *Sol. Energy Mater. Sol. Cells* 90, 1147 (2006).
78. S. J. Yoo, J. W. Lim, Y. E. Sung, Y. H. Jung, H. G. Choi, and D. K. Kim, *Appl. Phys. Lett.* 90, 173126 (2007).
79. K. C. Cheng, F. R. Chen, and J. J. Kai, *Sol. Energy Mater. Sol. Cells* 90, 1156 (2006).
80. C. Xiong, A. E. Aliev, B. Gnade, and K. J. Balkus, *ACS Nano* 2, 293 (2008).
81. P. G. Bruce, B. Scrosati, and J. M. Tarascon, *Angew. Chem. Int. Ed.* 47, 2930 (2008).
82. A. S. Arico, P. Bruce, B. Scrosati, J. M. Tarascon, and W. Van Schalkwijk, *Nat. Mater.* 4, 366 (2005).
83. B. Gao, A. Kleinhammes, X. P. Tang, C. Bower, L. Fleming, Y. Wu, and O. Zhou, *Chem. Phys. Lett.* 307, 153 (1999).
84. X. X. Wang, J. N. Wang, H. Chang, and Y. F. Zhang, *Adv. Funct. Mater.* 17, 3613 (2007).
85. C. Kim, K. S. Yang, M. Kojima, K. Yoshida, Y. J. Kim, Y. A. Kim, and M. Endo, *Adv. Funct. Mater.* 16, 2393 (2006).
86. A. N. Dey, *J. Electrochem. Soc.* 118, 1547 (1971).
87. W. Y. Li, L. N. Xu, and J. Chen, *Adv. Funct. Mater.* 15, 851 (2005).

88. N. Du, H. Zhang, B. Chen, J. B. Wu, X. Y. Ma, Z. H. Liu, Y. Q. Zhang, D. Yang, X. H. Huang, and J. P. Tu, *Adv. Mater.* 19, 4505 (2007).

89. K. T. Nam, D. W. Kim, P. J. Yoo, C. Y. Chiang, N. Meethong, P. T. Hammond, Y. M. Chiang, and A. M. Belcher, *Science* 312, 885 (2006).

90. Y. G. Li, B. Tan, and Y. Y. Wu, *Nano Lett.* 8, 265 (2008).

91. M. S. Park, Y. M. Kang, G. X. Wang, S. X. Doti, and H. K. Liu, *Adv. Funct. Mater.* 18, 455 (2008).

92. M. S. Park, G. X. Wang, Y. M. Kang, D. Wexler, S. X. Dou, and H. K. Liu, *Angew. Chem. Int. Ed.* 46, 750 (2007).

93. Z. Ying, Q. Wan, H. Cao, Z. T. Song, and S. L. Feng, *Appl. Phys. Lett.* 87 (2005).

94. C. Z. Wu, P. Yin, X. Zhu, C. Z. OuYang, and Y. Xie, *J. Phys. Chem. B* 110, 17806 (2006).

95. J. Chen, L. N. Xu, W. Y. Li, and X. L. Gou, *Adv. Mater.* 17, 582 (2005).

96. L. Taberna, S. Mitra, P. Poizot, P. Simon, and J. M. Tarascon, *Nat. Mater.* 5, 567 (2006).

97. K. X. Wang, M. D. Wei, M. A. Morris, H. S. Zhou, and J. D. Holmes, *Adv. Mater.* 19, 3016 (2007).

98. Z. W. Zhao, Z. P. Guo, D. Wexler, Z. F. Ma, X. Wu, and H. K. Liu, *Electrochem. Commun.* 9, 697 (2007).

99. G. Armstrong, A. R. Armstrong, P. G. Bruce, P. Reale, and B. Scrosati, *Adv. Mater.* 18, 2597 (2006).

100. Q. Wang, Z. H. Wen, and J. H. Li, *Inorg. Chem.* 45, 6944 (2006).

101. C. R. Sides, N. C. Li, C. J. Patrissi, B. Scrosati, and C. R. Martin, *MRS Bull.* 27, 604 (2002).

102. P. Meduri, C. Pendyala, V. Kumar, G. U. Sumanasekera, and M. K. Sunkara, *Nano Lett.* 9, 612 (2009).

103. J. P. Liu, Y. Y. Li, X. T. Huang, G. Y. Li, and Z. K. Li, *Adv. Funct. Mater.* 18, 1448 (2008).

104. B. Gao, S. Sinha, L. Fleming, and O. Zhou, *Adv. Mater.* 13, 816 (2001).

105. C. K. Chan, X. F. Zhang, and Y. Cui, *Nano Lett.* 8, 307 (2008).

106. H. Lee and J. Cho, *Nano Lett.* 7, 2638 (2007).

107. M. Green, E. Fielder, B. Scrosati, M. Wachtler, and J. S. Moreno, *Electrochem. Solid-State Lett.* 6, A75 (2003).

108. C. K. Chan, H. L. Peng, G. Liu, K. McIlwrath, X. F. Zhang, R. A. Huggins, and Y. Cui, *Nat. Nanotechnol.* 3, 31 (2008).

109. P. H. L. Notten, F. Roozeboom, R. A. H. Niessen, and L. Baggetto, *Adv. Mater.* 19, 4564 (2007).

110. A. Dobley, K. Ngala, S. F. Yang, P. Y. Zavalij, and M. S. Whittingham, *Chem. Mater.* 13, 4382 (2001).

111. S. Nordlinder, K. Edstrom, and T. Gustafsson, *Electrochem. Solid-State Lett.* 4, A129 (2001).

112. L. Q. Mai, W. Chen, Q. Xu, J. F. Peng, and Q. Y. Zhu, *Chem. Phys. Lett.* 382, 307 (2003).

113. S. Nordlinder, J. Lindgren, T. Gustafsson, and K. Edstrom, *J. Electrochem. Soc.* 150, E280 (2003).

114. C. J. Patrissi and C. R. Martin, *J. Electrochem. Soc.* 148, A1247 (2001).

115. N. C. Li, C. J. Patrissi, G. L. Che, and C. R. Martin, *J. Electrochem. Soc.* 147, 2044 (2000).

116. X. X. Li, F. Y. Cheng, B. Guo, and J. Chen, *J. Phys. Chem. B* 109, 14017 (2005).

117. Y. K. Zhou and H. L. Li, *J. Mater. Chem.* 12, 681 (2002).
118. F. Jiao, K. M. Shaju, and P. G. Bruce, *Angew. Chem. Int. Ed.* 44, 6550 (2005).
119. L. Q. Mai, B. Hu, W. Chen, Y. Y. Qi, C. S. Lao, R. S. Yang, Y. Dai, and Z. L. Wang, *Adv. Mater.* 19, 3712 (2007).
120. W. Y. Li, F. Y. Cheng, Z. L. Tao, and J. Chen, *J. Phys. Chem.* B 110, 119 (2006).
121. D. H. Park, S. T. Lim, S. J. Hwang, C. S. Yoon, Y. K. Sun, and J. H. Choy, *Adv. Mater.* 17, 2834 (2005).
122. Y. K. Sun, S. T. Myung, B. C. Park, and K. Amine, *Chem. Mater.* 18, 5159 (2006).
123. Y. Lee, M. G. Kim, and J. Cho, *Nano Lett.* 8, 957 (2008).
124. H. Chen, and C. P. Grey, *Adv. Mater.* 20, 2206 (2008).
125. P. Poizot, S. Laruelle, S. Grugeon, L. Dupont, and J. M. Tarascon, *Nature* 407, 496 (2000).
126. Y. Wang and J. Y. Lee, *J. Phys. Chem.* B 108, 17832 (2004).
127. X. W. Lou, Y. Wang, C. L. Yuan, J. Y. Lee, and L. A. Archer, *Adv. Mater.* 18, 2325 (2006).
128. D. W. Kim, I. S. Hwang, S. J. Kwon, H. Y. Kang, K. S. Park, Y. J. Choi, K. J. Choi, and J. G. Park, *Nano Lett.* 7, 3041 (2007).
129. M. S. Wu, P. C. J. Chiang, J. T. Lee, and J. C. Lin, *J. Phys. Chem.* B 109, 23279 (2005).
130. Y. G. Li, B. Tan, and Y. Y. Wu, *Chem. Mater.* 20, 567 (2008).
131. J. R. Li, Z. L. Tang, and Z. T. Zhang, *Chem. Mater.* 17, 5848 (2005).
132. Y. Xiong, Y. Xie, S. Chen, and Z. Li, *Chem. Eur. J.* 9, 4991 (2003).
133. C. Z. Wu, Y. Xie, L. Y. Lei, S. Q. Hu, and C. Z. OuYang, *Adv. Mater.* 18, 1727 (2006).
134. D. H. Park, S.H. Lee, T. W. Kim, S. T. Lim, S.J. Hwang, Y. S. Yoon, Y.H. Lee, and J. H. Choy, *Adv. Funct. Mater.* 17, 2949 (2007).
135. H. Ma, S. Zhang, W. Ji, Z. Tao, and J. Chen, *J. Am. Chem. Soc.* 130, 5361 (2008).
136. C. R. Sides and C. R. Martin, *Adv. Mater.* 17, 125 (2005).
137. Y. Wang, and G. Cao, *Chem. Mater.* 18, 2787 (2006).
138. M. K. Brendan, A. A. Harry, and S. L. Nathan, *J. Appl. Phys.* 97, 114302 (2005).
139. J. M. Tarascon and M. Armand, *Nature* 414, 359 (2001).

16

Other Applications

Besides the applications in electronics, optoelectronics, sensors, and energy production/storage devices, inorganic nanowires have been considered in several other fields as well. Field emission technology for flat panel displays and other related applications is an interesting opportunity for which carbon nanotubes (CNTs) have received a great deal of attention in the last decade. The literature also features reports on the use of some nanowires for this purpose, which is covered in this chapter. Nanowires of silicon, indium antimonide, and bismuth telluride find unique possibilities in the fabrication of efficient thermoelectric (TE) devices for both power generation and refrigeration.

16.1 Field Emission Devices

16.1.1 Background

When subjected to a high electric field, electrons near the Fermi level in some materials can overcome the energy barrier to escape to the vacuum level. This is commonly called field emission and the emitter itself is sometimes referred to as cold cathode to contrast it with thermionic emission. In the case of the latter, a source such as tungsten is heated to 1000°C or above to give the electrons the energy needed to overcome the surface potential barrier. This high-temperature operation may not always be advisable in any application. Common (field) emitter materials include silicon, molybdenum, diamond, and CNTs. Potential applications include all those which need an electron source and where thermionic emission currently dominates such as cathode ray lighting elements, flat panel displays, gas discharge tubes in telecom networks, electron guns in electron microscopy, and microwave amplifiers.

Field emission has long been pursued as an alternative to thermionic emission in many applications due to the low-temperature operation, anticipated high reliability, and low maintenance. In display applications, field emission is also amenable for large-scale production as witnessed in microelectronics and other competing display technologies. In the 1980s and 1990s, diamond received much attention as a field emitter material. However, the desire to obtain large field enhancement factor β (to be defined later) required etching high aspect ratio emitters out of patterned diamond thin films, which turned out to be difficult and expensive and as a result, commercialization efforts failed. Various forms of CNTs have been investigated in the last 5–7 years, driven by the desire to replace the liquid crystal and plasma displays in television applications. While the results in the literature are rather impressive for CNT displays, the current bottlenecks all appear to be related to manufacturing, particularly issues involving deposition of nanotubes on large-size glass substrates and ensuring uniform and high emission site intensity. Recently, there has been an increased focus in the literature on various nanowires also for producing field emission devices. The motivation arises from the fact that nanostructures with reduced dimensions exhibit superior field emission properties such as lower turn on voltages and higher emission currents. Nanostructures with high aspect ratios, lower work functions, high thermal and electrical conductivity, and high stability are potential field emitters. Field emission mainly depends on characteristics and geometry of the emitter. Thus the materials with high aspect ratio and sharp edges result in high emission currents and high field enhancement factor β [1].

Field emission in materials is described by the Fowler–Nordhiem equation given below:

$$I = aV^2 \exp\left[-b\phi^{1.5}/\beta V\right] \qquad (16.1)$$

where
   I is the emission current
   V is the voltage
   ϕ is the work function of the emitter
   β is a field enhancement factor
   a and *b* are constants

The measurements are usually presented in the form of $\ln(I/V^2)$ versus $(1/V)$ and in this mode of presentation, the plot is linear at low emission levels and in the high field region, the current usually saturates. The expectations from a good emitter include low threshold electric field (defined as the field required to yield $10\,mA/cm^2$), high current density, and high emission site density (for high resolution displays). Display applications demand current densities in the $1–10\,mA/cm^2$ range and microwave amplifiers require greater than $500\,mA/cm^2$. A low work function material is desirable to obtain a low threshold field. A large field enhancement factor is desirable, which depends on the geometry of the emitter, and is found to be proportional to $1/r$, where *r* is the tip radius of curvature.

16.1.2 Work Function (Φ)

Work function is the minimum amount of energy required to move the electrons from the Fermi level of a material into the vacuum. Figure 16.1 illustrates the band diagram for a typical semiconductor. Field emission from any

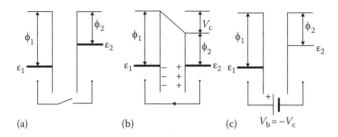

FIGURE 16.1
Schematic illustration of the contact potential between a material of interest with respect to a reference electrode: (a) relative band diagrams of sample and reference; (b) relative band diagrams under applied potential (V_b); and (c) relative band diagrams under null point ($V_c = -V_b$). (Courtesy of M. Meyyappan.)

material mainly depends on its work function. Work function of a material is the surface property, thus any change in the surface morphology or adsorption of oxygen, water vapor, and other impurities will strongly influence the work function value and in turn affect the field emission characteristics.

The work function values for many materials are reported for bulk materials with little or no measurements reported for nanowires or individual nanostructures. Thus it is necessary to determine the work function of nanowire-based materials to understand the resulting field emission characteristics. One can use a Kelvin probe to determine the work function value both under vacuum and ambient environments. In this approach, the surfaces of the sample of interest and a vibrating reference electrode (with work function Φ_1) are brought parallel to each other. When an AC signal is applied between the surfaces, the scanned amplitude varies linearly against the backing potential (V_b). At the null point ($-I = 0$ A), the contact potential difference (V_c) can be obtained as $V_c = -V_b$. The work function of the sample can be calculated as

$$\Phi_2 = \Phi_1 + V_c \tag{16.2}$$

where Φ_1 and Φ_2 are the work functions of the reference electrode and sample of interest.

16.1.3 Field Emission Testing

A field emission test apparatus is easy to construct in the laboratory and consists of a glass chamber or cell to house the cathode and anode under a high vacuum of 10^{-9} to 10^{-8} Torr. The cathode may consist of a glass or polytetrafluoroethylene substrate with metal-patterned lines and the emitter material may be transferred to the substrate or grown directly on it. The anode is located at 20–500 μm from the cathode and coated with a phosphor material. The test may be conducted in a diode or triode mode as shown in

Figure 16.2. In the triode configuration, a gate is used to modulate the current, independent of the acceleration voltage. In the diode configuration, a high voltage is typically needed and the gap, d, may need to be adjusted as well.

Figure 16.3 shows sample plot of current (I) versus electric field (E) for tungsten oxide nanowires for a gap of 12.5 μm, and (a) the plot of $\ln(I/E^2)$ versus ($1/E$), depicting a linear relationship indicative of field emission. The emission characteristics, in addition to being dependent on emitter material selection, ϕ, β, etc., are sensitive to the nature of material preparation, clean emitting sites versus adsorbates such as water vapor, oxygen, etc., film microstructure, and screening effect arising from interference from neighboring emission sites. Adsorbates can alter the material work function and affect the emission characteristics. In the long term, this will pose a reliability problem as well and that is the reason for using high vacuum in emission

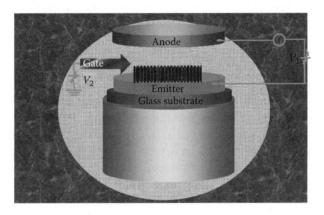

FIGURE 16.2
Schematic depicting the experimental set up for field emission measurements in a diode or triode mode. (Courtesy of M. Meyyappan.)

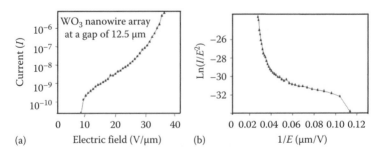

FIGURE 16.3
Sample plot of (a) current (I) versus electric field (E) for tungsten oxide nanowires at a gap (d) of 12.5 μm, and (b) the plot of $\ln(I/E^2)$ versus ($1/E$), depicting the linear relationship of the plot indicative of field emission. (Courtesy of M. Meyyappan.)

devices. The screening effect is illustrated with the aid of Figure 16.4, which shows simulated electric field surrounding the emitter. It is clear when emitters are placed close to each other, the electric field surrounding a given emitter changes significantly, which will have an impact on the turn-on field. Under otherwise identical conditions, an isolated individual emitter

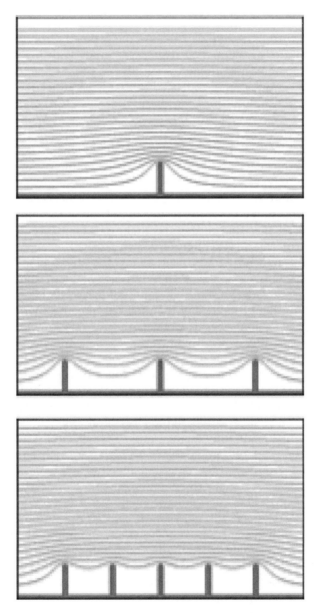

FIGURE 16.4
Field screening effect with an array of emitters. (Courtesy of C. Nguyen.)

will asymptotically have a lower turn-on field due to reduced field screening than the one with neighbors nearby. On the other hand, an array of emitters would be needed in most applications to increase the current density and therefore, a balance is required between emitter density and spacing.

16.1.4 Field Emission Characteristics of Nanowire-Based Materials

The emission current mainly depends on the geometry and the work function of the emitter as mentioned earlier. Materials with high aspect ratio, lower work functions, and aligned vertically tend to produce high emission currents, such as CuO nanoneedles [2] with a work function of 0.29–2.8 eV. Their high aspect ratio (from very sharp curvature tips) and well-aligned nature result in much lower turn-on voltages as low as 0.5 V/µm. A summary of CuO nanostructures field emission characteristics along with the results for other nanowires is given in Table 16.1. Not all the materials with low work function can be used as field emitters; for example, cesium with a work function of 1.87 eV poses serious problems related to stability and lifetime [1]. On the other hand ZnS nanobelts [3], with a work function of 7 eV, which is higher than other nanostructures, results in improved field emission properties such as lower turn-on voltages and higher field enhancement factor due to their high aspect ratio and alignment.

Different ZnO nanostructures have been widely studied for field emission (shown in Table 16.1) due to their rich morphology [1]. ZnO nanobelts of several millimeters in length yield high values of β, 1.4×10^4 [4]. Field emission also depends on the properties of the material in contact with the emitter. For example, ZnO nanowires on carbon cloth have very low turn-on voltages and higher β (4.0×10^4), which might be due to the extra emission sites from the carbon cloth [5].

GaN, as a semiconducting material with low electron affinity (2.7–3.3 eV) and strong chemical and mechanical properties, has gained attention for its field emission potential [6]. GaN nanorod films on a silicon substrate show very low turn-on voltages compared to CNTs and ZnO nanostructures due to their geometry, crystalline quality, and lower electron affinity [6]. AlN, with its stability at high temperatures and a very small electron affinity value of 0.25 eV results in high emission currents and superior field emission characteristics. A film of AlN nanowires with diameters of 15 nm show low threshold fields and high enhancement factor (8.2×10^4), attributed to smaller diameters of the wires [7]. In another study, Si-doped AlN nanoneedles show very low turn-on voltages (1.8 V/µm) and high enhancement factor (3.2×10^3), attributed to the high aspect ratio, small curvature of the tips and silicon doping, which lowers the resistance of AlN and increases the carrier concentration [8]. WO_3-based field emitters have received attention due to their large surface area and high aspect ratio. As seen in Table 16.1, quasialigned nanowires on tungsten foil have lower turn-on voltages of 2.6 V/µm [9], which is lower than most of the other nanostuctures.

TABLE 16.1

Summary of Structural and Field Emission Characteristics of Various Forms of Nanostructures Including CuO, Cu₂S, CoO, ZnO, GaN, GaAs, AlN, WO₃, Si, SiC, ZnS, and α-Fe₂O₃

Emitter	Morphology	Synthesis Method	Turn-On Field (V/µm)	Threshold Field (V/µm)	Field Enhancement Factor (β)	Work Function (Φ)	Reference
CuO	Vertically aligned high-density nanowire film on Cu substrate	Vapor–solid reaction on copper plates	3.5–4.5	0.5 mA/cm² at 7 V/µm	1570	2.5–2.8	[16]
	Vertical nanorods (low aspect ratio) with preferential orientation on a Cu substrate	Electrodeposition and self-catalytic mechanism of copper element	6–7	>0.2 mA/cm² at 11 V/µm	—		[17]
	Vertically aligned nanobelt film with individual lengths (10 µm) and width (20–150 nm) on Cu foil	Prepared using aqueous solution on copper foils	11	—	—		[18]
	High-density hybrid CuO and CuCO₃ (400 nm) nanostructures, 2.7 µm long and 160 nm diameter	Simple heating of copper-coated carbon paper in ambient	2.1	1 mA/cm² at 3.2 V/µm	2960		[19]
	Well-aligned, high-density CuO nanoneedles on a Cu substrate	Needles grown through thermal oxidation in air	0.5	1 mA/cm² at 21 V/µm	—	0.2–1.0	[2]
Cu₂S	Arrays consist of straight Cu₂S nanowires with dia. 50–70 nm on Cu foils	Gas solid reaction on copper foils		100 µA/cm² at 11 V/µm	—		[20]
CdS	Well-aligned nanowires (lengths of 700 nm and 30 nm dia.) on quartz plates	Hot wall metal organic chemical vapor deposition	7.8	1 mA/cm² at 12.2 V/µm	397	4.2	[21]

(continued)

TABLE 16.1 (continued)

Summary of Structural and Field Emission Characteristics of Various Forms of Nanostructures Including CuO, Cu$_2$S, CoO, ZnO, GaN, GaAs, AlN, WO$_3$, Si, SiC, ZnS, and α-Fe$_2$O$_3$

Emitter	Morphology	Synthesis Method	Turn-On Field (V/μm)	Threshold Field (V/μm)	Field Enhancement Factor (β)	Work Function (Φ)	Reference
Co$_3$O$_4$	Vertically aligned nanowires 2–10 μm long, 10–50 nm diameter on cobalt foil	Thermal oxidation of cobalt foil in air	6.4	50 μA/cm² at 7 V/μm	735	4.5	[22]
	Free standing nanowires of uniform dimensions (2.1 μm long, 20 nm at apex) and predetermined densities on silicon substrate	Electrodeposition into nanopores of track-etched polymer layers on silicon substrate	—	1 mA/cm² at 14 V/μm	211	5	[23]
	Free standing two-dimensional vertically aligned nanowalls	Thermal oxidation of cobalt foil in air	6	25 μA/cm² at 11 V/μm	1118 ± 41	4.5	[24]
ZnO	Quasialigned straight nanowire arrays uniform over the entire the Si substrate	Fabricated on Au-coated silicon wafer by heating a mixture of ZnO and graphite powders	7.4	1 mA/cm² at 13 V/μm	1028	5.3	[25]
	Medium-density nanowires (0.7–1 μm long, 30–40 nm dia.) vertically aligned with no bendings or interconnects between nanowires	Vapor–solid–liquid growth	8–10	20 μA/cm² at 10–13 V/μm	860	5.2	[26]
	Low-density, long nanowires (10 μm), vertically grown on carbon cloth	Carbothermal vapor transport and condensation approach (vapor–solid self-catalyzing mechanism)	—	1 mA/cm² at 0.7 V/μm	4.0×10⁴	5.3	[27]

Low-density, long nanowires (5–10 μm), on carbon cloth	Thermal evaporation (vapor–solid self-catalyzed mechanism)	0.1 μA/cm² at 0.2 V/μm	1 mA/cm² at 0.7 V/μm	4.1×10^4	5.4	[5]
Well-hexagonal faceted penholders (diameter 200 nm) with diameter of pen tips on the nanopencils in the range of 20–30 nm on a silicon wafer (randomly aligned)	Two-step pressure controlled thermal evaporation with no catalyst	3.7	1.3 mA/cm² at 4.6 V/μm	2300	5.3	[28]
High-density, vertically aligned nanonails uniformly grown on Si substrate (base—700 nm, cone shaped tip—300 nm)	Modified thermal evaporation process without using a catalyst or predeposited buffer layer	7.9	—	—	—	[29]
High-density, vertically aligned nanopencils 8 μm hexagonal stem with small diameter tip (50 nm) on Si substrate	Modified thermal evaporation process without using a catalyst or predeposited buffer layer	7.2	—	—	—	[29]
Arrays of nanopencils (5 μm long with tip diameter of 10–20 nm), uniformly distributed forming grass clusters on pyramidal Si substrate	Thermal evaporation method	3.8	1 mA/cm² at 5.8 V/μm	2776	5.3	[30]
Array of nanotowers (5 μm long) with needle tip (less than 10 nm) uniformly distributed on Si wafer	Thermal evaporation	3.6	1 mA/cm² at 6.2 V/μm	2.7×10^3	5.3	[31]

(continued)

TABLE 16.1 (continued)

Summary of Structural and Field Emission Characteristics of Various Forms of Nanostructures Including CuO, Cu_2S, CoO, ZnO, GaN, GaAs, AlN, WO_3, Si, SiC, ZnS, and $\alpha\text{-}Fe_2O_3$

Emitter	Morphology	Synthesis Method	Turn-On Field (V/μm)	Threshold Field (V/μm)	Field Enhancement Factor (β)	Work Function (Φ)	Reference
	Vertically well-aligned nanoneedles (0.9–1.0 μm long, 60–70 nm diameter tips) on Au/Ti/n-Si substrate	Metal organic chemical vapor deposition	0.1 μA/cm² at 0.9 V/μm	1 mA/cm² at 5.0 V/μm	8328	5.3	[32]
	Well-aligned arrays of nanoneedles (100 nm long, with apex 7 nm) on Si substrate	Chemical vapor deposition	2.4	2.4 mA/cm² at 7 V/μm	—	—	[33]
GaN	Patterned nanowires (1–1.5 μm long, 10–40 nm diameter) on silicon substrate	Pulsed laser ablation	8.4	1 mA/cm² at 10.8 V/μm	474	4.1	[34]
	Thick film of needlelike nanowires (tens of microns long and diameter decreasing from 200 to 10 nm along the wire axis) on Au-coated Si substrate	Thermal evaporation process	7.5	—	—	—	[35]
	Film of high-quality nanowires (tens of microns long with 10–50 nm in diameter) uniform on alumina substrate	Catalytic chemical vapor deposition	0.1 μA/cm² at 8.5 V/μm	0.2 mA/cm² at 17.5 V/μm	1170	4.1	[36]
	Nanorods (protrusions of 74.6 nm and diameter of 80 nm) with their *c*-axis perpendicular to the Si substrate	Radiofrequency plasma-enhanced molecular beam epitaxy	0.1 μA/cm² at 1.3 V/μm	2.5 mA/cm² at 2.5 V/μm	1270	—	[6]

GaAs	Large-area highly aligned nanowires (15–100 nm diameter) on GaAs wafer	H plasma etching	2.0	1 mA/cm² at 6.5 V/μm	3500	4.77	[12]
AlN	Film of hexagonal nanowires (tens of microns long, 15 nm diameter)	Extended VLS growth technique	—	1 mA/cm² at <1 V/μm	8.2×10^4	—	[7]
	Eiffel-tower-shape nanotips of 1 μm long (submicron base and shape tip of 10–100 nm) vertically assembled on Si substrate	Chemical vapor deposition	4.7	10 mA/cm² at 10.6 V/μm	1175.5	—	[37]
	Quasialigned nanotips (1.2 μm long, 100 nm at base, and 10 nm at tip) on Si substrate	Thermal chemical vapor deposition	6	0.2 A/cm² at 10 V/μm	—	4	[38]
	Flowerlike straight Si-doped nanoneedle (several microns long with base and tip diameters in the range of 50–150 nm and 5–30 nm) array on Si substrate	Chemical vapor deposition	1.8	10 mA/cm² at 4.6 V/μm	3271	3.7	[8]
	High density of nanoneedles (1 μm long with stem diameter of 100 nm and tip of 15 nm) on Si substrate	Vapor deposition method	3.1	4.7 mA/cm² at 9.9 V/μm	748	3.7	[39]
	Nanocones 2.5 μm long with tip sizes of 60 nm grown perpendicularly or slantingly forming quasiarrays on Si wafer	Chemical vapor deposition on Ni-coated Si wafer	17.8	—	1450	3.7	[40]

(continued)

TABLE 16.1 (continued)

Summary of Structural and Field Emission Characteristics of Various Forms of Nanostructures Including CuO, Cu$_2$S, CoO, ZnO, GaN, GaAs, AlN, WO$_3$, Si, SiC, ZnS, and α-Fe$_2$O$_3$

Emitter	Morphology	Synthesis Method	Turn-On Field (V/μm)	Threshold Field (V/μm)	Field Enhancement Factor (β)	Work Function (Φ)	Reference
	Well-aligned nanorods with high density of nanotips of 3–15 nm along the length, on a Si substrate	Vapor–solid process	3.8	1 mA/cm² at 7 V/μm	950	3.7	[41]
WO$_3$	Uniform and high density of straight nanowires (several microns long and diameter 70–400 nm) on tungsten substrate	Thermal evaporation process	4.8	—	—	—	[42]
	Film of quasialigned nanowires (W$_{18}$O$_{49}$) with diameters of 20–100 nm on tungsten foil	Infrared irradiation to heat tungsten foils	2.6 ± 0.1	10 mA/cm² at 6.2 V/μm	—	—	[9]
	Large-area quasialigned nanotips of 35 μm long on silicon substrate	Physical evaporation deposition process	2.0	10 mA/cm² at 4.4 V/μm	—	—	[43]
Si	High density of vertically aligned nanowires (1–1.2 μm long and 10–40 nm diameter) by top-down etching on Si wafer	Microwave plasma-enhanced chemical vapor deposition	0.8	1 mA/cm² at 5 V/μm	455	4	[44]
	Film of nanowires on Si wafer	Vapor–liquid–solid technique	—	10 mA/cm² at 7.8 ± 0.6 V/μm	500	4.5	[45]

	Description	Synthesis method					Ref.
	Low density of nanowires (10–20 μm long, 50 nm diameter) on carbon cloth	Vapor–liquid–solid reaction using silane gas	0.3	1 mA/cm² at 0.7 V/μm	6.1×10⁴	3.6	[10]
	Film of low density of nanowires (6 μm long, 5–10 nm diameter) Si substrate	High-temperature annealing of FeSi₂ nanodots on Si	6.3–7.3	10 mA/cm² at 9–10 V/μm	700–1000	3.6	[46]
	Film of low density of highly ordered Si nanowires (diameter of 25 nm) on indium tin oxide–coated glass substrate	Chemical vapor deposition method	7.4	10 mA/cm² at 9.9 V/μm	540	4.7	[47]
	Well-aligned and evenly distributed ensembles of nanowires (2 μm long with diameters of 20–50 nm) on Si wafer	Thermal evaporation process	7.3	—	424	3.6	[48]
	High-density arrays of needle-shaped carbon-coated nanocones (8.2 μm long with curvature radius of 4 nm) on porous Si wafer	Plasma etching in a hot filament chemical vapor deposition system	1.75 at 1 μA/cm²	109 μA/cm² at 8 V/μm	6350	4.7–2.7	[11]
	Islands of nanotip arrays of 10 nm long and diameters of 50 nm on a Si substrate	Si-based porous anodic alumina membrane as a mask	8.5	—	1100	4.7	[49]
SiC	Bunches of needle-shaped nanowires (1–1.2 μm long and 20–50 nm diameter and 2 nm sharp tip) on Si wafer	Thermal evaporation process	5	10 mA/cm² at 8.5 V/μm	839	4.4	[50]

(continued)

TABLE 16.1 (continued)

Summary of Structural and Field Emission Characteristics of Various Forms of Nanostructures Including CuO, Cu_2S, CoO, ZnO, GaN, GaAs, AlN, WO_3, Si, SiC, ZnS, and α-Fe_2O_3

Emitter	Morphology	Synthesis Method	Turn-On Field (V/μm)	Threshold Field (V/μm)	Field Enhancement Factor (β)	Work Function (Φ)	Reference
	Film of nanowires (several tens of microns long and 10–50nm diameter) on Si wafer	Metal organic chemical vapor deposition		1 mA/cm² at 4V/μm	2160	5	[13]
	Film of bamboo-like nanowires (10μm long, 300nm diameter)	Thermal evaporation process	10.1		—	—	[51]
	Nonaligned nanowires (8–20nm diameter) on activated carbon fibers, deposited on platinum film	High-temperature thermal evaporation process	3.1–3.5	60μA/cm² at 5.5V/μm	10^4	—	[15]
	Film of long nanowires (diameters of 14nm) on ceramic SiC substrate	Thermal evaporation process	3.3	10mA/cm² at 5.8V/μm	—	—	[52]
	High density of oriented nanowires (2mm long and diameters of 10–40nm) on a silica substrate	Chemical vapor deposition	0.7–1.5	10mA/cm² at 2.5–3.5V/μm	—	—	[14]
	Film of high density of carbon-coated nanowires (20nm core SiC and 3nm of carbon coat-layer) on NiO-catalyzed Si wafer	Heating NiO-catalyzed Si substrate along with carbothermal reduction of WO_3 by carbon	4.2	—	—	—	[53]

ZnS	Film of high-density nanowires (320–530nm at the base and 20–30nm sharp tip) on Si wafer	Vapor phase deposition	—	1 mA/cm² at 11.7 V/μm	522	—	[54]
	Film of nanobelts (widths of 5–30nm and peaked at 10–20nm)	Chemical vapor deposition	3.5	11.5 mA/cm² for 5.5 V/μm	2010	7.0	[3]
	Arrays of quasialigned nanobelts (tens to hundreds of bamboo-like nanowires (10 μm long, 300nm diameter)	Thermal evaporation process	3.6	14.5 mA/cm² at 5.5 V/μm	1.9×10^3	7.0	[55]
	Large-scale, well-aligned and oriented nanobelts (4 μm long, 30nm thick and several hundred nanometers in width) on zinc substrate	Solvothermal reaction	3.8	—	1812	5.4	[56]
$\alpha\text{-Fe}_2\text{O}_3$	Aligned arrays of nanoflakes (1 μm long and width of 100nm) with sharp tips of diameters 17nm on a AFM tip	Heating an iron-coated AFM tip	2.7–4.0	1.6 A/cm² at 6 V/μm	1.0×10^5 to 2.9×10^5	5.6	[57]

The interest in silicon nanowires for field emission is due to their well-understood electronic properties and their large-scale use in electronics industry. Silicon nanowires with controlled density grown on a carbon cloth require very low fields of $0.7\,V/\mu m$ to reach $1\,mA/cm^2$, the lowest ever reported value [10]. Field enhancement factors as high as 6.1×10^4 were achieved from controlled density of nanowires. Surface treatment of the nanostructures can also enhance the field emission properties. For example, silicon nano-cones synthesized on porous silicon show that an amorphous carbon coating on these structures increases field emission by lowering the work function from 4.15 to 2.37 eV [11]. In another example, highly aligned GaAs nanowires have high field enhancement factor ($\sim 10^3$) and low turn-on voltages, attributed to the very thin oxide layer/oxygen doping of these nanowires, which enhances the field emission by lifting the Fermi level and lowering the work function [12].

Silicon carbide nanowires with their high melting point, high thermal and chemical stability, high mechanical, and electrical breakdown combined with their high aspect ratio possess superior field emission characteristics [13]. Oriented silicon carbide nanowires on silica substrate present the lowest reported electric fields for various current densities while achieving continuous emission for about 24 h with low current fluctuations [14,15].

16.2 Thermoelectric Devices

In Chapter 11, bismuth telluride, indium antimonide, and a few other materials were referred to as the TE materials. TE devices using these materials convert heat into electricity using the principle of Seebeck effect, and vice versa by the Peltier effect. The Seebeck effect concerns the voltage generated across a junction of two dissimilar conducting or semiconducting materials when a temperature gradient is maintained. The Seebeck coefficient S is given by

$$S = dV/dT \qquad (16.3)$$

The Peltier effect states the opposite, that is, if a current is passed through a junction of two dissimilar materials, it will lead to a temperature difference. The Peltier coefficient Π is given by

$$Q = \Pi I \qquad (16.4)$$

where
  Q is the heat flux
  I is the current

A combination of high values for S and conductivity is needed to construct TE devices. The TE figure-of-merit ZT is defined as

$$ZT = S^2 \sigma T / k \qquad (16.5)$$

where
  σ is the electrical conductivity
  k thermal conductivity given by $k_e + k_p$ representing electron and phonon contributions respectively

The power factor is given by

$$PF = S^2 \sigma T \qquad (16.6)$$

which is just the numerator in the definition of ZT.

TE devices have a range of applications in waste heat recovery, cooling of instruments, chip cooling in semiconductor industry, auxiliary power in spacecraft and aircraft, commercial and home refrigeration, power for space vehicles, and others. Currently some upscale automobiles feature seat warmers/coolers using TE devices and personal soda/wine coolers are also available. NASA has employed TE devices in several missions to date. But large-scale applications in both refrigeration and power generation have not yet been realized since it has been difficult to achieve the desirable ZT values. A ZT value greater than 3 is the goal in solid state refrigeration development since that would give a performance comparable to a conventional refrigerator. However, the ZT of the best bulk materials to date is less than 1. It is clear from Equation 16.5 that high Seebeck coefficient, high electrical conductivity, and low thermal conductivity are desirable to boost ZT. But achieving these simultaneously is difficult since these are interdependent in bulk materials. Bulk metals exhibit high electrical conductivity but a low Seebeck coefficient. In contrast, bulk semiconductors exhibit high Seebeck coefficient and low thermal conductivity, but a low electrical conductivity due to low carrier concentration. However, the availability of electrons and holes for transport allows many n-type and p-type subunits connected in series. The best TE performance using a bulk material to date is from bismuth telluride, which shows a ZT value of close to 1 at room temperature. This has led to the investigation of quantum well superlattices and nanomaterials for TE devices, as they offer some inherent advantages. For example, the size effect on thermal conductivity discussed in the context of silicon nanowires in Section 6.2.8 is recalled here. SiNW thermal conductivities can be up to two orders of magnitude lower than for bulk silicon. Typically, a decrease in diameter is accompanied by a decrease in thermal conductivity. The electronic contribution to thermal conductivity is typically weaker than the phonon contribution in semiconductors. Boundary and interface scattering increases in nanostructures, serving to reduce the thermal conductivity.

Another effect of reduced dimensions is an increase in Seebeck coefficient due to a high density of states near Fermi level.

Theoretical studies indicate that nanowires of III–V and V–VI materials can indeed yield higher ZT values than the corresponding bulk counterparts [58,59]. Mingo shows that InSb nanowires stand out as the best choice for TE devices since they can reach a ZT of 6 with a wire diameter of 10 nm [58]. Achieving a ZT of 6 with GaAs would require an unrealistically small diameter of 1 nm. InP nanowire performance is even poorer than GaAs. Among the II–VI materials, Mingo's modeling studies find CdTe to be the best when compared with ZnTe, ZnSe, and ZnS but again requires an unrealistically small diameter of 1 nm to achieve a ZT of 6 [59].

Nanowires of silicon, InSb, bismuth telluride, and other materials have been studied for their TE potential [60–71]. In spite of the potential predicted by theoretical studies, the use of nanowire forms of appropriate TE candidate materials is limited to date [60–71] and no actual TE device capable of power generation or refrigeration using nanowires has been demonstrated yet. The references cited here basically demonstrate nanowire growth and occasional measurements of one or more parameters among k, σ, and S. Bi_2Te_3 nanowires of about 50 nm diameter grown using an anodized alumina template show slightly higher Seebeck coefficients than the bulk material, 270 and $-188\,\mu V/K$ for p- and n-type nanowires [60]. A nanowire array left intact inside an alumina template forms a composite and ZT of this hybrid is shown to be 0.12 [65]. Silicon nanowires have been investigated for their TE potential as well and Boukai et al. [69] demonstrated a 100-fold improvement in ZT for 10–20 nm nanowires compared to bulk silicon. The ZT value is about 1 at 200 K for 20 nm SiNWs. Hochbaum et al. [70] obtained a ZT value of 0.6 at room temperature for 50 nm SiNWs.

References

1. X. Fang, Y. Bando, U. K. Gautam, C. Ye, and D. Golberg, *J. Mater. Chem.* 18, 509 (2008).
2. Y. L. Liu, L. Liao, J. C. Li, and C. X. Pan, *J. Phys. Chem. C* 111, 5050 (2007).
3. X. S. Fang, Y. Bando, G. Z. Shen, C. H. Ye, U. K. Gautam, P. Costa, C. Y. Zhi, C. C. Tang, and D. Golberg, *Adv. Mater.* 19, 2593 (2007).
4. W. Z. Wang, B. Q. Zeng, J. Yang, B. Poudel, J. Y. Huang, M. J. Naughton, and Z. F. Ren, *Adv. Mater.* 18, 3275 (2006).
5. S. H. Jo, D. Banerjee, and Z. F. Ren, *Appl. Phys. Lett.* 85, 1407 (2004).
6. T. Yamashita, S. Hasegawa, S. Nishida, M. Ishimaru, Y. Hirotsu, and H. Asahi, *Appl. Phys. Lett.* 86, 082109 (2005).
7. Q. Wu, Z. Hu, X. Z. Wang, Y. N. Lu, K. F. Huo, S. Z. Deng, N. S. Xu, B. Shen, R. Zhang, and Y. Chen, *J. Mater. Chem.* 13, 2024 (2003).
8. Y. B. Tang, H. T. Cong, Z. M. Wang, and H. M. Cheng, *Appl. Phys. Lett.* 89, 253112 (2006).

9. Y. B. Li, Y. Bando, and D. Golberg, *Adv. Mater.* 15, 1294 (2003).

10. B. Q. Zeng, G. Y. Xiong, S. Chen, W. Z. Wang, D. Z. Wang, and Z. F. Ren, *Appl. Phys. Lett.* 90, 033112 (2007).

11. Q. Wang, J. J. Li, Y. J. Ma, X. D. Bai, Z. L. Wang, P. Xu, C. Y. Shi, B. G. Quan, S. L. Yue, and C. Z. Gu, *Nanotechnology* 16, 2919 (2005).

12. C. Y. Zhi, X. D. Bai, and E. G. Wang, *Appl. Phys. Lett.* 86, 213108 (2005).

13. D. W. Kim, Y. J. Choi, K. J. Choi, J. G. Park, J. H. Park, S. M. Pimenov, V. D. Frolov, N. P. Abanshin, B. I. Gorfinkel, N. M. Rossukanyi, and A. I. Rukovishnikov, *Nanotechnology* 19, 225706 (2008).

14. Z. W. Pan, H. L. Lai, F. C. K. Au, X. F. Duan, W. Y. Zhou, W. S. Shi, N. Wang, C. S. Lee, N. B. Wong, S. T. Lee, and S. S. Xie, *Adv. Mater.* 12, 1186 (2000).

15. W. M. Zhou, Y. J. Wu, E. S. W. Kong, F. Zhu, Z. Y. Hou, and Y. F. Zhang, *Appl. Surf. Sci.* 253, 2056 (2006).

16. Y. W. Zhu, T. Yu, F. C. Cheong, X. J. Xu, C. T. Lim, V. B. C. Tan, J. T. L. Thong, and C. H. Sow, *Nanotechnology* 16, 88 (2005).

17. H. Chien-Te, C. Jin-Ming, L. Hung-Hsiao, and S. Han-Chang, *Appl. Phys. Lett.* 83, 3383 (2003).

18. C. Jun, S. Z. Deng, N. S. Xu, Z. Weixin, W. Xiaogang, and Y. Shihe, *Appl. Phys. Lett.* 83, 746 (2003).

19. C. H. Teo, Y. W. Zhu, X. Y. Gao, A. T. S. Wee, and C. H. Sow, *Solid State Commun.* 145, 241 (2008).

20. J. Chen, S. Z. Deng, N. S. Xu, S. H. Wang, X. G. Wen, S. H. Yang, C. L. Yang, J. N. Wang, and W. K. Ge, *Appl. Phys. Lett.* 80, 3620 (2002).

21. Y. F. Lin, Y. J. Hsu, S. Y. Lu, and S. C. Kung, *Chem. Commun.* 22, 2391 (2006).

22. B. Varghese, T. C. Hoong, Z. Yanwu, M. V. Reddy, B. V. R. Chowdari, A. T. S. Wee, T. B. C. Vincent, C. T. Lim, and C. H. Sow, *Adv. Funct. Mater.* 17, 1932 (2007).

23. L. Vila, P. Vincent, L. Dauginet-De Pra, G. Pirio, E. Minoux, L. Gangloff, S. Demoustier-Champagne, N. Sarazin, E. Ferain, R. Legras, L. Piraux, and P. Legagneux, *Nano Letters* 4, 521 (2004).

24. T. Yu, Y. W. Zhu, X. J. Xu, Z. X. Shen, P. Chen, C.-T. Lim, J. T.-L. Thong, and C.-H. Sow, *Adv. Mater.* 17, 1595 (2005).

25. M.-K. Li, D.-Z. Wang, Y.-W. Ding, X.-Y. Guo, S. Ding, and H. Jin, *Mater. Sci. Eng. A* 452, 417 (2007).

26. X. D. Wang, J. Zhou, C. S. Lao, J. H. Song, N. S. Xu, and Z. L. Wang, *Adv. Mater.* 19, 1627 (2007).

27. D. Banerjee, S. H. Jo, and Z. F. Ren, *Adv. Mater.* 16, 2028 (2004).

28. R. C. Wang, C. P. Liu, J. L. Huang, S. J. Chen, Y. K. Tseng, and S. C. Kung, *Appl. Phys. Lett.* 87, 013110 (2005).

29. G. Z. Shen, Y. Bando, B. D. Liu, D. Golberg, and C.-J. Lee, *Adv. Funct. Mater.* 16, 410 (2006).

30. J. Xiao, Y. Wu, W. Zhang, X. Bai, L. G. Yu, S. Q. Li, and G. M. Zhang, *Appl. Surf. Sci.* 254, 5426 (2008).

31. J. Xiao, X. X. Zhang, and G. M. Zhang, *Nanotechnology* 19, 295706 (2008).

32. C. J. Park, D. K. Choi, J. Yoo, G. C. Yi, and C. J. Lee, *Appl. Phys. Lett.* 90, 083107 (2007).

33. Y. W. Zhu, H. Z. Zhang, X. C. Sun, S. Q. Feng, J. Xu, Q. Zhao, B. Xiang, R. M. Wang, and D. P. Yu, *Appl. Phys. Lett.* 83, 144 (2003).

34. D. K. T. Ng, M. H. Hong, L. S. Tan, Y. W. Zhu, and C. H. Sow, *Nanotechnology* 18, 375707 (2007).

35. B. D. Liu, Y. Bando, C. C. Tang, F. F. Xu, J. Q. Hu, and D. Golberg, *J. Phys. Chem. B* 109, 17082 (2005).
36. B. Ha, S. H. Seo, J. H. Cho, C. S. Yoon, J. Yoo, G. C. Yi, C. Y. Park, and C. J. Lee, *J. Phys. Chem. B* 109, 11095 (2005).
37. Y. B. Tang, H. T. Cong, Z. G. Chen, and H. M. Cheng, *Appl. Phys. Lett.* 86, 233104 (2005).
38. S. C. Shi, C. F. Chen, S. Chattopadhyay, K. H. Chen, and L. C. Chen, *Appl. Phys. Lett.* 87, 073109 (2005).
39. Q. Zhao, J. Xu, X. Y. Xu, Z. Wang, and D. P. Yu, *Appl. Phys. Lett.* 85, 5331 (2004).
40. C. Liu, Z. Hu, Q. Wu, X. Z. Wang, Y. Chen, W. W. Lin, H. Sang, S. Z. Deng, and N. S. Xu, *Appl. Surf. Sci.* 251, 220 (2005).
41. J. H. He, R. S. Yang, Y. L. Chueh, L. J. Chou, L. J. Chen, and Z. L. Wang, *Adv. Mater.* 18, 650 (2006).
42. Y. Baek and K. Yong, *J. Phys. Chem. C* 111, 1213 (2007).
43. J. Zhou, L. Gong, S. Z. Deng, J. Chen, J. C. She, N. S. Xu, R. S. Yang, and Z. L. Wang, *Appl. Phys. Lett.* 87, 223108 (2005).
44. J. C. She, S. Z. Deng, N. S. Xu, R. H. Yao, and J. Chen, *Appl. Phys. Lett.* 88, 013112 (2006).
45. N. N. Kulkarni, J. Bae, C. K. Shih, S. K. Stanley, S. S. Coffee, and J. G. Ekerdt, *Appl. Phys. Lett.* 87, 213115 (2005).
46. Y. L. Chueh, L. J. Chou, S. L. Cheng, J. H. He, W. W. Wu, and L. J. Chen, *Appl. Phys. Lett.* 86, 133112 (2005).
47. D. McClain, R. Solanki, L. F. Dong, and J. Jiao, *J. Vac. Sci. Technol. B* 24, 20 (2006).
48. X. S. Fang, Y. Bando, C. H. Ye, G. Z. Shen, U. K. Gautam, C. C. Tang, and D. Golberg, *Chem. Commun.* 40, 4093 (2007).
49. G. S. Huang, X. L. Wu, Y. C. Cheng, X. F. Li, S. H. Luo, T. Feng, and P. K. Chu, *Nanotechnology* 17, 5573 (2006).
50. Z. S. Wu, S. Z. Deng, N. S. Xu, J. Chen, and J. Zhou, *Appl. Phys. Lett.* 80, 3829 (2002).
51. G. Z. Shen, Y. Bando, C. H. Ye, B. D. Liu, and D. Golberg, *Nanotechnology* 17, 3468 (2006).
52. S. Z. Deng, Z. B. Li, W. L. Wang, N. S. Xu, J. Zhou, X. G. Zheng, H. T. Xu, J. Chen, and J. C. She, *Appl. Phys. Lett.* 89, 023118 (2006).
53. Y. Ryu, B. Park, Y. Song, and K. Yong, *J. Crystal Growth* 271, 99 (2004).
54. Y. Q. Chang, M. W. Wang, X. H. Chen, S. L. Ni, and W. J. Qiang, *Solid State Commun.* 142, 295 (2007).
55. X. S. Fang, Y. Bando, C. H. Ye, and D. Golberg, *Chem. Commun.* 29, 3048 (2007).
56. F. Lu, W. P. Cai, Y. G. Zhang, Y. Li, F. Q. Sun, S. H. Heo, and S. O. Cho, *Appl. Phys. Lett.* 89, 231928 (2006).
57. Y. W. Zhu, T. Yu, C. H. Sow, Y. J. Liu, A. T. S. Wee, X. J. Xu, C. T. Lim, and J. T. L. Thong, *Appl. Phys. Lett.* 87, 023103 (2005).
58. N. Mingo, *Appl. Phys. Lett.* 84, 2652 (2004).
59. N. Mingo, *Appl. Phys. Lett.* 85, 5986 (2004).
60. W. Wang, F. Jia, Q. Huang, and J. Zhang, *Microelectron. Eng.* 77, 223 (2005).
61. J. R. Lim, J. F. Whitacre, J. P. Fleurial, C. K. Huang, M. A. Ryan, and N. V. Myung, *Adv. Mater.* 17, 1488 (2005).
62. P. Wang, A. Bar-Cohen, B. Yang, G. L. Solbrekken, and A. Shakouri, *J. Appl. Phys.* 100, 014501 (2006).

63. J. Zhou, C. Jin, J.H. Seol, X. Li, and L. Shi, *Appl. Phys. Lett.* 87, 133109 (2005).
64. J.H. Seol, A.L. Moore, S.K. Saha, F. Zhou, L. Shi, Q.L. Ye, R. Scheffler, N. Mango, and T. Yamada, *J. Appl. Phys.* 101, 023706 (2007).
65. J. Keyani, A.M. Stacy, and J. Sharp, *Appl. Phys. Lett.* 89, 233106 (2006).
66. M.F. O'Dwyer, T.E. Humphrey, and H. Linke, *Nanotechnology* 17, S338 (2006).
67. M. S. Dresselhaus, G. Chen, M. Y. Tang, R. Yang, H. Lee, D. Wang, Z. Ren, J. P. Fleurial, and P. Gogna, *Adv. Mater.* 19, 1043 (2007).
68. E. Shapira, A. Tsukernik, and Y. Selzer, *Nanotechnology* 18, 485703 (2007).
69. A. I. Boukai, Y. Bunimovich, J. Tahir-Kheli, J. K. Yu, W. A. Goddard III, and J. R. Heath, *Nature* 451, 168 (2008).
70. A. I. Hochbaum, R. Chen, R. D. Delgado, W. Liang, E. C. Garnett, M. Najarian, A. Majumdar, and P. Yang, *Nature* 451, 163 (2008).
71. J. Lee, S. Farhangfar, J. Lee, L. Cagnon, R. Scholz, U. Gosele, and K. Nielsch, *Nanotechnology* 19, 365701 (2008).

Index

A

Aluminum nitride (AlN), 225, 231
Anodized alumina membrane (AAM)
  BiNWs, 171–172
  nanowire
    copper, 174
    silver, 173
    Zn, 179
  NiNW preparation, 176
Antimonide
  band gaps, InSb and GaSb, 257
  Bohr radius and AAM template
    approach, 259
  gallium/indium dissolution,
    257–258
  GaSb crystal nuclei, 258
  indium nanowire, 258–259
  IR absorption spectra, 259–260
Antoine equation, 46
Avalanche photodiode (APD), 300

B

Band-edge (BE) emission, 261
BiNWs, *see* Bismuth nanowires
Biosensor
  advantages and limitations, 337
  detection
    LDL cholesterol, 339, 341
    protein binding, 338–339
    PSA, 340, 341
    virus, 339–340
  NEA
    construction, 342–343
    structure, 341–342
    wafer fabrication, 343–344
Bismuth nanowires (BiNWs)
  electrochemical deposition,
    171–172
  energy overlap, 171
  melting points, 172–173
Bismuth telluride (Bi_2Te_3), 264
Bohr radius values, 249

Branching, nanowire
  hetero, 248–249
  homo, 247–248
  structure
    direct thermal evaporation, 206
    low-melting metals, 207

C

Cadmium selenide (CdSe), 43, 263,
  360, 366
Cadmium sulfide (CdS), 267–268
Cadmium telluride (CdTe), 27, 265, 267,
  350, 416
Carbon nanofibers (CNFs), 342–343
Carbon nanotubes (CNTs)
  GaN films, 404
  television applications, 400
Catalyst-assisted synthesis, oxide
    nanowire
  epitaxial arrays, 187–188
  lattice mismatches, 188
  metal oxide, 184
  metal source temperature, 189
  quartz tube reactor, 184–185
  temperature ramping stage, 186
  VLS mechanism, 185
  web-like morphology, 190
  zinc oxide, 185–186
Chemical beam epitaxy (CBE)
  role, 49
  species diffusion length, InAs, 93
Chemical sensor
  e-nose approach
    autoscaling purpose, 335–336
    CNTs and olfactory receptor
      genes, 332
    PCA, 334
    principal components, 336–337
    response pattern, NO_2 and Cl_2,
      334–335
    32-sensor chip and individual IDE
      array pattern, 332–333
    SWNTs and DMF, 333–334